OXFORD MATHEMATICAL MONOGRAPHS

Series Editors

I. G. MACDONALD R. PENROSE H. MCKEAN
J. T. STUART

OXFORD MATHEMATICAL MONOGRAPHS

A. Belleni-Morante: *Applied semigroups and evolution equations*
I. G. Macdonald: *Symmetric functions and Hall polynomials*
J. W. P. Hirschfeld: *Projective geometries over finite fields*
N. Woodhouse: *Geometric quantization*
A. M. Arthurs: *Complementary variational principles* Second edition
P. L. Bhatnagar: *Nonlinear waves in one-dimensional dispersive systems*
N. Aronszajn, T. M. Creese, and L. J. Lipkin: *Polyharmonic functions*
J. A. Goldstein: *Semigroups of linear operators*
M. Rosenblum and J. Rovnyak: *Hardy classes and operator theory*
J. W. P. Hirschfeld: *Finite projective spaces of three dimensions*
K. Iwasawa: *Local class field theory*
A. Pressley and G. Segal: *Loop groups*
J. C. Lennox and S. E. Stonehewer: *Subnormal subgroups of groups*
D. E. Edmunds and W. D. Evans: *Spectral theory and differential operators*
Wang Jianhua: *The theory of games*
S. Omatu and J. H. Seinfeld: *Distributed parameter system theory and applications*

Loop Groups

ANDREW PRESSLEY
King's College London

AND

GRAEME SEGAL
University of Oxford

CLARENDON PRESS · OXFORD

Oxford University Press, Walton Street, Oxford OX2 6DP
Oxford New York Toronto
Delhi Bombay Calcutta Madras Karachi
Petaling Jaya Singapore Hong Kong Tokyo
Nairobi Dar es Salaam Cape Town
Melbourne Auckland
and associated companies in
Berlin Ibadan

Oxford is a trade mark of Oxford University Press

Published in the United States
by Oxford University Press, New York

© Andrew Pressley and Graeme Segal, 1986

First published 1986
Reprinted (with corrections) 1988
First published in paperback (with corrections) 1988

All rights reserved. No part of this publication may be reproduced, stored in a retrieval system, or transmitted, in any form or by any means, electronic, mechanical, photocopying, recording, or otherwise, without the prior permission of Oxford University Press

This book is sold subject to the condition that it shall not, by way of trade or otherwise, be lent, re-sold, hired out or otherwise circulated without the publisher's prior consent in any form of binding or cover other than that in which it is published and without a similar condition including this condition being imposed on the subsequent purchaser

British Library Cataloguing in Publication Data
Pressley, Andrew
Loop groups.—(Oxford mathematical monographs)
1. Loops (Group theory)
I. Title   II. Segal, Graeme
512'.55   QA171
ISBN 0-19-853535-X
0-19-853561-9(Pbk)

Library of Congress Cataloging in Publication Data
Pressley, Andrew
Loop groups.
(Oxford mathematical monographs)
Bibliography: p.
Includes index.
1. Lie groups.   I. Segal, Graeme.   II. Title
III. Series
QA387.P725 1986   512'.55   86-5193
ISBN 0-19-853535-X
0-19-853561-9(Pbk)

Printed in Great Britain by
Antony Rowe Ltd, Chippenham, Wiltshire

CONTENTS

1 INTRODUCTION 1

PART I

2 FINITE DIMENSIONAL LIE GROUPS 11
  2.1 The Lie algebra 11
  2.2 Complex groups 12
  2.3 Compact Lie groups 13
  2.4 The root system 15
  2.5 Simply laced groups 17
  2.6 The Weyl group and the Weyl chambers: positive roots 18
  2.7 Irreducible representations and antidominant weights 18
  2.8 Complex homogeneous spaces 19
  2.9 The Borel–Weil theorem 21

3 GROUPS OF SMOOTH MAPS 24
  3.1 Infinite dimensional manifolds 24
  3.2 Groups of maps as infinite dimensional Lie groups 26
  3.3 Diffeomorphism groups 28
  3.4 Some group-theoretic properties of $\mathrm{Map}(X;G)$ 30
  3.5 Subgroups of LG: polynomial loops 32
  3.6 Maximal abelian subgroups of LG 34
  3.7 Twisted loop groups 36

4 CENTRAL EXTENSIONS 38
  4.1 Introduction 38
  4.2 The Lie algebra extensions 39
  4.3 The coadjoint action of LG on $\tilde{L}\mathfrak{g}$, and its orbits 43
  4.4 The group extensions when G is simply connected 46
  4.5 Circle bundles, connections, and curvature 50
  4.6 The group extensions when G is semisimple but not simply connected 55
  4.7 The basic central extension of LU_n 57
  4.8 The restriction of the extension to LT 60
  4.9 The inner product on $\mathbb{R} \oplus \tilde{L}\mathfrak{g}$ 64
  4.10 Extensions of $\mathrm{Map}(X;G)$; the Mickelsson–Faddeev extension 65
  4.11 Appendix: The cohomology of LG and its Lie algebra 67

5 THE ROOT SYSTEM: KAC–MOODY ALGEBRAS — 70

5.1 The root system and the affine Weyl group — 70
5.2 Generators and relations — 74
5.3 Kac–Moody Lie algebras — 76

6 LOOP GROUPS AS GROUPS OF OPERATORS IN HILBERT SPACE — 79

6.1 Loops as multiplication operators — 79
6.2 The restricted general linear group of Hilbert space — 80
6.3 The map $LGL_n(\mathbb{C}) \to GL_{\text{res}}(H^{(n)})$ — 82
6.4 Bott periodicity — 85
6.5 The isomorphism $H^{(n)} \cong H$ and the embedding $L\mathbb{T} \to LU_n$ — 85
6.6 The central extension of $GL_{\text{res}}(H)$ — 86
6.7 The central extension of $LGL_n(\mathbb{C})$ — 89
6.8 Embedding $\text{Diff}^+(S^1)$ in $U_{\text{res}}(H)$ — 91
6.9 Other polarizations of H: replacing the circle by the line, and the introduction of 'mass' — 93
6.10 Generalizations to other groups of maps — 98

7 THE GRASSMANNIAN OF HILBERT SPACE AND THE DETERMINANT LINE BUNDLE — 101

7.1 The definition of $\text{Gr}(H)$ — 101
7.2 Some dense submanifolds of $\text{Gr}(H)$ — 104
7.3 The stratification of $\text{Gr}(H)$ — 106
7.4 The cell decomposition of $\text{Gr}_0(H)$ — 108
7.5 The Plücker embedding — 109
7.6 The $\mathbb{C}^\times_{\leq 1}$-action — 111
7.7 The determinant bundle — 113
7.8 $\text{Gr}(H)$ as a Kähler manifold and a symplectic manifold — 117

8 THE FUNDAMENTAL HOMOGENEOUS SPACE — 120

8.1 Introduction: the factorization theorems — 120
8.2 Two applications of the Birkhoff factorization — 123
   Singularities of ordinary differential equations
   Holomorphic vector bundles on the Riemann sphere
8.3 The Grassmannian model of ΩU_n — 125
8.4 The stratification of $\text{Gr}^{(n)}$: the Birkhoff and Bruhat decompositions — 129
8.5 The Grassmannian model for the other classical groups — 136
8.6 The Grassmannian model for a general compact Lie group — 138

CONTENTS

8.7	The homogeneous space LG/T and the periodic flag manifold	143
8.8	Bott periodicity	146
8.9	ΩG as a Kähler manifold: the energy flow	147
8.10	ΩG and holomorphic bundles	152
8.11	The homogeneous space associated to a Riemann surface: the moduli spaces of vector bundles	154
8.12	Appendix: Scattering theory	160

PART II

9 REPRESENTATION THEORY — 165

9.1	General remarks about representations	167
9.2	The positive energy condition	171
9.3	The classification and main properties of representations of positive energy	176
9.4	The Casimir operator and the infinitesimal action of the diffeomorphism group	182
9.5	Heisenberg groups and their standard representations	188

10 THE FUNDAMENTAL REPRESENTATION — 195

10.1	Γ as an exterior algebra: the fermionic Fock space	196
10.2	The Hilbert space structure	200
10.3	The ring of symmetric polynomials	202
10.4	Γ as a sum of polynomial algebras	205
10.5	Jacobi's triple product identity	208
10.6	The basic representations of LU_n and LSU_n	211
10.7	Quantum field theory in two dimensions	214

11 THE BOREL–WEIL THEORY — 216

11.1	The space of holomorphic sections of a homogeneous line bundle	216
11.2	The decomposition of representations: complete reducibility	221
11.3	The existence of holomorphic sections	224
11.4	The smoothness condition	227

12 THE SPIN REPRESENTATION — 230

12.1	The Clifford algebra	230
12.2	Second construction of the spin representation	233
12.3	The spin representation as the sections of a holomorphic line bundle	242
12.4	The infinite dimensional spin representation	244
12.5	The basic representation of LO_{2n}	246
12.6	An analogy: the 'extra-special 2-group'	247

CONTENTS

13 'BLIPS' OR 'VERTEX OPERATORS' — 249
  13.1 The fermionic operators on $\mathcal{H}$ — 250
  13.2 The action of $U_{res}(H)$ and $O_{res}(H)$ on $\mathcal{H}$ — 260
  13.3 The representations of level 1 of LG when G is simply laced — 263
  13.4 The action of $\text{Diff}^+(S^1)$ on positive energy representations of loop groups — 268
  13.5 General remarks: the case of a general maximal abelian subgroup of LG — 271

14 THE KAC CHARACTER FORMULA AND THE BERNSTEIN–GELFAND–GELFAND RESOLUTION — 273
  14.1 General remarks about characters — 273
  14.2 Motivation of the character formula: Weyl's formula for compact groups — 276
  14.3 The character formula — 279
  14.4 The algebraic proof of the character formula — 286
  14.5 The Bernstein–Gelfand–Gelfand resolution — 290
  14.6 Applications of the resolution: the cohomology of $L\mathfrak{g}$ — 299

REFERENCES — 304

INDEX OF NOTATION — 311

INDEX — 315

1
INTRODUCTION

A loop group LG is the group of parametrized loops in another group G, i.e. the group of maps from the circle S^1 into G. Its composition law comes from pointwise composition in G. As far as this book is concerned G will always be either a compact Lie group or its complexification. There are a number of points of view from which one might be led to study these groups.

From a purely mathematical standpoint it is natural to ask whether the rich and highly developed theory of finite dimensional Lie groups can be extended to include infinite dimensional groups. In that light the groups Map$(X; G)$ of maps from a compact space X to a Lie group G are probably the most obvious examples of infinite dimensional Lie groups. Their behaviour is untypical in its simplicity. For example the exponential map $\mathfrak{g} \to G$ from the Lie algebra $\mathfrak{g}$ of G induces an exponential map Map$(X; \mathfrak{g}) \to$ Map$(X; G)$ which is locally one-to-one and onto: that is not a property one can expect of infinite dimensional Lie groups in general.

In quantum field theory groups of the form Map$(X; G)$—where X is usually physical space $\mathbb{R}^3$—arise in two slightly different ways, as gauge groups and as current groups. Loop groups therefore appear in the simplified model of quantum field theory in which space is taken to be one-dimensional, and also in 'string' models of elementary particles, where the particles are represented by one-dimensional extended objects. Some of the important facts in the representation theory of loop groups were first discovered by physicists, and in this book our whole approach to the subject has been coloured by quantum field theory.

Although it is intrinsically a very simple and natural group we know surprisingly little about Map$(X; G)$—particularly about its representations—except when X is a circle. In that case the situation is enormously easier, and has been very fully worked out. Loop groups turn out to behave like compact Lie groups to a quite remarkable extent. The aim of this book is to give a general exposition of what is known about them, concentrating on the global and analytical aspects of the theory rather than the algebraic ones. This has not been the usual approach to the subject, and deserves some comment. Usually loop groups have been approached by way of their Lie algebras, which are (essentially) examples of what are called Kac–Moody algebras. These are the Lie algebras which can be described in terms of generators and relations in the same way as the finite dimensional semisimple algebras. The classical theory of

Cartan and Killing shows how to associate an algebra to a finite integer matrix satisfying certain conditions which include positive definiteness. Omitting the positivity condition leads to the Kac–Moody algebras. If one merely weakens 'positive-definite' to 'positive-semidefinite' one obtains a subclass of the Kac–Moody algebras which are usually called *affine* algebras. These are (up to a one-dimensional central extension) the Lie algebras of loop groups and of the twisted versions of loop groups which we shall describe presently. For Kac–Moody algebras which are not affine no description is known other than the one in terms of generators and relations, and it is a mystery in what contexts the corresponding groups may arise. They are certainly not of the form Map$(X; G)$ with $\dim(X) > 1$. On the other hand when the theory of Kac–Moody algebras is developed algebraically it does not make very much difference—for many purposes at least—whether or not the algebra is affine.

A considerable stimulus was given to the theory of general Kac–Moody algebras by the discovery by Macdonald [107] in 1972 of a class of formal power series identities analogous to the Weyl denominator formula, but reducing in particular cases to such results as the Jacobi triple product identity. Kac pointed out that Macdonald's identities resulted immediately from a generalization of the Weyl character formula to Kac–Moody algebras, and the character formula has been much studied since then from various points of view. The characters turn out to be modular functions in a certain sense, though why that should be true remains another of the mysteries of the subject.

It is perfectly reasonable to conclude from all this that the groups we are studying are interesting not because they are loop groups but because they possess a certain very special combinatorial structure. From that point of view this book, which ignores all Kac–Moody algebras except the 'affine' ones, must seem rather perverse. The case for our approach is partly aesthetic. In studying loop groups we are pursuing geometry and analysis rather than algebra and combinatorics, and for some people the geometrical picture is more illuminating and attractive. On the other hand, while acknowledging that it would be very rash and optimistic to think that the theory of loop groups will tell us anything directly about more general groups of the form Map$(X; G)$, it is plain that the methods and constructions we use are very basic ones and belong to the mainstream of mathematics, especially in connection with quantum field theory. It does not seem unreasonable to think they may find application elsewhere.

The first main feature of our approach is to think of loop groups as groups of operators in Hilbert space. We regard an element of LG as a multiplication operator in the Hilbert space $H = L^2(S^1; V)$ of L^2 functions on the circle with values in some finite dimensional representation V

of G. We decompose H as $H_+ \oplus H_-$, where H_+ (resp. H_-) consists of the functions whose negative (resp. positive) Fourier coefficients vanish, and we study the way in which the multiplication operators behave with respect to this decomposition. The idea here comes from quantum field theory, where H is the space of solutions of a relativistic wave equation and H_+ and H_- are the solutions of positive and negative energy: we examine how an operator in H moves particles from states of positive energy to ones of negative energy and vice versa.

Our second main technique is the geometrical study of the 'fundamental homogeneous space' X of LG. This is LG/G, where G is regarded as the subgroup of LG consisting of constant loops. It can be identified with the space ΩG of *based* loops in G. It has two crucial properties. First it is a *complex* manifold, for it can be identified with the homogeneous space $LG_{\mathbb{C}}/L^+G_{\mathbb{C}}$, where $G_{\mathbb{C}}$ is the complexification of G and $L^+G_{\mathbb{C}}$ consists of the loops which are boundary values of holomorphic maps

$$\{z \in \mathbb{C}: |z| < 1\} \to G_{\mathbb{C}}.$$

The second property is that X has a stratification by complex manifolds of finite codimension, indexed by the conjugacy classes of homomorphisms $S^1 \to G$. This is precisely analogous to the subdivision of a Grassmann manifold into its Schubert cells, and also to the Bruhat decomposition of a complex semisimple group. Both properties together amount to a restatement of the Birkhoff factorization theorem of 1909, which asserts that a loop γ in $G_{\mathbb{C}}$ can be factorized $\gamma = \gamma_- \cdot \lambda \cdot \gamma_+$, where $\gamma_\pm$ are loops which extend holomorphically inside (resp. outside) the unit circle, and $\lambda: S^1 \to G$ is a homomorphism.

A more geometrical way of thinking of the stratification of $X = \Omega G$ is in terms of the *energy* function $\mathscr{E}: \Omega G \to \mathbb{R}$, defined by

$$\mathscr{E}(\gamma) = \frac{1}{4\pi} \int_0^{2\pi} \|\gamma'(\theta)\|^2 \, d\theta.$$

The critical points of $\mathscr{E}$ are the homomorphisms $\gamma: S^1 \to G$, and they fall into connected components according to their conjugacy class. The stratum corresponding to a conjugacy class consists of all loops which flow to it under the downwards gradient flow of $\mathscr{E}$.

Putting the two ideas together, we observe that because LG acts on the Hilbert space H it acts on the Grassmann manifold of closed subspaces of H. The orbit of the subspace H_+ under this action is a copy of the homogeneous space X. We shall constantly use this embedding of X as a subvariety of the Grassmannian.

The importance of the geometry of X in the theory of loop groups is partly as a tool in proving structural theorems such as the Birkhoff factorization theorem itself, but more fundamentally is because the irreducible representations of LG arise as spaces of holomorphic sections of line bundles on X. (Strictly speaking, we need here the closely related space $Y = LG/T$, where T is a maximal torus of G.)

The present book falls roughly into two halves, the first eight chapters being concerned with the groups and the rest with their representations. For an introduction to the representation theory we refer the reader to the beginning of Chapter 9. Indeed one can begin reading the book there, referring back to the earlier chapters only when they are needed: Chapter 8 in particular goes into rather more detail than many readers will find interesting.

We originally intended to devote part of the book to the applications of loop groups, but in the end we have not felt competent or energetic enough to do so. Loop groups arise in two dimensional quantum field theory, as has already been mentioned; more recently they have found extensive applications in connection with so-called 'completely integrable systems' of partial differential equations. The more combinatorial aspects of the theory arise in connection with the classification of certain kinds of singularities in algebraic geometry—a generalization of the classical correspondence between simple singularities and finite dimensional simple Lie algebras ([22], [140], [105]). There is the same kind of correspondence in the classification of systems of subspaces in linear algebra, generalizing Gabriel's theorem [53]. Finally, the character formula for representations of loop groups can be used as a fruitful source of combinatorial identities, as was originally pointed out by Macdonald [107].

Of all these applications only the first two, to quantum field theory and to differential equations, appear to involve the groups themselves rather than just their combinatorial presentation. The quantum field theory applications are hard to survey, as they consist of scattered instances of a number of different types. We refer the interested reader to Jacob [80], Dolan [38], Chau et al. [26], Witten [155]. The application to differential equations is more straightforward. It was first worked out by Sato [35], [126], although the ideas had been implicit in earlier writings on the subject (cf. especially Zakharov and Shabat [156]). There are now a number of accounts of the subject in the literature. For an expository account from the point of view of the present book we refer to [132]. The other applications are not really within our terms of reference, but we refer to Slodowy [141] for the classification of singularities, and to [85] for the generalizations of Gabriel's theory (sometimes called the theory of 'quivers'). The most extensive work in connection with combinatorial identities has been done by Lepowsky (cf. [43], [101], [103]).

Apart from applications in the strict sense we should mention one or two related matters. The central idea in the theory is the Birkhoff factorization theorem, which was found in the course of classifying singularities of ordinary differential equations (see Section 8.2 below). The related Riemann–Hilbert problem of finding a meromorphic function with prescribed monodromies about prescribed poles has recently been given a constructive solution using the kinds of method described in this book—notably 'vertex operators'. (Cf. [127].) In a completely different direction there is the recent work of Frenkel, Lepowsky, and Meurman [50] in connection with the 'monster' group in which once again the central point is the use of vertex operators. And vertex operators are also responsible for the boson–fermion correspondence which is important in two-dimensional quantum field theory (cf. Coleman [29], Mandelstam [111], Frenkel [45], and also Section 10.7 below).

The aim of this book is expository, and we have made no attempt to give a historical account of the evolution of the ideas and theorems we describe. It would in any case be rather hard to do so objectively, for many of the ideas have been worked out by a number of people independently in somewhat different contexts, and have often remained for some time 'well known to experts' without being written down. The best course for us seems to be to list a fairly representative sample of the different approaches to the subject. The definitive treatment of the Lie algebra theory is to be found in Kac's book [86], which contains an extensive list of references. Another algebraic approach is due to Garland and Lepowsky [54], [55], [56]. Goodman and Wallach [64] have approached the subject from the point of view of Banach Lie groups. Frenkel's paper [47] treats the character theory in terms of Wiener integration on orbits. We should also mention the important papers [49], [87], [100], and [88].

The representation theory of loop groups is closely related to that of the group Diff(S^1) of diffeomorphisms of the circle, which acts as a group of automorphisms of every loop group. It turns out that Diff$^+$(S^1)—the orientation preserving diffeomorphisms—acts projectively on all the representations of loop groups which we consider, and all the known representations of Diff$^+$(S^1) have been constructed in this way—cf. [131], [64], [61]. We shall not study Diff$^+$(S^1) systematically in this book, but we shall prove the crucial intertwining property.

We have already mentioned twisted loop groups. If α is an automorphism of G then the corresponding twisted loop group $L_{(\alpha)}G$ consists of the maps $\gamma : \mathbb{R} \to G$ such that

$$\gamma(\theta + 2\pi) = \alpha(\gamma(\theta))$$

for all θ. The theory of loop groups extends essentially without change to the twisted groups. We have made a few sporadic remarks about the twisted case, but for the most part it presents nothing new, and we have not pursued it.

Let us now outline the contents of the book systematically.

Chapter 2 is a survey of the results about finite dimensional Lie groups which we shall use. It is included simply to make the book more self-contained, and—we hope—more accessible to readers with a variety of different backgrounds.

Chapter 3 introduces infinite dimensional Lie groups, and considers what can be said about loop groups—and some related groups—from that point of view, without entering into their more specific and characteristic properties.

Chapter 4, on the other hand, is devoted to one of the most important and distinctive features of loop groups, the existence of a natural class of central extensions by the circle-group $\mathbb{T}$, or equivalently the fact that loop groups admit non-trivial projective representations. The extended loop groups play a bigger role in the theory than the loop groups themselves, and all the representations we shall construct are projective ones. In this chapter the extensions are constructed by differential–geometric methods. In fact we find all possible extensions of the group $\text{Map}(X; G)$ for any compact manifold X, though the result shows that only the case $X = S^1$ is important. We notice that the extensions exist only for groups of *smooth* (rather than continuous) loops.

Chapter 5 is a brief account of the theory of the Lie algebras of loop groups, about which we have said only what is needed for our purposes. We do give the definition of a Kac–Moody algebra, and explain how loop groups provide examples of them; but we do not mention the question of their classification.

Chapter 6 considers loop groups as groups of operators in Hilbert space. We introduce the restricted general linear group $GL_{\text{res}}(H)$ of a Hilbert space H with a polarization, i.e. a decomposition $H = H_+ \oplus H_-$. This group, which consists of the invertible operators in H whose off-diagonal blocks $H_\pm \to H_\mp$ are Hilbert–Schmidt, is central throughout the rest of the book. If $H_+ \oplus H_-$ is the space of positive and negative energy solutions of a relativistic wave equation then the elements of $GL_{\text{res}}(H)$ are precisely the transformations of H which are 'implementable' on the corresponding Fock space. (Cf. Shale [136].)

The group $GL_{\text{res}}(H)$ has a basic central extension by $\mathbb{C}^\times$ from which all the central extensions of loop groups are derived. Roughly speaking, the extension measures the extent to which the process of associating to an operator in H its $H_+ \to H_+$ component fails to be a homomorphism. We believe it becomes clear at this point that the direction to be followed if

1 INTRODUCTION

one wants to study the group Map$(X; G)$ when $\dim(X) > 1$ is the theory of Connes [32]. We have said a little about that in Section 6.10.

Chapter 7 introduces the Grassmannian Gr(H) of a polarized Hilbert space, another fundamental concept for our approach. Its most important property is that one can define a determinant line bundle on it. This is a holomorphic bundle which is homogeneous under the central extension of $GL_{res}(H)$: in fact it provides a definition of the central extension. The Grassmannian has a decomposition into Schubert varieties exactly like a finite dimensional Grassmannian: more precisely, it has a stratification by manifolds of finite codimension, and a dual decomposition into finite dimensional cells.

Chapter 8 is devoted to the geometry of the fundamental homogeneous space X of LG, and the theorems about the structure of LG which are obtained from it. Our basic tool, as we have said, is the embedding of X in the Grassmannian Gr(H), from which we see that X is a complex Kähler manifold. We derive the decomposition of X—and hence in particular the Birkhoff factorization theorems—from the Schubert decomposition of Gr(H). (We believe that this geometric point of view is more perspicuous than the usual treatment of Birkhoff's theorem by integral equations [63].) Our procedure works in the first instance for the loop group of U_n, and the general case is derived from that. Having obtained the decomposition by these methods we can then (in Section 8.9) consider the situation anew in the light of the gradient flow of the energy function. We show that the decomposition is precisely the Morse decomposition of X corresponding to the energy. In particular we obtain a very complete and appealing description of the energy flow (Proposition (8.9.8)).

In Section 8.10 we consider some of the properties X possesses simply as a complex manifold. Although it is an infinite dimensional homogeneous space it has many 'finite dimensional' properties. Not only is every holomorphic function on it constant on each connected component, but each connected component of the space of (based) maps from any compact complex manifold to X is of finite dimension. Our treatment here follows Atiyah [5].

Besides the homogeneous space $X = \Omega G$ of LG we shall see that there is a similar space X_M associated to any closed Riemann surface M. It has the homotopy type of the space of principal G-bundles on M.

The relationship between ΩG and the Grassmannian was first observed in 'scattering theory' in the sense of Lax and Phillips [99], as we explain in Section 8.12. It is also at the heart of the Bott periodicity theorem (see Section 8.8).

After Chapter 8 the remainder of the book is devoted to representation theory, and we refer to Chapter 9 for a survey of its contents. Our approach is on the one hand to imitate the finite dimensional Borel–Weil theory, and on the other hand to make use of the natural representations which are well-known from quantum field theory. The character formula is postponed until the last chapter, where we also show how the decomposition of X described in Chapter 8 leads directly to the Bernstein–Gelfand–Gelfand resolution of a representation of LG by Verma modules. This is the basis of all the more delicate algebraic and cohomological analysis of the representations into which we have refrained from entering.

This book has taken a long time to write. It began from lectures given by the second author at Berkeley at the beginning of 1982. A first draft of the book was written by the first author along the lines of the lectures, and was subsequently greatly expanded by the second author to produce the present work. Obviously we have profited greatly from the influence of many people, and we have tried to acknowledge our indebtedness at various places in the text. But we should like here to express our especial gratitude to Sir Michael Atiyah, who suggested the whole project and has constantly given us encouragement and advice, and to Daniel Quillen, who shaped our approach to the subject in 1978 by teaching us about the Grassmannian model of ΩG and its importance, and has been a continual influence since then.

We hope that some parts of the book, at least, will be of interest to readers who are not primarily concerned with loop groups. A few sections, such as the treatment of the spin representation, have indeed been written with that specifically in mind. We have tried within reason to make the sections of the book fairly independent, even at the cost of some repetition, in order to encourage readers to turn at once to the parts they find interesting. We have also deliberately written different sections with different levels of mathematical sophistication. We have tried hard to presuppose very little when dealing with the central parts of the theory.

PART I

2

FINITE DIMENSIONAL LIE GROUPS

The purpose of this chapter is to outline and assemble the basic facts about finite dimensional Lie groups which will be used in the course of the book, and to establish some notation and terminology. We make no attempt to provide proofs, referring the reader to any of a number of standard treatments, of which the nearest in spirit to our approach is probably Adams [1]. (Cf. also [20], [72], [76].) We have called attention especially, however to the complex homogeneous spaces and the Borel–Weil theorem—see Sections 2.8 and 2.9 below—as they are the basis of our approach to loop groups.

2.1 The Lie algebra

A Lie group G is a topological group which is locally like a Euclidean space $\mathbb{R}^n$. It is then automatically a differentiable manifold; in other words it can be covered by a collection of coordinate charts between which the transition functions are differentiable—in fact of class C^∞.

The tangent space $\mathfrak{g}$ to G at its identity element 1 is called the Lie algebra of G. For every vector ξ in $\mathfrak{g}$ there is a unique one-parameter subgroup $\{g_t\}_{t \in \mathbb{R}}$ in G which has ξ as its tangent vector

$$\left. \frac{\mathrm{d} g_t}{\mathrm{d} t} \right|_{t=0}$$

at the identity. The group element g_t is denoted by $\exp(t\xi)$; and the map $\xi \mapsto \exp(\xi)$ from $\mathfrak{g}$ to G is called the exponential map. It provides a one-to-one correspondence between a neighbourhood of 0 in $\mathfrak{g}$ and a neighbourhood of 1 in G, and is one of the preferred coordinate charts.

The group which will be central in this book is the unitary group U_n, which consists of all $n \times n$ complex matrices u such that $u^*u = 1$. (Here u^* denotes the transpose of the complex conjugate of u.) The Lie algebra of U_n is the vector space $\mathfrak{u}_n$ of skew-Hermitian matrices (i.e. those such that $\xi^* = -\xi$), and the exponential map is given by ordinary exponentiation of matrices.

For any Lie group G there is an operation $\mathfrak{g} \times \mathfrak{g} \to \mathfrak{g}$ defined on $\mathfrak{g}$, denoted by $(\xi, \eta) \mapsto [\xi, \eta]$ and called the 'bracket'. It is defined by

$$[\xi, \eta] = \lim_{s,t \to 0} \frac{1}{st} \exp^{-1}\{\exp(s\xi)\exp(t\eta)\exp(-s\xi)\exp(-t\eta)\}.$$

We see that it measures, in some sense, the group's failure to be commutative. The bracket is bilinear, and has the properties

$$[\xi, \eta] = -[\eta, \xi]$$

and

$$[\xi, [\eta, \zeta]] + [\eta, [\zeta, \xi]] + [\zeta, [\xi, \eta]] = 0.$$

(The latter is called the 'Jacobi identity'.) In the case of U_n, or indeed of any other matrix group, it is easy to see that the bracket is given in terms of matrix multiplication by

$$[\xi, \eta] = \xi\eta - \eta\xi.$$

An alternative description of the Lie algebra $\mathfrak{g}$ is as the space of left-invariant vector fields on the group G. For a tangent vector to G at the identity element defines by left-translation a tangent vector at each point of G, and hence a smooth vector field. Conversely, a left-invariant vector field is completely determined by its value at $1 \in G$. From this point of view the bracket operation $\mathfrak{g} \times \mathfrak{g} \to \mathfrak{g}$ is the usual bracket of vector fields: if ξ and η are expressed in some local coordinate chart as

$$\xi = \sum_i \xi_i \frac{\partial}{\partial x_i} \quad \text{and} \quad \eta = \sum_i \eta_i \frac{\partial}{\partial x_i},$$

then in the same chart $[\xi, \eta]$ is given by

$$\sum_{i,j} \left(\xi_j \frac{\partial \eta_i}{\partial x_j} - \eta_j \frac{\partial \xi_i}{\partial x_j} \right) \frac{\partial}{\partial x_i}.$$

In other words if ξ and η are regarded as differentiation operators acting on smooth functions on G then $[\xi, \eta] = \xi\eta - \eta\xi$.

If the group G is connected then the Lie algebra $\mathfrak{g}$, together with its bracket operation, determines G completely, up to the possibility of replacing it by a locally isomorphic group. The group U_n, for example, has the same Lie algebra as its simply connected covering group $\tilde{U}_n$, which is the subgroup of the product $U_n \times \mathbb{R}$ consisting of pairs (u, t) such that $\det(u) = e^{it}$. (In other words, an element of $\tilde{U}_n$ is an element of U_n together with a choice of the logarithm of its determinant.) For any finite dimensional Lie algebra $\mathfrak{g}$ there is always a simply connected group G whose Lie algebra is $\mathfrak{g}$, and any homomorphism of Lie algebras $\mathfrak{g} \to \mathfrak{g}'$, where $\mathfrak{g}'$ is the Lie algebra of a group G', arises from a homomorphism of groups $G \to G'$.

2.2 Complex groups

If a Lie group G is a complex manifold and the composition law $G \times G \to G$ is a holomorphic map then G is called a complex Lie group.

The most obvious example is $GL_n(\mathbb{C})$, the group of all invertible $n \times n$ complex matrices. The Lie algebra of a complex Lie group is a complex vector space, and the bracket $\mathfrak{g} \times \mathfrak{g} \to \mathfrak{g}$ is complex-bilinear. Conversely, such a complex Lie algebra always arises from a complex Lie group.

Any Lie algebra $\mathfrak{g}$ has a complexification $\mathfrak{g}_\mathbb{C} = \mathfrak{g} \otimes_\mathbb{R} \mathbb{C}$. If $\mathfrak{g}$ is the Lie algebra of a group G then a complex group $G_\mathbb{C}$ corresponding to $\mathfrak{g}_\mathbb{C}$ which contains G as a subgroup is called a *complexification* of G. Such a complexification need not exist.

Example. The group $SL_2(\mathbb{R})$ of real 2×2 matrices of determinant one has the complexification $SL_2(\mathbb{C})$. Now $SL_2(\mathbb{C})$ is simply connected, while the fundamental group of $SL_2(\mathbb{R})$ is $\mathbb{Z}$. (A loop in $SL_2(\mathbb{R})$ has a winding number which is the winding number of its first column, which is a non-zero vector in $\mathbb{R}^2$.) Thus there is a simply connected covering group G of $SL_2(\mathbb{R})$ such that the kernel of $G \to SL_2(\mathbb{R})$ is $\mathbb{Z}$. If G possessed a complexification it would necessarily be a covering group of $SL_2(\mathbb{C})$, which is impossible because $SL_2(\mathbb{C})$ is simply connected. We can express the same thing by saying that the kernel of any homomorphism from G to a complex group must contain the subgroup $\mathbb{Z}$. (In particular, G cannot possess any faithful finite dimensional representation $G \to GL_n(\mathbb{C})$.)

If G is a compact group, however, then it does possess a complexification $G_\mathbb{C}$. For G can be embedded in a unitary group U_n, and $G_\mathbb{C}$ can be realized as a subgroup of the complexification $GL_n(\mathbb{C})$ of U_n. This group $G_\mathbb{C}$ is unique up to isomorphism, and we shall always refer to it as *the* complexification of G. Thus the complexification of $\mathbb{T}$ will always mean $\mathbb{C}^\times$: other possible complexifications such as $\mathbb{C}/\mathbb{Z}^2 \cong \mathbb{T} \times \mathbb{T}$ cannot arise as complex subgroups of a general linear group.

2.3 Compact Lie groups

The most important fact about compact Lie groups is the Peter–Weyl theorem, which is essentially the assertion that any compact Lie group is isomorphic to a subgroup of some unitary group U_n. This is deduced fairly easily from the most basic property of compact groups, the existence of Haar measure, a probability measure on the group which is invariant under both left and right translations.

A more obvious application of the existence of Haar measure is to show that when a compact group G acts linearly on a finite dimensional vector space V there is always a positive definite inner product on V which is invariant under G. (One can take an arbitrary inner product on V and make it invariant by averaging it with respect to the action of G.) The existence of an invariant inner product, in turn, implies that V is the orthogonal direct sum of subspaces on which G acts irreducibly.

Let us apply the preceding remark when V is the Lie algebra $\mathfrak{g}$ of G. The group G acts on $\mathfrak{g}$ by the *adjoint representation*: the adjoint action of $g \in G$ is defined as the derivative of the map $x \mapsto gxg^{-1}$ at the identity element $x = 1$. We find

$$\mathfrak{g} = \mathfrak{g}_1 \oplus \ldots \oplus \mathfrak{g}_k, \tag{2.3.1}$$

where G acts irreducibly on each $\mathfrak{g}_i$. It is immediate that the $\mathfrak{g}_i$ are sub-Lie-algebras, and that $[\mathfrak{g}_i, \mathfrak{g}_j] = 0$ when $i \neq j$. If G_i is the subgroup of G corresponding to $\mathfrak{g}_i$ then G is locally isomorphic to the product $G_1 \times \ldots \times G_k$.

The groups G_i into which we have decomposed G clearly have no non-trivial connected normal subgroups. Apart from the circle group $\mathbb{T}$ such groups are usually called *simple* groups—although the terminology is not ideal as the groups can possess finite normal subgroups (necessarily contained in the centre). Thus any compact Lie group is locally isomorphic to a product of circles and simple groups. If there are no circles in the decomposition then the group is called *semisimple*.

The simple compact groups have been classified. They consist of the special unitary groups SU_n, the special orthogonal groups SO_n, the symplectic groups Sp_n, and five exceptional groups—and, of course, groups locally isomorphic to these. The traditional notation for the Lie algebras is: $\mathfrak{su}_n = A_{n-1}$, $\mathfrak{so}_{2n+1} = B_n$, $\mathfrak{sp}_n = C_n$, and $\mathfrak{so}_{2n} = D_n$. The exceptional groups are called G_2, F_4, E_6, E_7, and E_8. In all cases the subscript gives the rank (see Section 2.4).

In Chapter Four we shall need to use the following simple result whose proof can conveniently be given here.

Proposition (2.3.2). *If $\mathfrak{g}$ is the Lie algebra of a compact semisimple group G then any $\mathbb{C}$-bilinear G-invariant map*

$$B: \mathfrak{g}_\mathbb{C} \times \mathfrak{g}_\mathbb{C} \to \mathbb{C}$$

is necessarily symmetric.

Proof. Such a map B can be equivalently regarded as a $\mathbb{C}$-linear map $\tilde{B}$ from $\mathfrak{g}_\mathbb{C}$ to its dual $\mathfrak{g}_\mathbb{C}^*$ which commutes with the action of G.

First suppose G is simple. Then $\mathfrak{g}_\mathbb{C}$ is irreducible as a complex representation of G, so by Schur's lemma ([1] 3.22) any two choices of $\tilde{B}$ differ only by multiplication by a complex number. On the other hand there is a choice of $\tilde{B}$ which corresponds to a symmetric map B, for $\mathfrak{g}$ possesses an invariant inner product. It follows that any choice of B is symmetric.

In general we can decompose $\mathfrak{g}$ as in (2.3.1). The factors $\mathfrak{g}_i$ are obviously all non-isomorphic as representations of G, and the same applies to their complexifications. So by Schur's lemma the map

$$\tilde{B}: \oplus \mathfrak{g}_{i,\mathbb{C}} \to \oplus \mathfrak{g}_{i,\mathbb{C}}^*$$

must be of the form $\oplus \tilde{B}_i$, where $\tilde{B}_i: \mathfrak{g}_{i,\mathbb{C}} \to \mathfrak{g}_{i,\mathbb{C}}^*$. By the preceding argument each $\tilde{B}_i$ must be symmetric, and so $\tilde{B}$ is symmetric.

2.4 The root system

In studying the structure of a compact connected Lie group G one begins by choosing a maximal torus T. Any maximal connected abelian subgroup of G is necessarily a torus, i.e. a product $\mathbb{T}^\ell$ of circles, and any two such subgroups are conjugate in G (cf. [1] 2.32, 4.22). The dimension ℓ of a maximal torus is called the *rank* of G. In the case of the unitary group U_n the diagonal matrices form a maximal torus $\mathbb{T}^n$. (In this case it is elementary that any two maximal tori are conjugate, as any set of commuting matrices can be simultaneously diagonalized. The general case, however, is less simple.)

The group G acts linearly on its Lie algebra $\mathfrak{g}$ by the adjoint representation; this action induces, of course, a complex-linear action of G on $\mathfrak{g}_\mathbb{C}$. The crucial step in analysing the structure of G is to decompose the vector space $\mathfrak{g}_\mathbb{C}$ under the action of the maximal torus T. Any complex representation of a compact abelian group such as T breaks up as a direct sum of one-dimensional representations on each of which T acts by means of a homomorphism

$$\alpha: T \to \mathbb{T} \subset \mathbb{C}^\times.$$

Of course T acts trivially on its own complexified Lie algebra $\mathfrak{t}_\mathbb{C}$, but no other vectors in $\mathfrak{g}_\mathbb{C}$ are left fixed by T because T is a maximal abelian subgroup. We can write

$$\mathfrak{g}_\mathbb{C} = \mathfrak{t}_\mathbb{C} \oplus \bigoplus_\alpha \mathfrak{g}_\alpha, \qquad (2.4.1)$$

where $\mathfrak{g}_\alpha$ is the vector subspace of $\mathfrak{g}_\mathbb{C}$ on which T acts by the homomorphism $\alpha: T \to \mathbb{T}$. The homomorphisms α which occur in this decomposition are called the *roots* of G. They form a finite subset of the character group $\hat{T}$ of T. A homomorphism $\alpha: T \to \mathbb{T}$ is determined by its derivative $\dot\alpha$ at the identity, which lies in the vector space $\mathfrak{t}^*$ of linear maps $\mathfrak{t} \to \mathbb{R}$ and is such that

$$\alpha(\exp \xi) = e^{i\dot\alpha(\xi)}.$$

It is usual to think of $\hat{T}$ as a lattice in $\mathfrak{t}^*$, and to write it additively. In other words we shall usually not distinguish in notation between α and $\dot\alpha$. Notice that if α is a root then so is $-\alpha$, for $\mathfrak{g}_{-\alpha} = \bar{\mathfrak{g}}_\alpha$. The lattice $\hat{T}$ is called the lattice of *weights* of G.

In the case of the unitary group U_n the Lie algebra $\mathfrak{g}_{\mathbb{C}}$ consists of all $n \times n$ complex matrices, and the roots α_{ij} are indexed by the ordered pairs (i, j) with $i \neq j$ and $1 \leq i, j \leq n$. The space $\mathfrak{g}_{\alpha_{ij}}$ consists of the matrices which are zero except in the $(i, j)^{\text{th}}$ place, and α_{ij} maps the diagonal matrix with entries $(u_1, \ldots, u_n)$ to $u_i u_j^{-1} \in \mathbb{T}$, and is identified with the linear map $(\xi_1, \ldots, \xi_n) \mapsto \xi_i - \xi_j$ on $\mathbb{R}^n$.

The centre of G is contained in every maximal torus, and is obviously the intersection of the kernels of all the roots $\alpha: T \to \mathbb{T}$. From this it follows that if G is semisimple and so has a finite centre then the roots span the vector space $\mathfrak{t}^*$.

It turns out—cf. [1] 5.5—that the subspaces $\mathfrak{g}_\alpha$ in (2.4.1) are always one-dimensional. This fact enables one to give a simple description of the Lie algebra $\mathfrak{g}_{\mathbb{C}}$ in terms of generators and relations.

Let us choose for each root α a non-zero vector e_α in $\mathfrak{g}_\alpha$. We shall assume that $e_{-\alpha} = \bar{e}_\alpha$. It is easy to see that the bracket

$$h_\alpha = -i[e_\alpha, e_{-\alpha}]$$

belongs to $\mathfrak{t}$, and cannot be zero.

It follows that the three vectors $\{e_\alpha, e_{-\alpha}, h_\alpha\}$ span a sub-Lie-algebra of $\mathfrak{g}_{\mathbb{C}}$ which is isomorphic to the complexification of the Lie algebra of SU_2. If we normalize e_α so that the relations take the form

$$[h_\alpha, e_\alpha] = 2ie_\alpha$$
$$[h_\alpha, e_{-\alpha}] = -2ie_{-\alpha}$$
$$[e_\alpha, e_{-\alpha}] = ih_\alpha,$$

corresponding to those of the matrices

$$e = \begin{pmatrix} 0 & 1 \\ 0 & 0 \end{pmatrix} \quad f = \begin{pmatrix} 0 & 0 \\ -1 & 0 \end{pmatrix} \quad h = \begin{pmatrix} i & 0 \\ 0 & -i \end{pmatrix},$$

then h_α is canonically determined by α, and $2\pi h_\alpha$ belongs to the kernel of the exponential map $\exp: \mathfrak{t} \to T$.

The element h_α is called the *coroot* corresponding to α. For any root $\beta: \mathfrak{t} \to \mathbb{R}$ the number $\beta(h_\alpha)$ is an integer (because $\exp(2\pi h_\alpha) = 1$), and $\alpha(h_\alpha) = 2$. The coroot h_α defines a homomorphism $\eta_\alpha: \mathbb{T} \to T$ by

$$\eta_\alpha(e^{i\theta}) = \exp(\theta h_\alpha); \qquad (2.4.2)$$

this extends canonically to a homomorphism $i_\alpha: SU_2 \to G$. We shall usually think of the lattice $\check{T}$ of all homomorphisms $\mathbb{T} \to T$ as contained in the vector space $\mathfrak{t}$, just as we regard the lattice $\hat{T} = \text{Hom}(T; \mathbb{T})$ as contained in $\mathfrak{t}^*$; in other words we shall usually not distinguish between the coroot h_α and the homomorphism η_α. The lattice $\check{T}$ is canonically dual to $\hat{T}$ over

the integers by the composition
$$\check{T} \times \hat{T} \to \mathrm{Hom}(\mathbb{T}; \mathbb{T}) = \mathbb{Z}.$$
It is an important fact that for a simply connected group G the coroots h_α generate the lattice $\check{T}$. ([1] 5.47)

The Lie algebra $\mathfrak{g}_\mathbb{C}$ is clearly generated additively by the root vectors e_α together with the elements of $\mathfrak{t}$, and the relations are necessarily of the form

$$[e_\alpha, e_\beta] = n_{\alpha\beta} e_{\alpha+\beta} \quad \text{if } \alpha + \beta \text{ is a root,}$$
$$= i h_\alpha \quad \text{if } \alpha + \beta = 0,$$
$$= 0 \quad \text{otherwise;} \tag{2.4.3}$$
$$[h, e_\alpha] = i\alpha(h) e_\alpha.$$

So far the elements e_α have been fixed only up to multiplication by complex numbers of modulus one. It turns out that they can be chosen so that the numbers $n_{\alpha\beta}$ are integers—in fact all $n_{\alpha\beta}$ are non-zero. Cf. [20] Chapter 8 Section 2.4.

2.5 Simply laced groups

There is a class of groups for which the relations (2.4.3) take an especially simple form: they are the *simply laced* groups. A group G is called simply laced if there is a G-invariant inner product $\langle\ ,\ \rangle$ on $\mathfrak{g}$ for which all the coroots h_α have the same length. In that case we shall normalize the inner product so that $\langle h_\alpha, h_\alpha \rangle = 2$. The resulting identification $\mathfrak{t} \cong \mathfrak{t}^*$ makes h_α correspond to α. The unitary group U_n and the orthogonal group SO_{2n} are simply laced: the preferred inner product is

$$\langle A, B \rangle = -\mathrm{trace}(AB)$$

for U_n and

$$\langle A, B \rangle = -\tfrac{1}{2} \mathrm{trace}(AB)$$

for SO_{2n}, when the Lie algebras are identified with the algebras of skew-Hermitian (resp. skew) matrices. The exceptional groups of type E are also simply laced. In general, a compact group is simply laced if its Lie algebra has no simple factors of types B, C, F, or G.

For a simply laced group the inner product on $\mathfrak{t}$ induces an inner product on $\mathfrak{t}^*$ which is integral on the lattice $\hat{T}$:

$$\langle\ \rangle : \hat{T} \times \hat{T} \to \mathbb{Z}.$$

Furthermore $\langle \lambda, \lambda \rangle$ is *even* for each $\lambda \in \hat{T}$. Let us choose a bilinear form

$$B : \hat{T} \times \hat{T} \to \mathbb{Z}/2$$

such that

$$B(\lambda, \lambda) \equiv \tfrac{1}{2}\langle \lambda, \lambda \rangle \quad (\text{mod } 2).$$

(This form B cannot be symmetric.) Then

(i) the roots of G are precisely the set of all vectors α in $\hat{T}$ such that $\langle \alpha, \alpha \rangle = 2$, and

(ii) the first relation in the set (2.4.3) can be taken to be

$$[e_\alpha, e_\beta] = (-1)^{B(\alpha, \beta)} e_{\alpha+\beta}. \qquad (2.5.1)$$

It is easy to see that different choices of B amount simply to changing the signs of some of the e_α.

2.6 The Weyl group and the Weyl chambers: positive roots

For a compact Lie group G with maximal torus T the group of all automorphisms of T which are obtained by conjugating by elements of G is called the *Weyl group* W. Thus $W \cong N(T)/T$, where $N(T)$ is the normalizer of T in G. For the unitary group U_n the Weyl group is the symmetric group S_n, which acts on the diagonal matrices by permuting their entries.

The Weyl group is a finite group of isometries of $\mathfrak{t}$: it preserves the lattice $\check{T}$, and permutes the set of roots in $\hat{T}$. For each root α it contains an element s_α of order two represented by $\exp \tfrac{1}{2}\pi(e_\alpha + e_{-\alpha})$ in $N(T)$. The action of s_α on $\mathfrak{t}$ is given by

$$s_\alpha(\xi) = \xi - \alpha(\xi) h_\alpha; \qquad (2.6.1)$$

it is the reflection in the hyperplane H_α of $\mathfrak{t}$ whose equation is $\alpha(\xi) = 0$. The reflections s_α generate W.

The elements of $\mathfrak{t}$ which do not belong to any of the root hyperplanes H_α are called *regular*. They fall into a number of connected components, called *Weyl chambers*, which are permuted simply transitively by W. It is customary to select one of the chambers C and call it the *positive* Weyl chamber. Then the roots α are classified as positive or negative according as they take positive or negative values on C, and a positive root α is called *simple* if the hyperplane H_α is a wall of C. For a semisimple group of rank ℓ there are ℓ simple roots $\alpha_1, \ldots, \alpha_\ell$, and the 3ℓ elements e_{α_i}, $e_{-\alpha_i}$, h_{α_i} generate the Lie algebra $\mathfrak{g}_\mathbb{C}$.

In the case of U_n we can take the positive roots to be the α_{ij} with $i < j$, and the simple roots to be $\alpha_{i,i+1}$ for $1 \leq i < n$.

2.7 Irreducible representations and antidominant weights

Every irreducible representation of a compact group is finite dimensional. If G acts on a finite dimensional complex vector space V then one can find a basis $\{\varepsilon_i\}$ of V with respect to which the operation of the maximal torus T is diagonal. The torus then acts on ε_i by a homomorphism $\lambda_i : T \to \mathbb{T}$ called the *weight* of ε_i. The set of weights is a finite subset of $\hat{T}$ which is invariant under W.

If the representation V is irreducible then it possesses a unique basis vector ε_1 whose weight λ_1 is dominated by the other λ_i's: one says that λ dominates μ if $\lambda - \mu$ is positive on the positive chamber. The association of λ_1 to V defines a one-to-one correspondence between the equivalence classes of irreducible representations of G and the set $\hat{T}_-$ of antidominant weights; a weight $\lambda \in \hat{T}$ is called *antidominant* if λ is dominated by $w \cdot \lambda$ for each $w \in W$, or equivalently (from (2.6.1)) if $\lambda(h_\alpha) \leq 0$ for each positive root α. We can identify $\hat{T}_-$ with the set $\hat{T}/W$ of orbits of W on $\hat{T}$.

It is with considerable hesitation that we have decided, with loop groups in mind, to describe representations in terms of *lowest* weights rather than highest weights as is usual. That has led us to use the unattractive compromise term 'antidominant'. Of course λ is antidominant if and only if $-\lambda$ is dominant in the usual sense.

One method of associating an irreducible representation V_λ to a weight $\lambda \in \hat{T}_-$ is described in Section 2.9.

2.8 Complex homogeneous spaces

Much of our study of loop groups will be based on the consideration of their complex homogeneous spaces. We shall outline here the main facts about the complex homogeneous spaces of compact groups, beginning with the unitary group U_n.

The complex algebraic homogeneous spaces for U_n are the Grassmannians and flag manifolds. For each ordered partition $\mathbf{k}$ of n, i.e. $\mathbf{k} = (k_1, k_2, \ldots, k_r)$ with $k_i > 0$ and $\Sigma k_i = n$, we define the flag manifold $Fl_\mathbf{k}$ as the space of r-tuples $\mathbf{E} = (E_1, \ldots, E_r)$ of subspaces of $\mathbb{C}^n$ such that $E_1 \subset E_2 \subset \ldots \subset E_r$ and $\dim(E_i) = k_1 + \ldots + k_i$. If $\mathbf{k} = (k, n-k)$ then $Fl_\mathbf{k}$ is the *Grassmannian* $\mathrm{Gr}_k(\mathbb{C}^n)$ of all k-dimensional subspaces of $\mathbb{C}^n$. If $\mathbf{k} = (1, 1, \ldots, 1)$ we shall write $Fl(\mathbb{C}^n)$ for $Fl_\mathbf{k}$.

The space $Fl_\mathbf{k}$ is a homogeneous space under the action of U_n, and the isotropy group of its natural base-point—the flag $\mathbf{E}$ such that E_i is the subspace $\mathbb{C}^{k_1 + \ldots + k_i}$ spanned by the first $k_1 + \ldots + k_i$ vectors of the standard basis of $\mathbb{C}^n$—is $U_\mathbf{k} = U_{k_1} \times \ldots \times U_{k_r} \subset U_n$. So $Fl_\mathbf{k}$ can be identified with $U_n/U_\mathbf{k}$. On the other hand $Fl_\mathbf{k}$ is also a complex algebraic variety, and is a homogeneous space of the complex group $GL_n(\mathbb{C})$. Thus

$$Fl_\mathbf{k} \cong U_n/U_\mathbf{k} \cong GL_n(\mathbb{C})/P_\mathbf{k},$$

where $P_\mathbf{k}$ is the group of upper echelon matrices of type $\mathbf{k}$. In particular

$Fl(\mathbb{C}^n) \cong U_n/T \cong GL_n(\mathbb{C})/B^+$, where T is the standard torus of U_n and B^+ is the group of upper triangular matrices.

The spaces $Fl_\mathbf{k}$ are, up to isomorphism, the only homogeneous spaces of U_n which are complex algebraic varieties, and they are the only compact homogeneous spaces of $GL_n(\mathbb{C})$. The subgroups $P_\mathbf{k}$ are the only subgroups of $GL_n(\mathbb{C})$ which contain B^+. One of the important properties of $Fl_\mathbf{k}$ is that it possesses a canonical decomposition into complex cells, i.e. subspaces isomorphic to some $\mathbb{C}^r$. The cells are simply the orbits of B^+ on $Fl_\mathbf{k}$. Their closures are usually called 'Schubert varieties' ([116] §6, [68] p. 196). For example, $Gr_k(\mathbb{C}^n)$ is the union of $\binom{n}{k}$ cells $C_\mathbf{m}$ indexed by the sequences $\mathbf{m} = (m_1, \ldots, m_k)$ such that $1 \leq m_1 < m_2 < \ldots < m_k \leq n$. In fact

$$C_\mathbf{m} = \{W \subset \mathbb{C}^n : \dim(W \cap \mathbb{C}^j) = i \text{ when } m_i \leq j < m_{i+1}\}, \quad (2.8.1)$$

and it has dimension $\Sigma (m_i - i)$.

The situation just described has a precise analogue for any compact Lie group G. The subgroup of the complexification $G_\mathbb{C}$ which plays the role of the upper triangular matrices is the standard *Borel subgroup* B^+ whose Lie algebra is spanned by $\mathbf{t}_\mathbb{C}$ and the root vectors e_α corresponding to the positive roots α. We have $B^+ \cap G = T$, and $G/T \cong G_\mathbb{C}/B^+$.

Proposition (2.8.2). *There is a one-to-one correspondence between*
  (i) *complex algebraic homogeneous spaces for G,*
  (ii) *compact Kähler homogeneous spaces for $G_\mathbb{C}$,*
  (iii) *subgroups of $G_\mathbb{C}$ containing B^+, and*
  (iv) *subsets of the set of simple roots of G.*

For the proof see Wang [152] and Serre [133], and also [20] Chapter 4 Section 2.5. To a subset A of the set of simple roots there corresponds the homogeneous space $G_\mathbb{C}/P_A$, where P_A is the subgroup of $G_\mathbb{C}$ whose Lie algebra is generated by the Lie algebra of B^+ and by the elements $e_{-\alpha}$ for $\alpha \in A$.

Subgroups of $G_\mathbb{C}$ which are conjugate to one of the P_A are called *parabolic*. We have $G_\mathbb{C}/P_A = G/(P_A \cap G)$, and each such space has a canonical decomposition into complex cells which are the orbits of B^+: this is called the *Bruhat decomposition* ([20] Chapter 6 §2, [72] Chapter 9 §1).

In the case of $G_\mathbb{C}/B^+ = G/T$ one can think of the Weyl group $W = N(T)/T$ as a subset of G/T, and there is exactly one element of W in each cell: in other words the cell decomposition of G/T is $\{C_w\}_{w \in W}$, where $C_w = B^+ w$. The dimension of C_w is the *length* of w, which is defined as the number of positive roots α such that $w \cdot \alpha$ is negative.

There is also a dual cell decomposition of G/T provided by the orbits of the opposite Borel subgroup B^-, which is the complex conjugate† of B^+. The cells B^+w and B^-w have complementary dimensions, and intersect transversally in the single point w.

When $G = O_{2n}$ the space G/T has a description as a flag manifold like that for U_n/T. It consists of all flags $E_1 \subset E_2 \subset \ldots \subset E_n \subset \mathbb{C}^{2n}$ such that $\dim(E_i) = i$ and each E_i is *isotropic* for the standard bilinear form on $\mathbb{C}^{2n}$.

2.9 The Borel–Weil theorem

The importance of the complex homogeneous spaces of G arises from their role in constructing the irreducible representations. In fact only the largest one $G/T \cong G_\mathbb{C}/B^+$ is needed. Every homomorphism $\lambda: T \to \mathbb{T}$ extends uniquely to a holomorphic homomorphism $\lambda: B^+ \to \mathbb{C}^\times$. It therefore defines a homogeneous holomorphic line bundle $L_\lambda = G_\mathbb{C} \times_{B^+} \mathbb{C}_\lambda$ on $G_\mathbb{C}/B^+$. (The notation $G_\mathbb{C} \times_{B^+} \mathbb{C}_\lambda$ means the quotient of $G_\mathbb{C} \times \mathbb{C}$ by the equivalence relation which identifies (gb, ξ) with $(g, \lambda(b)\xi)$ for all $b \in B^+$.) The group $G_\mathbb{C}$ acts on the line bundle L_λ, and hence acts on its holomorphic cross-sections.

The Borel–Weil theorem (cf. Bott [15]) is

Theorem (2.9.1).
 (i) *The line bundle L_λ has no non-zero holomorphic sections unless λ is an antidominant weight.*
 (ii) *If λ is an antidominant weight then the space of holomorphic sections of L_λ is an irreducible representation of G with lowest weight λ.*

It may be worth explaining briefly why the space Γ_λ of holomorphic sections of L_λ is an irreducible representation. We first observe that if Γ_λ is expressed as a sum of irreducible representations then each component contains an element of lowest weight. Now an element of lowest weight is invariant under the subgroup N^- whose Lie algebra is spanned by the e_α with $\alpha < 0$ (for acting on it with such an e_α would give an element of lower weight). So it is enough to show that L_λ cannot have two linearly independent N^--invariant sections. But N^- acts on the base $G_\mathbb{C}/B^+$ with an open dense orbit—the orbit of the base-point. So if s_1 and s_2 were two N^--invariant sections their ratio would have to be constant on the open orbit, and so it would be constant on all of $G_\mathbb{C}/B^+$.

We should also mention the relation between holomorphic line bundles on a manifold X and holomorphic maps from X to complex projective space.

---

† 'Complex conjugation' here means the involution of $G_\mathbb{C}$ whose fixed points are G: thus for $GL_n(\mathbb{C})$ it means $A \mapsto (\bar{A}')^{-1}$.

In one direction, suppose we have a holomorphic map $f:X\to P(V)$, where $P(V)$ denotes the projective space of rays in a vector space V. Then we can define a holomorphic line bundle L_f on X whose fibre at x is the line $f(x)$ of V. Thus L_f is a subspace of $X\times V$, and there is a map $\pi:L_f\to V$ which is linear on each fibre. A linear form $\alpha:V\to\mathbb{C}$ therefore defines by composition with π a section of the dual line bundle L_f^*, whose fibre at x is the dual of the fibre of L_f at x, and so we have a linear map $V^*\to\Gamma(L_f^*)$, where $\Gamma(L_f^*)$ denotes the space of holomorphic sections of L_f^*.

In the other direction, suppose that L is a line bundle on X, and suppose that for each $x\in X$ there is a section of L which does not vanish at x. Then there is a canonical map $f_L:X\to P(\Gamma^*)$, where Γ is the space of sections of L. One defines $f_L(x)$ as the map $\Gamma\to\mathbb{C}$ given by evaluating sections at the point x—one must choose an identification of the fibre of L at x with $\mathbb{C}$, but the choice affects $f_L(x)$ only up to multiplication by an element of $\mathbb{C}^\times$.

In the light of this reinterpretation the proof of another part of the Borel–Weil theorem is almost obvious. To prove that an irreducible representation of $G_\mathbb{C}$ on V arises as the sections of a line bundle on $G_\mathbb{C}/B^+$ it is enough to show that there is a ray Ω in V^* which is stable under B^+. For then considering the orbit of Ω gives us an equivariant map $f:G_\mathbb{C}/B^+\to P(V^*)$, and hence a line bundle L_f and a map $V\to\Gamma(L_f^*)$. Any vector of highest weight in V^* defines a ray stable under B^+.

Example. As an example of the Borel–Weil theorem let us consider the irreducible representation of U_n on the kth exterior power $\Lambda^k(\mathbb{C}^n)$.

Perhaps the most obvious of all holomorphic line bundles is the determinant bundle Det on the Grassmannian $\text{Gr}_k(V)$ of k-dimensional subspaces of a finite dimensional vector space V. This is the bundle whose fibre at a subspace $W\subset V$ is the top exterior power $\Lambda^k(W)$. It has no non-zero holomorphic sections, but its dual Det*, whose fibre at W is the dual line $\Lambda^k(W)^*$, does possess sections. The following well-known fact will be crucial for us in Chapter 10, so we shall give a proof of it here.

Proposition (2.9.2). *The space of holomorphic sections of* Det* *on* $\text{Gr}_k(V)$ *is naturally isomorphic to* $\Lambda^k(V^*)$.

Proof. A holomorphic section of Det* is the same thing as a holomorphic map $s:\text{Det}\to\mathbb{C}$ which is linear on each fibre. A typical point of Det can be represented in the form $\lambda v_1\wedge\ldots\wedge v_k$, where $\lambda\in\mathbb{C}$ and $\{v_1,\ldots,v_k\}$ is a basis for some $W\in\text{Gr}_k(V)$. We can therefore define a

2.9 THE BOREL-WEIL THEOREM

map
$$\Lambda^k(V^*) \to \Gamma(\text{Det}^*) \qquad (2.9.3)$$
by
$$\alpha_1 \wedge \ldots \wedge \alpha_k \mapsto \{\lambda v_1 \wedge \ldots \wedge v_k \mapsto \langle \alpha_1 \wedge \ldots \wedge \alpha_k, \lambda v_1 \wedge \ldots \wedge v_k \rangle\},$$
where $\langle \alpha_1 \wedge \ldots \wedge \alpha_k, v_1 \wedge \ldots \wedge v_k \rangle$ is the determinant of the matrix $(\langle \alpha_i, v_j \rangle)$.

It is clear that the map (2.9.3) is injective. To prove it is surjective, let U denote the open subspace of V^k consisting of k-tuples of linearly independent vectors. There is a natural map $\pi : U \to \text{Det}$. If $s : \text{Det} \to \mathbb{C}$ is a section of Det* then what we must show is that the composite $s \circ \pi$ extends to a multilinear map $V^k \to \mathbb{C}$. To prove that, consider $s(\pi(v_1, v_2, \ldots, v_k))$ as a function of v_1 for fixed $v_2, \ldots, v_k$. The resulting holomorphic function f is defined on the complement of the subspace $\langle v_2, \ldots, v_k \rangle$ of V, and satisfies $f(\lambda v_1) = \lambda f(v_1)$ for every $\lambda \in \mathbb{C}^\times$. The subspace $\langle v_2, \ldots, v_k \rangle$ has codimension greater than 1 (for we can assume that $k < n$), and so f extends to a holomorphic function on all of V by Hartogs's theorem ([68] page 7). We can then expand f in a Taylor's series at the origin, and because of the condition $f(\lambda v_1) = \lambda f(v_1)$ we find that f must be a linear function on V. Treating the other variables in the same way shows that $s \circ \pi$ is multilinear.

3
GROUPS OF SMOOTH MAPS

3.1 Infinite dimensional manifolds

Before discussing infinite dimensional Lie groups we must make clear what we mean by an infinite dimensional smooth manifold, if only to emphasize that there is nothing esoteric involved in the idea. For an excellent short treatment of the subject we refer the reader to Milnor [115]. We shall follow his approach closely. A more detailed account can be found in Hamilton [70].

The manifolds we consider will be paracompact topological spaces X 'modelled on' some topological vector space E, in the sense that X is covered by an atlas of open sets $\{U_\alpha\}$ each of which is homeomorphic to an open set E_α of E by a given homeomorphism $\phi_\alpha : U_\alpha \to E_\alpha$. The vector space E will always be locally convex and complete. The transition functions between charts

$$\phi_\alpha(U_\alpha \cap U_\beta) \xrightarrow{\phi_\alpha^{-1}} U_\alpha \cap U_\beta \xrightarrow{\phi_\beta} \phi_\beta(U_\alpha \cap U_\beta)$$

are assumed to be smooth, i.e. infinitely differentiable. The meaning of 'infinitely differentiable' is as follows.

A map $f : U \to E$, where U is an open set of E, is continuously differentiable (or C^1) if the limit

$$Df(u; v) = \lim_{t \to 0} t^{-1}(f(u + tv) - f(u))$$

exists for all $u \in U$ and $v \in E$, and is continuous as a map $Df : U \times E \to E$. (Of course Df is linear in its second variable.) The second derivative, if it exists, is then the map

$$D^2f : U \times E \times E \to E$$

defined by

$$D^2f(u; v, w) = \lim_{t \to 0} t^{-1}(Df(u + tw; v) - Df(u; v)),$$

and so on.

We shall collect here some remarks about calculus on infinite dimensional manifolds.

3.1 INFINITE DIMENSIONAL MANIFOLDS

Complex manifolds

If E is a complex topological vector space and the transition functions are holomorphic, then we have a *complex* manifold. To say that $f: U \to E$ is holomorphic, where U is an open set of E, means that f is smooth and that $Df: U \times E \to E$ is complex-linear in the second variable.

Differential forms

If U is an open set of E then a differential form of degree p on U is a smooth map
$$U \times \underbrace{E \times \ldots \times E}_{p} \to E$$

which is multilinear and alternating in the last p variables. Differential forms can then be defined in the usual way on a smooth manifold, and the usual definition of the exterior derivative, and the proof of the Poincaré lemma, apply without modification.

In order to make serious use of differential forms, however, one needs to know that for each open covering of the manifold there is a subordinate smooth partition of unity. That is true providing the following two conditions are satisfied.

(I) The vector space E has enough smooth functions, in the sense that for each open set U of E there is a non-vanishing smooth function $E \to \mathbb{R}$ which vanishes outside U.

(II) The manifold is Lindelöf, i.e. each open covering has a countable refinement.

Both of these conditions are satisfied for all the manifolds we shall consider.

De Rham's theorem holds for any manifold X which has smooth partitions of unity. The usual proof applies. In particular, if a cohomology class $c \in H^p(X; \mathbb{R})$ is represented by a Čech cocycle $\{c_{\alpha_0 \ldots \alpha_p}\}$ with respect to an open covering $\{U_\alpha\}$ of X, then c is also represented by the differential form

$$\sum_{\alpha_0, \ldots, \alpha_p} c_{\alpha_0 \ldots \alpha_p} \lambda_{\alpha_0} \, d\lambda_{\alpha_1} \wedge \ldots \wedge d\lambda_{\alpha_p}, \qquad (3.1.1)$$

where $\{\lambda_\alpha\}$ is a partition of unity subordinate to $\{U_\alpha\}$.

Vector fields

There is no difficulty in defining smooth vector fields, or the bracket of two vector fields; and vector fields act as differentiation operators on functions in the usual way. One must beware, however, that vector fields on infinite dimensional manifolds do *not* in general have trajectories. We shall meet an interesting example of this phenomenon when we discuss the gradient flow of the energy function on a loop space in Chapter 8.

3.2 Groups of maps as infinite dimensional Lie groups

An infinite dimensional Lie group is a group Γ which is at the same time an infinite dimensional smooth manifold, and is such that the composition law $\Gamma \times \Gamma \to \Gamma$ and the operation of inversion $\Gamma \to \Gamma$ are given by smooth maps. The tangent space to Γ at the identity element is its Lie algebra, the bracket being defined by identifying tangent vectors at the identity element with left-invariant vector fields on Γ. If for each element ξ of the Lie algebra there is a unique one-parameter subgroup $\gamma_\xi : \mathbb{R} \to \Gamma$ such that $\gamma'_\xi(0) = \xi$, then the exponential map is defined. This is the case in all known examples.

For infinite dimensional Lie groups modelled on Banach spaces there is a well-developed theory ([20] Chapter 3) which is closely parallel to the theory of finite dimensional Lie groups. For groups modelled on more general topological vector spaces there is no such theory, and most of the standard theorems about Lie groups do not hold. We shall meet interesting examples of Lie algebras which do not correspond to any Lie group and of Lie groups whose exponential maps are not locally bijective. We hope that it will emerge all the same that the concept of a general infinite dimensional Lie group is a useful one.

Probably the simplest and most immediate example of an infinite dimensional Lie group is the group $\text{Map}_{\text{cts}}(X; G)$ of all continuous maps from a compact space X to a finite dimensional Lie group G. (The group law, of course, is pointwise composition in G.) The natural topology on $\text{Map}_{\text{cts}}(X; G)$ is the topology of uniform convergence. We see that it is a smooth manifold as follows.

If U is an open neighbourhood of the identity element in G which is homeomorphic by the exponential map to an open set $\check{U}$ of the Lie algebra $\mathfrak{g}$ of G, then $\mathcal{U} = \text{Map}_{\text{cts}}(X; U)$ is an open neighbourhood of the identity in $\text{Map}_{\text{cts}}(X; G)$ which is homeomorphic to the open set $\check{\mathcal{U}} = \text{Map}_{\text{cts}}(X; \check{U})$ of the Banach space $\text{Map}_{\text{cts}}(X; \mathfrak{g})$. If f is any element of $\text{Map}_{\text{cts}}(X; G)$, then $\mathcal{U}_f = \mathcal{U}.f$ is a neighbourhood of f which is also homeomorphic to $\check{\mathcal{U}}$. The sets $\mathcal{U}_f$ provide an atlas which makes $\text{Map}_{\text{cts}}(X; G)$ into a smooth manifold, and in fact into a Lie group: there is no difficulty at all in checking that the transition functions are smooth, or that multiplication and inversion are smooth maps.

In this book, however, we shall be concerned not with groups of continuous maps but with groups of smooth maps.

Suppose now that X is a finite dimensional compact smooth manifold, and let $\text{Map}(X; G)$ denote the group of *smooth* maps $X \to G$. The case we are primarily interested in is when X is the circle S^1; then $\text{Map}(X; G)$ is the *loop group* of G, which is denoted by LG. We shall think of the circle as consisting interchangeably of real numbers θ modulo 2π or of complex numbers $z = e^{i\theta}$ of modulus one.

3.2 GROUPS OF MAPS

Defining the atlas $\{\mathcal{U}_f\}$ for $\mathrm{Map}(X; G)$ just as in the continuous case, we find that the set $\tilde{\mathcal{U}}$ is an open set in the vector space $E = \mathrm{Map}(X; \mathfrak{g})$ of all smooth maps $X \to \mathfrak{g}$. The simplest way to define the topology of $\mathrm{Map}(X; G)$ is to prescribe that the sets $\mathcal{U}_f$ are open and homeomorphic to the open set $\tilde{\mathcal{U}}$ of E. The standard topology on E is the topology of uniform convergence of the functions and all their partial derivatives of all orders [70]. It makes E into a complete separable metrizable topological vector space, but not a Banach space. We shall not describe it in detail here. But when X is the circle the convergence of a sequence $\{f_k\}$ in E to f means that $(d^n f_k/d\theta^n)$ converges uniformly to $d^n f/d\theta^n$ for each n. Again there is no difficulty in seeing that $\mathrm{Map}(X; G)$ is an infinite dimensional Lie group.

For most of the purposes of this book it would make no difference if we considered, instead of smooth maps, maps of a given finite degree r of differentiability. $\mathrm{Map}(X; G)$ would then be a Banach Lie group. (We should have to interpret C^r maps in the Sobolev sense [144], otherwise the manifold would not have enough smooth functions.) No practical advantage would be gained by the change, however, and so we shall keep to smooth maps, which seem aesthetically more appealing. Thus $\mathrm{Map}(X; G)$ will always denote smooth maps, and LG will denote the smooth loop group. In the case of diffeomorphism groups, as we shall see, there is no choice but to work with smooth maps.

The Lie algebra of $\mathrm{Map}(X; G)$ is obviously $\mathrm{Map}(X; \mathfrak{g})$, and the exponential map

$$\exp : \mathrm{Map}(X; \mathfrak{g}) \to \mathrm{Map}(X; G)$$

is defined, and is a local homeomorphism near the identity. One of our themes is that the loop group of a compact group G behaves surprisingly like a compact group itself, but we shall begin by pointing out a slight difference. In a compact group G every element in the identity component G^0 lies on a one-parameter subgroup, i.e. the exponential map $\mathfrak{g} \to G^0$ is surjective. This property is not inherited by $\mathrm{Map}(X; G)$.

Example. Consider LG, where $G = SU_2$. Then G is simply connected, so LG is connected. The element γ of LG defined by

$$z \mapsto \begin{pmatrix} z & 0 \\ 0 & z^{-1} \end{pmatrix}$$

does not lie on any one-parameter subgroup. For if γ is $\exp(\xi)$ for some $\xi \in L\mathfrak{g}$ then ξ must commute with γ and hence must be diagonal: but there is no smooth function θ on the circle such that $e^{i\theta} = z$. Notice that this example is precisely analogous to the non-surjectivity of exp for finite dimensional non-compact groups: the element

$$\begin{pmatrix} -2 & 0 \\ 0 & -\tfrac{1}{2} \end{pmatrix}$$

of $SL_2(\mathbb{R})$ does not lie on a one-parameter subgroup. It is easy to see, however, that when G is compact the image of the exponential map is dense in the identity component of LG. That is not true in groups like $SL_2(\mathbb{R})$.

Another obvious but important remark about groups of maps is that when G has a complexification $G_\mathbb{C}$ then Map$(X; G)$ has the complexification Map$(X; G_\mathbb{C})$. It is clear that the last group is a complex Lie group.

3.3 Diffeomorphism groups

The group of diffeomorphisms of the circle will play only a peripheral role in this book, but it is a very interesting example of an infinite dimensional Lie group.

First let us remark that for any finite dimensional compact smooth manifold X the group Diff(X) of all smooth diffeomorphisms $X \to X$ is a Lie group (cf. [70], [115]). Its Lie algebra is the vector space Vect(X) of all smooth vector fields on X, with the usual bracket operation, and the exponential map

$$\exp: \text{Vect}(X) \to \text{Diff}(X)$$

assigns to a vector field the unique flow that it generates. For a finite k, however, the group of k-times continuously differentiable diffeomorphisms obviously does *not* form a Lie group, for left translation is not a differentiable map. (Still more obviously, the bracket of two vector fields of class C^k is only of class C^{k-1}.)

Although the exponential map is defined for Diff(X), it is far from being a local homeomorphism. There are diffeomorphisms arbitrarily close to the identity which are not on any one-parameter subgroup, and others which are on many. The following discussion is based on Omori [120]. (Cf. also Milnor [115].)

Proposition (3.3.1). *The map* exp: Vect$(S^1) \to$ Diff(S^1) *is neither locally one-to-one nor locally surjective.*

Proof.

(i) Consider the rotation $R_{2\pi/n}$ through the angle $2\pi/n$. This belongs to the subgroup $\mathbb{T}$ of all rigid rotations in Diff(S^1). The centralizer of $R_{2\pi/n}$ is the subgroup H of all diffeomorphisms which are periodic with period $2\pi/n$. So $R_{2\pi/n}$ lies on all of the one-parameter subgroups $\phi \mathbb{T} \phi^{-1}$ for $\phi \in H$. This shows that exp is not locally one-to-one.

(ii) In seeing that exp is not locally surjective, the essential point is that a one-parameter subgroup of Diff(S^1) which has no stationary points is conjugate to the subgroup $\mathbb{T}$. Granting this, and observing that a diffeomorphism conjugate to a rotation has no fixed points unless it is the

3.3 DIFFEOMORPHISM GROUPS

identity, we see that a diffeomorphism cannot be on a one-parameter subgroup if
  (a) it has no fixed point,
  (b) it has a point of finite order n, and
  (c) it is not of order n.
Such diffeomorphisms ϕ are very plentiful, and can be chosen arbitrarily close to the identity. For example one can define $\phi(\theta) = \theta + \pi$ for $0 \leq \theta \leq \pi$, and then extend ϕ to the remainder of the circle in any way at all which does not make $\phi = R_\pi$. Alternatively, one can define

$$\phi(\theta) = \theta + \frac{2\pi}{n} + \varepsilon \sin n\theta,$$

where ε is small. Then $\phi^n(0) = 0$, but ϕ^n is not the identity, as its derivative there is $(1 + n\varepsilon)^n$.

To see that any one-parameter subgroup with no stationary points is conjugate to a group of rigid rotations it is enough to observe that any nowhere-vanishing vector field $v(\theta)\, d/d\theta$ can be conjugated to a constant vector field. The conjugating diffeomorphism ψ is given by

$$\psi(\theta) = k \int_0^\theta v(t)^{-1}\, dt,$$

where k is chosen so that $\psi(2\pi) = 2\pi$.

Before leaving diffeomorphism groups we should point out another way in which they differ from the loop groups. The complexification of the Lie algebra $\text{Vect}(X)$ does not correspond to any Lie group. This is intuitively unsurprising, for complex vector fields on S^1 generate paths in the space of maps $S^1 \to \mathbb{C}$, and these do not form a group. A proof that there is no Lie group corresponding to $\text{Vect}_\mathbb{C}(S^1)$ can be given as follows.

Proposition (3.3.2). *Any homomorphism from* $\text{Diff}^+(S^1)$ *to a complex Lie group is trivial.*

Proof. The group $PSL_2(\mathbb{R})$ is contained in $\text{Diff}^+(S^1)$, for S^1 can be regarded as the real projective line. Consider the n-fold covering map $\pi : S^1 \to S^1$ given by $z \mapsto z^n$. Let G_n denote the group of diffeomorphisms ϕ which are n-fold coverings of elements $\psi \in PSL_2(\mathbb{R})$, i.e. such that $\pi \circ \phi = \psi \circ \pi$. It is easy to see that G_n is isomorphic to the n-fold covering group of $PSL_2(\mathbb{R})$: its centre consists of the rotations $R_{2\pi k/n}$ for $k = 0, 1, \ldots, n-1$. But we have pointed out in Section 2.2 that any homomorphism from G_n into a complex Lie group must factorize through $SL_2(\mathbb{R})$ or $PSL_2(\mathbb{R})$. The kernel of any homomorphism from $\text{Diff}^+(S^1)$ into a complex group must therefore contain all rotations through $2\pi k/n$ with n odd, and so must contain all rotations. Being a normal subgroup, the kernel is therefore the whole of $\text{Diff}^+(S^1)$, in view of the following result.

Proposition (3.3.3). $\text{Diff}^+(S^1)$ *is a simple group.*

The proof of this result, due to Herman [74], is surprisingly difficult, and we shall not give it here.

3.4 Some group-theoretic properties of Map(X; G)

In this section G will be a compact connected Lie group, and X will be a compact smooth manifold. For brevity we shall write MG for the group of smooth maps $\text{Map}(X; G)$.

If G is semisimple then it is perfect, i.e. equal to its commutator subgroup $[G, G]$. We shall show that then the identity component $M_0 G$ is also perfect. (We cannot expect MG itself to be perfect: for example in the case of LG the group of connected components is the fundamental group $\pi_1(G)$, which is abelian.)

Proposition (3.4.1). *If G is semisimple then $M_0 G$ is perfect, and in fact $[G, M_0 G] = M_0 G$.*

Proof. Let us first consider the case $G = SU_2$. If

$$e_1 = \begin{pmatrix} 0 & 1 \\ -1 & 0 \end{pmatrix}, \quad e_2 = \begin{pmatrix} 0 & i \\ i & 0 \end{pmatrix}, \quad e_3 = \begin{pmatrix} i & 0 \\ 0 & -i \end{pmatrix}$$

is the usual basis for the Lie algebra of G and $\mathbb{T}_1, \mathbb{T}_2, \mathbb{T}_3$ are the circle subgroups they generate, then the multiplication $\mathbb{T}_1 \times \mathbb{T}_2 \times \mathbb{T}_3 \to G$ is surjective. The multiplication

$$M\mathbb{T}_1 \times M\mathbb{T}_2 \times M\mathbb{T}_3 \to MG$$

is therefore surjective in a neighbourhood of the identity; and so it is enough (because the subgroups $\mathbb{T}_1, \mathbb{T}_2, \mathbb{T}_3$ are conjugate) to prove that every element of the identity component of $M\mathbb{T}_3$ belongs to $[G, M_0 G]$. This last statement is true because

$$\begin{pmatrix} \phi & 0 \\ 0 & \phi^{-1} \end{pmatrix} = \left[\begin{pmatrix} 0 & 1 \\ -1 & 0 \end{pmatrix}, \begin{pmatrix} \phi^{-\frac{1}{2}} & 0 \\ 0 & \phi^{+\frac{1}{2}} \end{pmatrix} \right].$$

(The bracket here denotes the group-theoretic commutator $[x, y] = xyx^{-1}y^{-1}$, not the Lie bracket; and $\phi^{\frac{1}{2}}$ is defined because ϕ has winding number zero.)

The result for a general semisimple group G follows at once from the particular case of SU_2. For, as has been explained in Section 2.4, one can find a finite number of homomorphisms $i_1, \ldots, i_n : SU_2 \to G$,

3.4 SOME GROUP THEORETIC PROPERTIES OF MAP(X: G)

corresponding to the positive roots of G, such that the multiplication map

$$\Pi i_k : (SU_2)^n \to G,$$

and hence the induced map $(MSU_2)^n \to MG$, is locally surjective.

We shall now discuss the group of automorphisms of MG.

The group of diffeomorphisms of X acts on MG as a group of automorphisms. Apart from that, there are obvious pointwise automorphisms of MG arising from smooth maps $X \to A$, where A is the group of automorphisms of G. If G is simple there are essentially no others.

Proposition (3.4.2). *If G is simple then the group of automorphisms of M_0G is the semidirect product* $\text{Diff}(X) \tilde{\times} MA$.

Proof. Suppose that $\alpha : M_0G \to M_0G$ is an automorphism. Composing it with $\varepsilon_x : MG \to G$, the evaluation map at $x \in X$, gives a homomorphism $\alpha_x : M_0G \to G$. The restriction of α_x to the subgroup G of constant maps in MG must be an automorphism a_x of G, for if it were trivial then α_x would be trivial because $M_0G = [G, M_0G]$. Clearly $x \mapsto a_x$ is an element a of MA, and by replacing α with $a^{-1} \circ \alpha$ we may as well assume that a_x is the identity for each x. Then the crucial step is to see that $\alpha_x = \varepsilon_x \circ \alpha = \varepsilon_y$ for some $y \in X$. For this it is enough to consider the derivative of α_x, which is a homomorphism of Lie algebras $\dot{\alpha}_x : M\mathfrak{g} \to \mathfrak{g}$.

If U is an open set of X, let $M_U\mathfrak{g}$ denote the ideal of $M\mathfrak{g}$ consisting of elements with support in U. Because $\mathfrak{g}$ is simple and $\dot{\alpha}_x$ is surjective, $\dot{\alpha}_x | M_U$ must be either trivial or surjective. It follows that when $\dot{\alpha}_x$ is regarded as a distribution on X its support consists of a single point y: for if y and y' were distinct points of the support, and U and U' are disjoint neighbourhoods of y and y', then the commuting ideals $M_U\mathfrak{g}$ and $M_{U'}\mathfrak{g}$ would each map surjectively on to the non-abelian algebra $\mathfrak{g}$, which is impossible. Thus the kernel of $\dot{\alpha}_x$ contains the ideal $J_{y,k}$ of all elements of $M\mathfrak{g}$ which vanish to some order k at y. But

$$[\ldots [[J_{y,1}, J_{y,1}], J_{y,1}], \ldots, J_{y,1}] = J_{y,k}$$

(where $J_{y,1}$ occurs k times on the left), so the kernel must contain $J_{y,1}$. As $M\mathfrak{g}/J_{y,1} \cong \mathfrak{g}$, this proves that $\dot{\alpha}_x$ is evaluation at y.

If $\phi : X \to X$ is defined by $\alpha_x = \varepsilon_x \circ \alpha = \varepsilon_{\phi(x)}$, then $\alpha : M_0G \to M_0G$ is given by $\alpha(f)(x) = f(\phi(x)) = (\phi^*f)(x)$. The map ϕ must be smooth because ϕ^* takes smooth functions to smooth functions, and it must be a diffeomorphism because α is an automorphism.

Remarks.

(i) The preceding result obviously does not hold if G is not simple, but the method enables one to describe all automorphisms of MG when G is semisimple. If G has a torus factor then MG contains a large vector space as a factor, and the automorphism group contains its general linear group.

(ii) The proof of (3.4.2), as has been pointed out to us by P. de la Harpe, actually proves the following result.

Proposition (3.4.3). *If G is simple then the maximal normal subgroups of $M_0 G$ are precisely the kernels of the evaluation maps $M_0 G \to G$ at the points of X.*

To conclude this section let us return briefly to loop groups. The identity component A^0 of the group A of automorphisms of G consists of inner automorphisms, and $\pi_0(A) = A/A^0$ is the finite group of classes of outer automorphisms. Now LA acts as a group of automorphisms of LG, and again its identity component $(LA)^0$ consists of inner automorphisms, for any null-homotopic loop in A can be lifted to G. In fact a loop can be lifted precisely when its homotopy class belongs to the image of $\pi_1(G) \to \pi_1(A)$. The cokernel of this homomorphism is the centre Z of G, so we have

Proposition (3.4.4). *The semidirect product $\pi_0(A) \tilde{\times} Z$ is a subgroup of the group of outer automorphism classes of LG.*

The action of the centre Z is the important point. For any $g \in Z$ one can choose a smooth map $\eta : \mathbb{R} \to G$ such that $\eta(\theta + 2\pi) = g \cdot \eta(\theta)$, and then conjugation by η is the associated outer automorphism of LG.

3.5 Subgroups of LG: polynomial loops

From time to time we shall want to mention a number of subgroups of LG. The most obvious of these is the group $L_{\mathrm{an}} G$ of real-analytic loops. If G is embedded in a unitary group U_n, so that a loop γ in G is a matrix-valued function and can be expanded in a Fourier series

$$\gamma(z) = \sum_{k=-\infty}^{\infty} \gamma_k z^k, \qquad (3.5.1)$$

then the real-analytic loops are those such that the series converges in some annulus $r \leq |z| \leq r^{-1}$ with $r < 1$, i.e. such that $\|\gamma_k r^{-|k|}\|$ is bounded for all k for some $r < 1$. The natural topology on $L_{\mathrm{an}} G$ is got by regarding it as the direct limit of the Banach Lie groups $L_{\mathrm{an},r} G$ consisting of functions holomorphic in $r \leq |z| \leq r^{-1}$; the group $L_{\mathrm{an},r} G$ has the topology of uniform convergence. There is no difficulty in seeing that $L_{\mathrm{an}} G$ is a Lie group with Lie algebra $L_{\mathrm{an}} \mathfrak{g}$. (The choice of the embedding $G \subset U_n$ was immaterial, and indeed was not really used: it was introduced above only for concreteness.)

A slightly smaller subgroup is the group $L_{\mathrm{rat}} G$ of rational loops, i.e.

3.5 SUBGROUPS OF LG: POLYNOMIAL LOOPS

loops which, when regarded as matrix-valued functions, have entries which are rational functions of z with no poles on $|z|=1$. (A rational function means the quotient of two polynomials.) We shall not pursue the question of the appropriate topology to be put on $L_{\text{rat}}G$: let us notice only that it is a dense subgroup of LG.

The smallest subgroup we shall consider is $L_{\text{pol}}G$: the group of loops whose matrix entries are finite Laurent polynomials in z and z^{-1}, i.e. loops of the form (3.5.1) where only finitely many of the matrices γ_k are non-zero. This group is the union of the subsets $L_{\text{pol},N}G$ consisting of the loops (3.5.1) for which $\gamma_k = 0$ when $|k| > N$. Each of these subsets is naturally a compact space, and we give $L_{\text{pol}}G$ the direct limit topology. It is associated with the Lie algebra $L_{\text{pol}}\mathfrak{g}$ of all finite series

$$\sum_{k=-N}^{N} \xi_k z^k \qquad (3.5.2)$$

where ξ_k belongs to the complexified Lie algebra $\mathfrak{g}_\mathbb{C}$ and $\xi_{-k} = \bar{\xi}_k$. This vector space is the direct limit of its finite dimensional subspaces $L_{\text{pol},N}\mathfrak{g}$, and has the direct limit topology. There is of course no exponential map $L_{\text{pol}}\mathfrak{g} \to L_{\text{pol}}G$, for the exponential of a finite series (3.5.2) is usually not a finite series.

The group $L_{\text{pol}}G$ has the complexification $L_{\text{pol}}G_\mathbb{C}$ which consists of the loops in $G_\mathbb{C}$ which, *together with their inverses*, are given by finite Laurent polynomials (3.5.1). (In the case of $L_{\text{pol}}G$ we did not need to say 'together with their inverses' because for $\gamma \in LG$ we have $\gamma^{-1} = \gamma^*$, and so the inverse of a polynomial loop is automatically polynomial.) If $G = U_n$ then $L_{\text{pol}}G_\mathbb{C}$ is just $GL_n(\mathbb{C}[z, z^{-1}])$. In general, if G is thought of as an algebraic group, then $L_{\text{pol}}G_\mathbb{C}$ is the 'points of G with values in $\mathbb{C}[z, z^{-1}]$' in the sense of algebraic geometry.

It is *not* always true that $L_{\text{pol}}G$ is dense in LG. For example, if $G = \mathbb{T}$ then the only elements of $L_{\text{pol}}G$ are the loops uz^k, with $u \in \mathbb{T}$; i.e. the identity component of $L_{\text{pol}}G$ is simply the constant loops. (For the inverse of a non-constant polynomial cannot be a polynomial.) The following result is therefore a little surprising.

Proposition (3.5.3). *If G is semisimple, then $L_{\text{pol}}G$ is dense in LG.*

Proof. Let H be the closure of $L_{\text{pol}}G$ in LG, and let V be the subset of $L\mathfrak{g}$ formed by the tangent vectors ξ such that the corresponding one-parameter subgroup γ_ξ belongs to H. The essential observation is that V is a vector space. To see that it is closed under addition one uses the formula

$$\gamma_{\xi+\eta}(t) = \lim_{t \to \infty} (\gamma_\xi(t/n)\gamma_\eta(t/n))^n,$$

which holds in LG because, for a suitable neighbourhood U of the

identity in G, the sequence of maps $f_n: U \times U \to G$ defined by
$$f_n(x, y) = (x^{1/n} y^{1/n})^n$$
converges in the C^∞ topology.

It is clear that V is a closed subspace of $L\mathfrak{g}$. To prove (3.5.3) it is enough to show (because the exponential map is locally surjective in LG) that $V = L\mathfrak{g}$.

Consider first the case $G = SU_2$. Then the elements
$$\xi_n = \begin{pmatrix} 0 & z^n \\ -z^{-n} & 0 \end{pmatrix} \quad \text{and} \quad \eta_n = \begin{pmatrix} 0 & iz^n \\ iz^{-n} & 0 \end{pmatrix}$$
belong to V, as the corresponding one-parameter subgroups lie in $L_{\text{pol}}G$. (For $\xi_n^2 = \eta_n^2 = -1$.) By linearity then, and because it is closed, V contains every element of the form
$$f \begin{pmatrix} 0 & 1 \\ -1 & 0 \end{pmatrix} + g \begin{pmatrix} 0 & i \\ i & 0 \end{pmatrix},$$
where f and g are smooth real-valued functions on the circle. But V is invariant under conjugation by constant elements of SU_2, so we must have $V = L\mathfrak{g}$.

The general case follows in the usual way, because for any semisimple G there are a finite number of homomorphisms $SU_2 \to G$ for which the images of $\mathfrak{su}_2$ in $\mathfrak{g}$ span $\mathfrak{g}$. (The argument proves that the closure of $L_{\text{pol}}G$ contains the identity component of LG; the proof is completed by observing that $L_{\text{pol}}G$ contains at least one element from each connected component of LG.)

3.6 Maximal abelian subgroups of LG

We shall show that there is a conjugacy class of maximal abelian subgroups of LG associated naturally to each conjugacy class in the Weyl group of G.

If A is any abelian subgroup of LG then for any point θ of the circle the subgroup $A(\theta)$ of G got by evaluating the loops in A at θ is abelian, and so is contained in a maximal torus of G. Thus the most obvious maximal abelian subgroup of LG is LT, where T is a maximal torus of G. More generally, if λ is a map which assigns a maximal torus $T_{\lambda(\theta)}$ of G smoothly to each point θ of the circle, then the subgroup
$$A_\lambda = \{\gamma \in LG : \gamma(\theta) \in T_{\lambda(\theta)} \text{ for all } \theta\}$$
is a maximal abelian subgroup. As all maximal tori are conjugate, the space of maximal tori can be identified with G/N, where N is the normalizer of a fixed torus T. Thus λ is a smooth map $\lambda: S^1 \to G/N$.

3.6 MAXIMAL ABELIAN SUBGROUPS OF LG

The conjugacy class of A_λ depends only on the homotopy class of λ. This follows easily from the homotopy lifting property of the fibration $N \to G \to G/N$: for example if λ is contractible it can be lifted to $\tilde\lambda : S^1 \to G$, and then $T_{\lambda(\theta)} = \tilde\lambda(\theta) . T . \tilde\lambda(\theta)^{-1}$ and $A_\lambda = \tilde\lambda . LT . \tilde\lambda^{-1}$. The fundamental group of G/N is the Weyl group $W = N/T$, for W acts freely on the simply connected space G/T, and $G/N = (G/T)/W$. The set of homotopy classes of maps $S^1 \to G/N$, with no basepoints, is therefore the set of conjugacy classes of W (see Spanier [143] p. 379), and we shall think of λ as representing such a class. An element $w \in W$ defines an automorphism α_w of T by conjugation, and the corresponding A_λ can be described as follows.

Proposition (3.6.1). *If λ corresponds to $w \in W$ then A_λ is isomorphic to the group of smooth maps $\gamma : \mathbb{R} \to T$ such that*

$$\gamma(\theta + 2\pi) = \alpha_w^{-1}(\gamma(\theta)) \tag{3.6.2}$$

for all $\theta \in \mathbb{R}$.

Proof. Suppose w is represented by $n \in N$, and let ω be an element of the Lie algebra of G such that $\exp(2\pi\omega) = n$. Then we can take

$$A_{\lambda(\theta)} = \exp(\theta\omega) . T . \exp(-\theta\omega).$$

If $\tilde\gamma$ belongs to A_λ then $\gamma : \mathbb{R} \to T$, defined by

$$\gamma(\theta) = \exp(-\theta\omega)\tilde\gamma(\theta)\exp(\theta\omega),$$

satisfies (3.6.2) and conversely.

By using the description (3.6.1) of A_λ, and considering the exact sequence of groups

$$\Omega T \to A_\lambda \to T,$$

where $A_\lambda \to T$ is evaluation at $\theta = 0$, it is easy to prove

Proposition (3.6.3). *The group of connected components $\pi_0(A_\lambda)$ of A_λ, and its fundamental group $\pi_1(A_\lambda)$, are the cokernel and kernel of the homomorphism*

$$w_* - 1 : \check{T} \to \check{T},$$

where $\check{T}$ is the lattice $\pi_1(T)$, and w_* is the action of w on $\check{T}$.

The maximal abelian subgroups A_λ we have described are not the only ones, for example if T_0 and T_1 are two different maximal tori in G then the subgroup consisting of loops γ such that $\gamma(\theta) \in T_0$ for $0 \le \theta \le \pi$, and $\gamma(\theta) \in T_1$ for $\pi \le \theta \le 2\pi$, is clearly maximal. It seems likely, however, that the A_λ exhaust all the maximal abelian subgroups in the group of real-analytic loops.

We shall make considerable use in Part II of this book of the subgroup A_λ of LU_n corresponding to the *Coxeter element* w of the Weyl group. The Weyl group of U_n is the symmetric group S_n, which permutes the entries of the diagonal matrices which form the maximal torus T of U_n. The Coxeter element is the cyclic permutation $(12\ldots n)$.

Proposition (3.6.4). *The maximal abelian subgroup of LU_n corresponding to the Coxeter element is isomorphic to $L\mathbb{T}$.*

Proof. This follows from (3.6.1). For if $\gamma: \mathbb{R} \to T$ satisfies (3.6.2) when w is the Coxeter element then each diagonal element of γ is a function $\gamma_i : \mathbb{R} \to \mathbb{T}$ which is periodic with period $2\pi n$, and the γ_i differ from each other only by translation by multiples of 2π.

We shall describe this embedding of $L\mathbb{T}$ in LU_n in a somewhat different way in Section 6.5. Its importance was first recognized by Lepowsky and Wilson [102]. (Cf. also [87].) Its Lie algebra—or, strictly, a central extension of it—is sometimes referred to as a 'principal Heisenberg subalgebra' of $L\mathfrak{u}_n$.

Remark. The abelian subgroup corresponding to a general element w of S_n is easily seen to be a product of copies of $L\mathbb{T}$, one for each cycle in the permutation w.

3.7 Twisted loop groups

The abelian subgroups A_λ of LG which we have just described are examples of what are called *twisted loop groups*. If α is any automorphism of a group G then one can define

$$L_{(\alpha)}G = \{\gamma: \mathbb{R} \to G \text{ such that } \gamma(\theta + 2\pi) = \alpha(\gamma(\theta))\}. \quad (3.7.1)$$

The group $L_{(\alpha)}G$ depends (up to isomorphism) only on the class of α modulo inner automorphisms. For if

$$\beta(g) = c\alpha(g)c^{-1}$$

for some $c \in G$, then we can choose a smooth map $\lambda: \mathbb{R} \to G$ such that $\lambda(\theta + 2\pi) = c \cdot \alpha(\lambda(\theta))$, and then the map $\gamma \mapsto \tilde{\gamma}$, where

$$\tilde{\gamma}(\theta) = \lambda(\theta)\gamma(\theta)\lambda(\theta)^{-1},$$

defines an isomorphism $L_{(\alpha)}G \to L_{(\beta)}G$. This means that if G is semi-simple one may as well think of α as belonging to the finite group of outer automorphism classes of G; in particular one can assume that α has finite order.

An alternative description of $L_{(\alpha)}G$ is as the group of cross-sections of a fibre bundle on S^1 with fibre G. The bundle is the quotient space of

3.7 TWISTED LOOP GROUPS

$G \times \mathbb{R}$ by the equivalence relation which identifies (g, θ) with $(\alpha(g), \theta + 2\pi)$.

The theory of twisted loop groups is exactly analogous to that of loop groups, but we shall not pursue it in this book (cf. Section 5.3). The position would be different if we had anything significant to say about groups of the form Map$(X; G)$ for spaces X other than the circle: in that case the analogue of the twisted groups would include the groups of automorphisms of principal fibre bundles on X with structure group G—so called *gauge groups*.

4
CENTRAL EXTENSIONS

For the remainder of this book G will always denote a compact connected Lie group.

4.1 Introduction

A fundamental property of the loop group LG is the existence of interesting central extensions

$$\mathbb{T} \to \tilde{L}G \to LG$$

of LG by the circle $\mathbb{T}$. (In other words, $\tilde{L}G$ is a group containing $\mathbb{T}$ in its centre and such that the quotient group $\tilde{L}G/\mathbb{T}$ is LG.) The $\tilde{L}G$ are analogous to the finite-sheeted covering groups of a finite dimensional Lie group, in that any projective unitary representation of LG comes from a genuine representation of some $\tilde{L}G$: we recall that a projective unitary representation of a group L on a Hilbert space H is the assignment to each $\lambda \in L$ of a unitary operator $U_\lambda : H \to H$ so that

$$U_\lambda U_{\lambda'} = c(\lambda, \lambda') U_{\lambda\lambda'}$$

holds for all $\lambda, \lambda' \in L$, where $c(\lambda, \lambda')$ is a complex number of modulus 1. The function $c : L \times L \to \mathbb{T}$ is called the 'projective multiplier' or 'cocycle' of the representation.

As topological spaces the $\tilde{L}G$ are fibre bundles over LG with the circle as fibre. Except for the product extension $LG \times \mathbb{T}$ they are non-trivial fibre bundles: that is to say $\tilde{L}G$ is not homeomorphic to the cartesian product $LG \times \mathbb{T}$, and there is no continuous cross-section $LG \to \tilde{L}G$. In fact the group extension $\tilde{L}G$ is completely determined by its topological type as a fibre bundle, and every circle bundle on LG can be made into a group extension. It is interesting that the behaviour of $\text{Map}(X; G)$ when $\dim(X) > 1$ is completely different. There are often non-trivial circle bundles on $\text{Map}(X; G)$, but if X is simply connected only the flat ones can be made into groups. (That follows from Propositions (4.2.8) and (4.5.6) below.)

When G is a simple and simply connected group there is a *universal* central extension among the $\tilde{L}G$, i.e. one of which all the others are quotient groups. This is analogous to the universal covering group of a finite dimensional group. Any central extension E of LG by any abelian group A arises from the universal extension $\tilde{L}G$ by a homomorphism

$\theta: \mathbb{T} \to A$, in the sense that $E = \tilde{L}G \times_\mathbb{T} A$. (The last notation denotes the quotient group of $\tilde{L}G \times A$ by the subgroup consisting of all elements $\{(z, -\theta(z)): z \in \mathbb{T}\}$.) When G is simply connected but not simple there is still a universal central extension, but, as we shall see, it is an extension of LG by the homology group $H_3(G; \mathbb{T})$, a torus whose dimension is the number of simple factors in G.

The group LG has a complexification $LG_\mathbb{C}$. The extensions $\tilde{L}G$ also have complexifications $\tilde{L}G_\mathbb{C}$, which are extensions of $LG_\mathbb{C}$ by $\mathbb{C}^\times$. We shall postpone the construction of the complexifications, however, until Chapter 6.

It is worth noticing that the central extensions of LG are closely related to its natural affine action on the space of *connections* in the trivial principal G-bundle on the circle. (See (4.3.3).)

This chapter ends with an appendix discussing the cohomology of the space LG and of the Lie algebra $L\mathfrak{g}$.

4.2 The Lie algebra extensions

On the level of Lie algebras the extensions can be defined and classified very simply: they correspond precisely to invariant symmetric bilinear forms on $\mathfrak{g}$. As a vector space $\tilde{L}\mathfrak{g}$ is $L\mathfrak{g} \oplus \mathbb{R}$, and the bracket is given by

$$[(\xi, \lambda), (\eta, \mu)] = ([\xi, \eta], \omega(\xi, \eta)) \qquad (4.2.1)$$

for $\xi, \eta \in L\mathfrak{g}$ and $\lambda, \mu \in \mathbb{R}$, where $\omega: L\mathfrak{g} \times L\mathfrak{g} \to \mathbb{R}$ is the bilinear map

$$\omega(\xi, \eta) = \frac{1}{2\pi} \int_0^{2\pi} \langle \xi(\theta), \eta'(\theta) \rangle \, d\theta \qquad (4.2.2)$$

and $\langle \, , \, \rangle$ is a symmetric invariant form on the Lie algebra $\mathfrak{g}$. Let us recall that if $\mathfrak{g}$ is semisimple then *every* invariant bilinear form on $\mathfrak{g}$ is symmetric. (See (2.3.2).)

For the formula (4.2.1) to define a Lie algebra, ω must be skew—which is clear by integrating by parts in (4.2.2)—and must satisfy the 'cocycle condition'

$$\omega([\xi, \eta], \zeta) + \omega([\eta, \zeta], \xi) + \omega([\zeta, \xi], \eta) = 0. \qquad (4.2.3)$$

This condition follows from the Jacobi identity in the Lie algebra $L\mathfrak{g}$ and the fact that the inner product on $\mathfrak{g}$ is invariant:

$$\langle [\xi, \eta], \zeta \rangle = \langle \xi, [\eta, \zeta] \rangle.$$

One of the first things to notice about the cocycle ω is that it is invariant under the action of the group $\text{Diff}^+(S^1)$ of orientation-preserving diffeomorphisms of the circle, i.e.

$$\omega(f^*\xi, f^*\eta) = \omega(\xi, \eta),$$

for $f \in \mathrm{Diff}^+(S^1)$. (Here $f^*\xi(\theta)$ denotes $\xi(f(\theta))$.) This means that $\mathrm{Diff}^+(S^1)$ acts as a group of automorphisms of the extended Lie algebra. We shall see later that it also acts on the group extension. It is important that the extension singles out a particular orientation of S^1: orientation-reversing diffeomorphisms can act on $\tilde{L}\mathfrak{g}$ only by reversing the kernel $\mathbb{R}$.

There are essentially no other cocycles on $L\mathfrak{g}$ than the ω given by (4.2.2). To make this precise, notice that ω is invariant under conjugation by constant loops, i.e. $\omega(\xi, \eta) = \omega(g\xi, g\eta)$ for $g \in G$, where $g\xi, g\eta$ are the adjoint action of g on ξ, η. There is no point in considering cocycles which are not invariant in this way. Indeed, for any cocycle α, the cocycle $g \cdot \alpha$ defines the same extension as α, where $g \cdot \alpha$ is defined by $g \cdot \alpha(\xi, \eta) = \alpha(g^{-1}\xi, g^{-1}\eta)$. So the extension defined by α is also given by the invariant cocycle

$$\int_G g \cdot \alpha \, dg$$

obtained by averaging α over the compact group G. (Notice that the cocycle identity (4.2.3) expresses precisely that the cohomology class of the cocycle does not change under an infinitesimal conjugation.)

Then we have

Proposition (4.2.4). *If $\mathfrak{g}$ is semisimple then the only continuous G-invariant cocycles on the Lie algebra $L\mathfrak{g}$ are those given by (4.2.2).*

Remark. One cannot omit 'semisimple' here. For example, if $G = \mathbb{T}$ then *any* skew bilinear form on the vector space $L\mathfrak{g} = L\mathbb{R}$ is a cocycle. But if we require the cocycles to be invariant under $\mathrm{Diff}^+(S^1)$ then 'semisimple' is not needed, for $L\mathbb{R}/\mathbb{R}$ is an irreducible representation of $\mathrm{Diff}^+(S^1)$, and so it is easy to see that the only bilinear form on $L\mathbb{R}$ which is invariant under $\mathrm{Diff}^+(S^1)$ is, up to a scalar multiple,

$$(\xi, \eta) \mapsto \int_{S^1} \xi \, d\eta.$$

Proof of (4.2.4). Any cocycle $\alpha: L\mathfrak{g} \times L\mathfrak{g} \to \mathbb{R}$ can be extended to a complex bilinear map $\alpha: L\mathfrak{g}_\mathbb{C} \times L\mathfrak{g}_\mathbb{C} \to \mathbb{C}$. An element $\xi \in L\mathfrak{g}_\mathbb{C}$ can be expanded in a Fourier series $\sum \xi_k z^k$, with $\xi_k \in \mathfrak{g}_\mathbb{C}$. By continuity α is completely determined by its values on elements of the form $\xi_k z^k$. Let us write $\alpha_{p,q}(\xi, \eta) = \alpha(\xi z^p, \eta z^q)$ for $\xi, \eta \in \mathfrak{g}_\mathbb{C}$. Then $\alpha_{p,q}$ is a G-invariant bilinear map $\mathfrak{g}_\mathbb{C} \times \mathfrak{g}_\mathbb{C} \to \mathbb{C}$, which is necessarily symmetric, and $\alpha_{p,q} = -\alpha_{q,p}$.

The cocycle identity (4.2.3) translates into the statement

$$\alpha_{p+q,r} + \alpha_{q+r,p} + \alpha_{r+p,q} = 0 \tag{4.2.5}$$

for all p, q, r. Putting $q = r = 0$ we find $\alpha_{p,0} = 0$ for all p. Putting

4.2 THE LIE ALGEBRA EXTENSIONS

$r = -p - q$ we find

$$\alpha_{p+q,-p-q} = \alpha_{p,-p} + \alpha_{q,-q},$$

whence

$$\alpha_{p,-p} = p\alpha_{1,-1}.$$

Putting $r = n - p - q$ in (4.2.5) we find

$$\alpha_{n-p-q,p+q} = \alpha_{n-p,p} + \alpha_{n-q,q},$$

whence

$$\alpha_{n-k,k} = k\alpha_{n-1,1}.$$

This imples that $\alpha_{p,q} = 0$ if $p + q \neq 0$, for

$$n\alpha_{n-1,1} = \alpha_{0,n} = 0.$$

Returning to $\xi = \sum \xi_p z^p$ and $\eta = \sum \eta_q z^q$, we have

$$\alpha(\xi, \eta) = \sum p\alpha_{1,-1}(\xi_p, \eta_{-p})$$

$$= \frac{i}{2\pi} \int_0^{2\pi} \alpha_{1,-1}(\xi(\theta), \eta'(\theta)) \, d\theta,$$

which is of the form (4.2.2).

Proposition (4.2.4) determines the universal central extension of $L\mathfrak{g}$. We can reformulate it in the following way. For any finite dimensional Lie algebra $\mathfrak{g}$ there is a universal invariant symmetric bilinear form

$$\langle \, , \, \rangle_K : \mathfrak{g} \times \mathfrak{g} \to K \qquad (4.2.6)$$

from which every $\mathbb{R}$-valued form arises by a unique linear map $K \to \mathbb{R}$. (Of course K is simply the dual of the space of all $\mathbb{R}$-valued forms.) The cocycle ω_K given by

$$\omega_K(\xi, \eta) = \frac{1}{2\pi} \int_0^{2\pi} \langle \xi(\theta), \eta'(\theta) \rangle_K \, d\theta \qquad (4.2.7)$$

defines an extension of $L\mathfrak{g}$ by K, which by Proposition (4.2.4) is the universal central extension of $L\mathfrak{g}$ when $\mathfrak{g}$ is semisimple. For semisimple groups K can be identified with $H_3(\mathfrak{g}; \mathbb{R})$, because a bilinear form $\langle \, , \, \rangle$ on $\mathfrak{g}$ gives rise to an invariant skew 3-form

$$(\xi, \eta, \zeta) \mapsto \langle \xi, [\eta, \zeta] \rangle,$$

and all elements of $H^3(\mathfrak{g}; \mathbb{R})$ are so obtained. When $\mathfrak{g}$ is simple then $K = \mathbb{R}$.

Extensions of Map$(X; \mathfrak{g})$

Before leaving the subject of Lie algebra extensions, it is worth pointing out that very little extra work is needed to determine all central extensions of Map$(X; \mathfrak{g})$ for any smooth manifold X. We shall indicate briefly a proof of the following result, which is a very simple case of a general theory of Loday and Quillen [104] relating the cohomology of Lie algebras to Connes's cohomology [33]. We shall content ourselves with the case of a simple algebra $\mathfrak{g}$. There is then an essentially unique inner product $\langle \, , \, \rangle$.

Proposition (4.2.8). *If $\mathfrak{g}$ is simple then the kernel of the universal central extension of* Map$(X; \mathfrak{g})$ *is the space* $K = \Omega^1(X)/d\Omega^0(X)$ *of 1-forms on X modulo exact 1-forms. The extension is defined by the cocycle*

$$(\xi, \eta) \mapsto \langle \xi, d\eta \rangle. \tag{4.2.9}$$

Equivalently, the extensions of Map$(X; \mathfrak{g})$ *by $\mathbb{R}$ correspond to the one-dimensional closed currents C on X, the cocycle being given by integrating* (4.2.9) *over C.*

Before proving this let us remark that from one point of view it is a disappointing result, as it tells us that there are no 'interesting' extensions of Map$(X; \mathfrak{g})$ when $\dim(X) > 1$. More precisely, if $f: S^1 \to X$ is any smooth loop in X one can always obtain an extension of Map$(X; \mathfrak{g})$ by pulling back the universal extension of $L\mathfrak{g}$ by f. Proposition (4.2.8) asserts that any extension is a weighted linear combination of extensions of this form. The first 'interesting' cohomology class of Map$(X; \mathfrak{g})$, for a compact $(n-1)$-dimensional manifold X, is in dimension n, and is defined by the cocycle

$$(\xi_1, \ldots, \xi_n) \mapsto P(\xi_1, d\xi_2, \ldots, d\xi_n),$$

where P is an invariant polynomial of degree n on $\mathfrak{g}$.

Proof of (4.2.8). Let us write Map$(X; \mathfrak{g})$ as $A \otimes \mathfrak{g}$, where A is the ring of smooth functions on X. Any G-invariant real-valued bilinear form on $A \otimes \mathfrak{g}$ must be of the form

$$(f \otimes \xi, g \otimes \eta) \mapsto \alpha(f \otimes g)\langle \xi, \eta \rangle,$$

where $\alpha: A \otimes A \to \mathbb{R}$ is linear. Such an α can be identified with a distribution with compact support on $X \times X$. The cocycle condition translates into the statement that α vanishes on functions of the form

$$fg \otimes h + gh \otimes f + hf \otimes g, \tag{4.2.10}$$

where f, g, h are smooth functions on X. This means that $\alpha(f \otimes g) = 0$ when f and g have disjoint support, for then $fg = 0$ and one can find h so that $fh = f$ and $gh = 0$. Thus the distribution α has support along the

diagonal. Proposition (4.2.8) is the assertion that $\alpha(f \otimes g)$ depends only on the 1-form fdg. This in turn reduces to two facts:
(i) $\alpha(f \otimes 1) = 0$ for all f; and
(ii) $\alpha \mid I^2 = 0$, where I is the ideal of functions in $A \otimes A$ which vanish on the diagonal.

Both of these facts follow directly from (4.2.10), for I is generated additively by functions of the form $f \otimes g - fg \otimes 1$.

Extensions of Vect(S^1)

Another calculation that fits in very naturally at this point is that for the Lie algebra Vect(S^1) of smooth vector fields on the circle, i.e. the Lie algebra of the group Diff(S^1). A complex-linear 2-cocycle

$$\alpha : \text{Vect}_{\mathbb{C}}(S^1) \times \text{Vect}_{\mathbb{C}}(S^1) \to \mathbb{C},$$

where $\text{Vect}_{\mathbb{C}}(S^1) = \text{Vect}(S^1) \otimes \mathbb{C}$, is determined by the numbers $\alpha_{p,q} = \alpha(L_p, L_q)$, where $L_n = e^{in\theta}(d/d\theta)$. We have

$$[L_n, L_m] = i(m - n)L_{n+m}.$$

The cocycle identity for (L_0, L_p, L_q) shows that the cohomology class of α is not changed by rotation, and so we can (by averaging) assume that α is itself invariant. Then $\alpha_{p,q} = 0$ unless $p + q = 0$. If we write $\alpha_{p,-p} = \alpha_p$, and notice that $\alpha_{-p} = -\alpha_p$, then the cocycle identity gives

$$(p + 2q)\alpha_p - (2p + q)\alpha_q = (p - q)\alpha_{p+q}.$$

This determines all the α_p in terms of α_1 and α_2. The general solution is $\alpha_p = \lambda p^3 + \mu p$. But $\alpha_p = p$ is a coboundary, so the value of μ is unimportant. We have proved

Proposition (4.2.11). *The most general central extension of Vect(S^1) by $\mathbb{R}$ is described by the cocycle α, where*

$$\alpha\left(e^{in\theta}\frac{d}{d\theta}, e^{im\theta}\frac{d}{d\theta}\right) = i\lambda n(n^2 - 1) \quad \text{if} \quad n + m = 0,$$

$$= 0 \quad \text{if} \quad n + m \neq 0,$$

for some $\lambda \in \mathbb{R}$.

The representing cocycle given here is characterized by the fact that it is invariant under rotation and vanishes on the subalgebra $\mathfrak{sl}_2(\mathbb{R})$ of Vect(S^1).

4.3 The coadjoint action of LG on $\tilde{L}\mathfrak{g}$, and its orbits

In this section $\tilde{L}\mathfrak{g}$ denotes the extension of $L\mathfrak{g}$ by $\mathbb{R}$ associated to an invariant bilinear form $\langle \, , \, \rangle$ on $\mathfrak{g}$ by the formula (4.2.2). We shall use the form $\langle\ ,\ \rangle$, to define a form on $L\mathfrak{g}$, again denoted $\langle\ ,\ \rangle$, by

$$\langle \xi, \eta \rangle = \frac{1}{2\pi} \int_0^{2\pi} \langle \xi(\theta), \eta(\theta) \rangle \, d\theta.$$

Because we are dealing with a central extension, the adjoint action of $\tilde{L}\mathfrak{g}$ on itself is really an action of $L\mathfrak{g}$, given by

$$\eta \cdot (\xi, \lambda) = ([\eta, \xi], \omega(\eta, \xi)). \tag{4.3.1}$$

Proposition (4.3.2). *The adjoint action of $L\mathfrak{g}$ on $\tilde{L}\mathfrak{g}$ comes from an action of LG given by*

$$\gamma \cdot (\xi, \lambda) = (\gamma \cdot \xi, \lambda - \langle \gamma^{-1}\gamma', \xi \rangle).$$

Here $\gamma \cdot \xi$ denotes the adjoint action of $\gamma \in LG$ on $\xi \in L\mathfrak{g}$.

Proof. One has only to check that the desired formula does define a group action, and that its derivative at $\gamma = 1$ is given by (4.3.1). Both verifications are straightforward.

Now let us consider the dual $(\tilde{L}\mathfrak{g})^*$ of $\tilde{L}\mathfrak{g}$. This fits into the exact sequence

$$(L\mathfrak{g})^* \to (\tilde{L}\mathfrak{g})^* \to \mathbb{R},$$

on which the group LG acts. If we identify $(\tilde{L}\mathfrak{g})^*$ with $(L\mathfrak{g})^* \oplus \mathbb{R}$, and use the form $\langle\ ,\ \rangle$ on $L\mathfrak{g}$ to map $L\mathfrak{g}$ into $(L\mathfrak{g})^*$, then we have

Proposition (4.3.3). *The coadjoint action of LG on $(\tilde{L}\mathfrak{g})^*$ is given by*

$$\gamma \cdot (\phi, \lambda) = (\gamma \cdot \phi + \lambda \gamma' \gamma^{-1}, \lambda).$$

The reason for being interested in this coadjoint action is the heuristic principle due to Kirillov [92] that the irreducible unitary representations of a group Γ correspond roughly to the orbits of its coadjoint action. A little more precisely, the correspondence is with those orbits which satisfy an integrality condition (C) which is described below.

Let us assume that the inner product on $\mathfrak{g}$ is positive-definite. Then $L\mathfrak{g}$ is identified with a dense subspace of $(L\mathfrak{g})^*$ which we shall call the 'smooth part' of the dual. We can describe the orbits of the action of LG on this in the following way.

For each smooth element $(\phi, \lambda) \in (\tilde{L}\mathfrak{g})^*$ with $\lambda \neq 0$ we can find a unique smooth path $f: \mathbb{R} \to G$ by solving the differential equation

$$f'f^{-1} = -\lambda^{-1}\phi \tag{4.3.4}$$

with the initial condition $f(0) = 1$. Because ϕ is periodic in θ we have

$$f(\theta + 2\pi) = f(\theta) \cdot M_\phi,$$

where $M_\phi = f(2\pi)$. If (ϕ, λ) is transformed by $\gamma \in LG$ then f is changed to $\tilde{f}$, where

$$\tilde{f}(\theta) = \gamma(\theta)f(\theta)\gamma(0)^{-1}. \tag{4.3.5}$$

Thus M_ϕ is changed to $\gamma(0)M_\phi\gamma(0)^{-1}$. In fact (4.3.4) defines a bijection between $L\mathfrak{g} \times \{\lambda\}$ and the space of maps f such that $f(0) = 1$ and $f(\theta + 2\pi) = f(\theta) \cdot M$ for some $M \in G$. From this we can read off

Proposition (4.3.6).

 (i) *If G is simply connected and $\lambda \neq 0$ then the orbits of LG on the smooth part of $(L\mathfrak{g})^* \times \{\lambda\} \subset (\tilde{L}\mathfrak{g})^*$ correspond precisely to the conjugacy classes of G under the map $(\phi, \lambda) \mapsto M_\phi$.*

 (ii) *The stabilizer of (ϕ, λ) in LG is isomorphic to the centralizer Z_ϕ of M_ϕ in G by the map $\gamma \mapsto \gamma(0)$; and γ stabilizes (ϕ, λ) if and only if $\gamma(\theta) = f(\theta)\gamma(0)f(\theta)^{-1}$.*

According to Kirillov's idea, the irreducible unitary representations of a group Γ correspond to the coadjoint orbits Ω with the property

(C) if the stabilizer of $\Phi \in \Omega$ is the subgroup H of Γ then Φ is the derivative of a character of the identity component of H.

To apply the principle in our case we need to know that $\tilde{L}\mathfrak{g}$ is the Lie algebra of an extension $\tilde{L}G$ of LG by the circle $\mathbb{T}$. The conditions under which $\tilde{L}G$ exists will be determined in the following sections. Granting its existence for the present, we find at once that if an orbit belonging to $(L\mathfrak{g})^* \times \{\lambda\}$ is allowable then λ must be an *integer*. Then an orbit in the smooth part of the dual corresponds to the conjugacy class of an element $g \in G$, which we can assume to belong to a given maximal torus T. If we choose

$$\xi \in \mathfrak{t} \subset \mathfrak{g} \subset L\mathfrak{g} \subset (L\mathfrak{g})^*$$

so that $\exp(\lambda^{-1}\xi) = g$ then (ξ, λ) belongs to the orbit. If g is sufficiently generic then its centralizer in G is T, and the condition (C) clearly amounts to the requirement that $\xi \in \mathfrak{t} \subset \mathfrak{t}^*$ belongs to the lattice $\hat{T}$. In fact it is not hard to check that this is true in any case. On the other hand (ξ, λ) and $(\tilde{\xi}, \lambda)$ belong to the same orbit if $\tilde{\xi} = w \cdot \xi + \lambda\eta$ for some $\eta \in \check{T}$ and some w in the Weyl group W of G. Thus we have

Proposition (4.3.7). *If λ is a non-zero integer then the coadjoint orbits in the smooth part of $(L\mathfrak{g})^* \times \{\lambda\}$ which satisfy the condition (C) correspond to the orbits of the affine Weyl group† $W_{\text{aff}} = W \ltimes \check{T}$ on the lattice $\hat{T}$, where $(w, \eta) \in W_{\text{aff}}$ acts on $\hat{T}$ by*

$$\xi \mapsto w \cdot \xi + \lambda\eta.$$

We shall see later—in Chapters 9 and 11—that if $\lambda > 0$ these orbits

† For a discussion of the affine Weyl group we refer to Section 5.1.

correspond exactly to the representations of $\tilde{L}G$ of positive energy. Furthermore it is worth observing that an orbit belongs to the smooth part of $(\tilde{L}\mathfrak{g})^*$ if it is stable under the rotation action of $\mathbb{T}$ on $(\tilde{L}\mathfrak{g})^*$: for if (ϕ, λ) belongs to such a stable orbit then one must have

$$\frac{d\phi}{d\theta} = [\eta, \phi] + \lambda\eta'$$

for some $\eta \in L\mathfrak{g}$, which implies that ϕ is smooth. That fits well with Kirillov's viewpoint, because representations of positive energy are stable under rotations.

4.4 The group extensions when G is simply connected

The Lie algebra extensions that we have been describing do not all correspond to Lie groups. For that to be true, a certain integrality condition must be satisfied. The Lie algebra cocycle ω is a skew form on the tangent space to LG at the identity; it therefore defines a left-invariant 2-form ω on LG, and the cocycle condition (4.2.3) translates into the fact that this differential form is closed.

Theorem (4.4.1).

(i) *If G is simply connected then the Lie algebra extension*

$$\mathbb{R} \to \tilde{L}\mathfrak{g} \to L\mathfrak{g}$$

defined by a cocycle ω corresponds to a group extension

$$\mathbb{T} \to \tilde{L}G \to LG$$

if and only if the differential form $\omega/2\pi$ represents an integral *cohomology class on LG, i.e. its integral over every 2-cycle in LG is an integer.*

(ii) *In that case the group extension $\tilde{L}G$ is completely determined by ω, and there is a unique action of $\mathrm{Diff}^+(S^1)$ on $\tilde{L}G$ which covers its action on LG.*

(iii) *If $\lambda\omega$ is not integral for any non-zero real number λ, then $\tilde{L}\mathfrak{g}$ does not correspond to any Lie group.*

(iv) *The cocycle ω defined by the formula (4.2.2) satisfies the integrality condition if and only if $\langle h_\alpha, h_\alpha \rangle$ is an even integer for each coroot h_α of G. (See Section 2.4.)*

Remark. The part of the extension $\tilde{L}G$ over the subgroup G of constant loops is *canonically* isomorphic to $G \times \mathbb{T}$, as there are no non-trivial homomorphisms $G \to \mathbb{T}$. We shall therefore often think of G as a subgroup of $\tilde{L}G$.

Let us notice at once that (4.4.1) (iii) is an immediate consequence of

4.4 GROUP EXTENSIONS WHEN G IS SIMPLY CONNECTED

(4.4.1) (i). For if there is any Lie group at all which corresponds to $\tilde{L}\mathfrak{g}$ it will be an extension of LG by either $\mathbb{R}$ or $\mathbb{T}$, and an extension by $\mathbb{R}$ gives rise to an extension by $\mathbb{T}$. The multiplier λ corresponds to the different ways of identifying the Lie algebra of $\mathbb{T}$ with $\mathbb{R}$. This result, together with Theorem (4.4.1) (iv), provides us with a class of Lie algebras which do not correspond to any Lie group, for as soon as G has more than one simple factor a generic invariant inner product on $\mathfrak{g}$ is not a multiple of one which satisfies the integrality conditions.

The 'if' part of Theorem (4.4.1) (i) is deduced from the following quite general result, which will be proved in the next section. We shall return to the 'only if' part in Proposition (4.5.6).

Proposition (4.4.2). *Suppose that a Lie group Γ acts smoothly on a connected and simply connected manifold X, leaving invariant an integral closed 2-form $\omega/2\pi$ on X. (Both Γ and X may be infinite dimensional.) Then there is an extension $\tilde{\Gamma}$ of Γ by $\mathbb{T}$ canonically associated to (X, ω), and for any point $x \in X$ the associated extension of Lie algebras can be represented by the cocycle*

$$(\xi, \eta) \mapsto \omega(\xi_x, \eta_x),$$

where ξ_x denotes the tangent vector at $x \in X$ corresponding to the action of the infinitesimal element ξ of Γ.

The group $\tilde{\Gamma}$ in this proposition can be described quite explicitly. The integral closed 2-form ω allows us to associate to each piecewise smooth loop ℓ in X an element $C(\ell)$ of $\mathbb{T}$ by

$$C(\ell) = \exp i \int_\sigma \omega,$$

where σ is a piece of surface in X bounded by ℓ. (If σ and σ' are two such surfaces then $\int_\sigma \omega$ and $\int_{\sigma'} \omega$ differ by a multiple of 2π because $\omega/2\pi$ is integral; so $C(\ell)$ is well defined.) The assignment $\ell \mapsto C(\ell)$ has the three properties:

(H1) *independence of parametrization*, i.e. $C(\ell) = C(\ell \circ \phi)$ when $\phi : S^1 \to S^1$ is any piecewise smooth map of degree 1;

(H2) *additivity*, i.e. $C(p * r^{-1}) = C(p * q^{-1})C(q * r^{-1})$, when p, q, r are three paths from x_0 to x_1, and $p * q^{-1}$ denotes the loop obtained by performing p followed by the reverse of q; and

(H3) *Γ-invariance*, i.e. $C(\gamma \cdot \ell) = C(\ell)$ for any $\gamma \in \Gamma$.

Any map $\ell \mapsto C(\ell)$ with these three properties defines a central extension $\tilde{\Gamma}$ of Γ by $\mathbb{T}$, as follows. We choose a base-point x_0 in X. Then an element of $\tilde{\Gamma}$ is represented by a triple (γ, p, u), where $\gamma \in \Gamma$, $u \in \mathbb{T}$, and p is a path in X from x_0 to $\gamma \cdot x_0$. Two triples (γ, p, u) and

(γ', p', u') are regarded as equivalent if $\gamma = \gamma'$ and $u = C(p' * p^{-1}) \cdot u'$. The composition in $\tilde{\Gamma}$ is given by

$$(\gamma_1, p_1, u_1) \cdot (\gamma_2, p_2, u_2) = (\gamma_1\gamma_2, p_1 * \gamma_1 \cdot p_2, u_1 u_2).$$

It is easy to check that $\tilde{\Gamma}$ is a well-defined group.

The description of $\tilde{\Gamma}$ just given is very convenient and we shall often make use of it in this chapter. It is not, however, the best way to see that $\tilde{\Gamma}$ is a Lie group, or to understand its global topology; for that reason we shall give a different proof of (4.4.2) in the next section.

To obtain the desired central extensions of LG we can apply Proposition (4.4.2) with $\Gamma = X = LG$, for if G is simply connected then LG is simply connected too. Indeed as a space LG is the product $G \times \Omega G$, where G is the subgroup of constant loops and ΩG is the subgroup of loops γ such that $\gamma(1) = 1$. Thus

$$\pi_1(LG) \cong \pi_1(G) \oplus \pi_1(\Omega G) \cong \pi_1(G) \oplus \pi_2(G).$$

It is a classical theorem that $\pi_2(G) = 0$ for any compact Lie group: we shall assume this for the present, but a proof will be given in Section 8.6.

The explicit construction of $\tilde{L}G$ makes clear that $\text{Diff}^+(S^1)$ acts on it, for $\text{Diff}^+(S^1)$ acts on $X = LG$ preserving the 2-form ω and leaving fixed the identity element. That proves part of (4.4.1) (ii). We shall postpone to the next section the proof that there is no other extension of LG with the Lie algebra cocycle ω.

Now let us turn to (4.4.1) (iv). There is a so-called 'transgression' homomorphism

$$\tau : H^3(G) \to H^2(\Omega G), \tag{4.4.3}$$

where the cohomology has either real or integer coefficients, defined as the composite

$$H^3(G) \to H^3(S^1 \times \Omega G) \to H^2(\Omega G),$$

where the first map is induced by the evaluation $S^1 \times \Omega G \to G$, and the second is integration over S^1. (Cf. [18] p. 247.) When G is simply connected the transgression τ is an isomorphism: it reduces to the transpose of the obvious isomorphism $\pi_2(\Omega G) \to \pi_3(G)$ when one uses the Hurewicz isomorphisms $\pi_2(\Omega G) \cong H_2(\Omega G)$ and $\pi_3(G) \cong H_3(G)$. Thus (4.4.1) (iv) is obtained by putting together the following two results.

Proposition (4.4.4). *Let σ denote the left-invariant 3-form on G whose value at the identity element is given by*

$$\sigma(\xi, \eta, \zeta) = \langle [\xi, \eta], \zeta \rangle.$$

Then the transgression $\tau(\sigma)$ is cohomologous to the invariant form $\omega/2\pi$ on ΩG.

4.4 GROUP EXTENSIONS WHEN G IS SIMPLY CONNECTED

Proposition (4.4.5). *The skew form σ of (4.4.4) defines an integral cohomology class on the simply connected group G if and only if $\langle h_\alpha, h_\alpha \rangle \in 2\mathbb{Z}$ for each coroot h_α of G.*

Proof of (4.4.5). For each root α there is a homomorphism $i_\alpha : SU_2 \to G$ which on the diagonal matrices in SU_2 induces the coroot h_α—see Section 2.4. When the form σ is pulled back to $SU_2 \cong S^3$ by i_α one obtains $\frac{1}{2}\langle h_\alpha, h_\alpha \rangle \sigma_0$, where σ_0 is the invariant 3-form on SU_2 with integral 1. Thus (4.4.5) follows from the fact that the maps i_α generate $\pi_3(G)$. It is enough to prove this for simple groups G, and we can also replace G by any locally isomorphic group. One has then to show that for suitable α the homogeneous space $G/i_\alpha(SU_2)$ is 3-connected. That is obvious for the classical groups. We refer to Bott [14] for a simple proof by Morse theory which works in all cases. In fact α can always be taken to be the highest root.

Proof of (4.4.4). When the form $2\pi\sigma$ is pulled back to $S^1 \times \Omega G$ its value at the point (θ, γ) on the triple of tangent vectors $(\delta\theta, \delta_1\gamma, \delta_2\gamma)$ is

$$\frac{1}{4\pi} \langle \gamma(\theta)^{-1}\gamma'(\theta), [\xi_1(\theta), \xi_2(\theta)] \rangle \delta\theta,$$

where $\xi_i(\theta) = \gamma(\theta)^{-1}\delta_i\gamma(\theta)$. This is to be integrated over S^1 and compared with

$$\omega_\gamma(\delta_1\gamma, \delta_2\gamma) = \frac{1}{2\pi} \int_0^{2\pi} \langle \xi_1(\theta), \xi_2'(\theta) \rangle \, d\theta.$$

Consider the 1-form β on ΩG given by

$$\beta_\gamma(\delta\gamma) = \frac{1}{4\pi} \int_0^{2\pi} \langle \gamma(\theta)^{-1}\gamma'(\theta), \gamma(\theta)^{-1}\delta\gamma(\theta) \rangle \, d\theta.$$

A simple calculation shows that $d\beta = \tau(2\pi\sigma) - \omega$.

If $\mathfrak{g}$ is a simple algebra then all invariant inner products on it are proportional, and so there is a smallest one satisfying the integrality condition (4.4.5). We shall call this the *basic inner product*, and the associated extension the *basic central extension* of LG. If G is simply laced then the basic inner product is the one discussed in Section 2.5 for which $\langle h_\alpha, h_\alpha \rangle = 2$ for every coroot. In general it is characterized by the property that $\langle h_\alpha, h_\alpha \rangle = 2$ when α is the highest root. The Killing form [1] on $\mathfrak{g}$ satisfies the integrality condition, so it is an integer multiple of the basic form. We shall obtain a formula for the integer in Section 14.5. When G is simply laced it is the Coxeter number ([20] Chapter 6, Section 1.11) of G.

The basic central extension is universal.

Proposition (4.4.6). *If G is simply connected and simple then the extension $\tilde{L}G$ associated to the basic inner product is itself simply connected. It is the unique simply connected extension of LG by $\mathbb{T}$, and it is the universal central extension in the category of Lie groups. Furthermore $\pi_2(\tilde{L}G) = 0$.*

Proof. We calculate $\pi_1(\tilde{L}G)$ and $\pi_2(\tilde{L}G)$ from the homotopy exact sequence of the circle bundle $\mathbb{T} \to \tilde{L}G \to LG$. This gives us

$$0 \to \pi_2(\tilde{L}G) \to \pi_2(LG) \to \pi_1(\mathbb{T}) \to \pi_1(\tilde{L}G) \to 0.$$

The map $\pi_2(LG) \to \pi_1(\mathbb{T}) = \mathbb{Z}$ is defined by evaluating the first Chern class of the bundle on 2-spheres in LG; and the basic extension is defined so that the first Chern class generates H^2 of the 2-sphere corresponding to the highest root. Thus $\pi_1(\tilde{L}G) = 0$; and because we know $\pi_2(LG) \cong \pi_3(G) \cong \mathbb{Z}$ it follows that $\pi_2(\tilde{L}G) = 0$ too.

To prove the universality, let $A \to E \to LG$ be an arbitrary central extension. The corresponding Lie algebra extension can be defined by a skew form

$$\omega_A : L\mathfrak{g} \times L\mathfrak{g} \to \mathfrak{a},$$

where $\mathfrak{a}$ is the Lie algebra of A. Because $\tilde{L}\mathfrak{g}$ is universal (see (4.2.4)) we know that $\omega_A = \phi \circ \omega$, where ω is the basic cocycle and $\phi : \mathbb{R} \to \mathfrak{a}$ is some map. It follows that if we pull E back to $\tilde{L}G$, i.e. we form $\tilde{E} = \tilde{L}G \times_{LG} E$, the subgroup of $\tilde{L}G \times E$ consisting of pairs (x, y) such that x and y have the same image in LG, then the resulting Lie algebra extension of $\tilde{L}\mathfrak{g}$ by $\mathfrak{a}$ is trivial. But we shall see in the next section that an extension of a simply connected group such as $\tilde{L}G$ is trivial if its Lie algebra cocycle is trivial. So $\tilde{E}$ is a trivial extension of $\tilde{L}G$, and its splitting map gives us the desired homomorphism $\tilde{L}G \to E$.

Let us notice that the induced homomorphism $\mathbb{T} \to A$ can be determined as the image of the generator of $\pi_2(LG)$ in $\pi_1(A)$, for $\pi_1(A) \cong \text{Hom}(\mathbb{T}; A)$.

4.5 Circle bundles, connections, and curvature

The main object of this section is to prove Proposition (4.4.2), but we shall begin by summarizing the facts about circle bundles, connections, and curvature that we shall make use of. This material is all well known (cf. [27], [68]), at least in the finite dimensional case. At the end of the summary we shall give brief proofs of the essential points.

Suppose that $\pi : Y \to X$ is a smooth principal fibre bundle whose fibre is a circle and whose base X is a possibly infinite dimensional manifold. This means that the group $\mathbb{T}$ acts freely on Y, the fibres are its orbits, and X is the orbit-space $Y/\mathbb{T}$. A *connection* in the bundle is a prescription

4.5 CIRCLE BUNDLES, CONNECTIONS, AND CURVATURE

which decomposes the tangent space $T_y Y$ at a point $y \in Y$ as

$$\mathbb{R} \oplus T_y^{\text{horiz}} Y,$$

where $\mathbb{R}$ is the tangent space along the fibre and $T_y^{\text{horiz}} Y$—the 'horizontal' tangent vectors—is a replica of $T_{\pi(y)} X$. The decomposition is required to be invariant under the action of $\mathbb{T}$ on Y.

A connection tells one how a path in X can be lifted to a horizontal path in Y with a prescribed starting point. If one lifts a closed path in X the lifted path in Y will in general fail to close. The gap between its ends corresponds to an element of $\mathbb{T}$ called the *holonomy* around the path. The *curvature* of the connection measures the holonomy around infinitesimally small closed paths: it is the closed 2-form ω on X whose value on a pair of tangent vectors ξ, η at a point of X is the infinitesimal holonomy around the parallelogram spanned by ξ and η. The curvature defines an element of the cohomology group $H^2(X; \mathbb{R})$ which depends only on the topological type of the bundle. Moreover $\omega/2\pi$ is an *integral* class—i.e. its integral over any 2-cocycle in X is an integer—and it comes from a well-defined element of $H^2(X; \mathbb{Z})$ called the (first) *Chern class* of the bundle. The Chern class describes the topological type of the circle bundle completely, and any element of $H^2(X; \mathbb{Z})$ arises from a bundle. If X is simply connected then the natural map

$$i : H^2(X; \mathbb{Z}) \to H^2(X; \mathbb{R})$$

is injective, and the topological type is completely determined by the class of $\omega/2\pi$. In general the kernel of i corresponds to the *flat* bundles, i.e. those which can be given a connection with curvature zero.

Analytically a connection can be described in three ways.

(i) One can give the map $\xi \mapsto \tilde{\xi}$ which to each vector field ξ on X assigns the corresponding horizontal $\mathbb{T}$-invariant vector field $\tilde{\xi}$ on Y. From this point of view the curvature is given by

$$\omega(\xi, \eta) = [\tilde{\xi}, \tilde{\eta}] - [\xi, \eta]^{\sim}. \tag{4.5.1}$$

(The right hand side of this equation is a $\mathbb{T}$-invariant vertical vector field on Y; but we can identify it with a real-valued function on X.)

(ii) One can give the $\mathbb{T}$-invariant 1-form α on Y which assigns to a tangent vector to Y its vertical component, a real number. The restriction of α to each fibre is the standard 1-form $d\theta$. The derivative $d\alpha$ is $\mathbb{T}$-invariant and vanishes on vertical vectors, so $d\alpha = \pi^* \omega$ for a unique closed 2-form ω on X, which is the curvature.

(iii) One can introduce local trivializations of Y. That is, X is covered by open sets $\{U_a\}$, and the part of Y over U_a is identified with $U_a \times \mathbb{T}$. Then the connection is described in U_a by the 1-form $\alpha_a = s_a^* \alpha$ which is obtained by pulling back the form α described above by any section

$s_a : U_a \to Y$ which is constant in terms of the local trivialization $U_a \times \mathbb{T}$. In U_a the curvature ω is described by $d\alpha_a = \omega$. If the transition functions of the bundle are

$$f_{ab} : U_a \cap U_b \to \mathbb{T}$$

(i.e. the point $(x, \rho) \in U_a \times \mathbb{T}$ is the same point as $(x, f_{ab}(x)\rho) \in U_b \times \mathbb{T}$), then

$$\alpha_b = \alpha_a + i f_{ab}^{-1} \, df_{ab}. \tag{4.5.2}$$

That completes our summary. We shall now prove the essential result.

Proposition (4.5.3). *Let X be a connected and simply connected manifold.*

(i) If ω is a closed 2-form on X such that $\omega/2\pi$ represents an integral cohomology class then there is a circle bundle on X with a connection whose curvature is ω.

(ii) If Y and Y' are circle bundles on X with connections α and α' which have the same curvature ω then there is an isomorphism $\psi : Y \to Y'$ such that $\psi^*\alpha' = \alpha$. Furthermore ψ is unique up to composition with the action of an element of $\mathbb{T}$.

Proof.

(i) One way of expressing the condition that $\omega/2\pi$ is integral is to say that there is an integral Čech cocycle $\{v_{abc}\}$ defined with respect to an open covering $\{U_a\}$ of X such that $\omega = d\beta + 2\pi v$, for some 1-form β, where v is the 2-form associated to $\{v_{abc}\}$ by means of a smooth partition of unity $\{\lambda_a\}$ subordinate to $\{U_a\}$—see (3.1.1). (Thus v_{abc} is an integer defined when $U_a \cap U_b \cap U_c$ is non-empty, and we can assume that it is skew with respect to changing the order of a, b, c.)

Let us construct a bundle Y on X by means of the transition functions $\{f_{ab}\}$, where

$$f_{ab}(x) = \exp 2\pi i \sum_c v_{abc} \lambda_c.$$

The coherence of the f_{ab} follows from the cocycle condition

$$v_{bcd} - v_{acd} + v_{abd} - v_{abc} = 0.$$

A connection in this bundle is defined by the 1-forms

$$\alpha_a = 2\pi \sum_{b,c} v_{abc} \lambda_b \, d\lambda_c,$$

its curvature is the 2-form $2\pi v$:

$$2\pi v = 2\pi \sum_{a,b,c} v_{abc} \lambda_a \, d\lambda_b \wedge d\lambda_c.$$

4.5 CIRCLE BUNDLES, CONNECTIONS, AND CURVATURE

To obtain a connection with curvature ω we simply add the 1-form β to each α_a.

(ii) Suppose that Y and Y' are defined with respect to the same open covering $\{U_a\}$ of X by transition functions $\{f_{ab}\}$ and $\{f'_{ab}\}$. A map $\psi: Y \to Y'$ will be given locally by functions $\psi_a: U_a \to \mathbb{T}$ such that

$$\psi_b(x) f_{ab}(x) = f'_{ab}(x) \psi_a(x) \quad \text{for} \quad x \in U_a \cap U_b. \tag{4.5.4}$$

The condition $\psi^* \alpha' = \alpha$ is expressed by

$$\alpha'_a = \alpha_a + i \psi_a^{-1} \, d\psi_a. \tag{4.5.5}$$

It is permissible to assume that each set U_a is contractible, and each intersection $U_a \cap U_b$ connected. Then we can find a function $\phi_a: U_a \to \mathbb{R}$ such that $d\phi_a = \alpha'_a - \alpha_a$. From (4.5.2) we then find that

$$d(e^{-i\phi_b} f_{ab}) = d(f'_{ab} e^{-i\phi_a})$$

in $U_a \cap U_b$. In other words

$$e^{-i\phi_b} f_{ab} = f'_{ab} e^{-i\phi_a} e^{i\mu_{ab}},$$

where $\mu_{ab} \in \mathbb{R}$ is constant. Because X is simply connected (and $\{\mu_{ab}\}$ is a Čech 1-cocycle) we can find numbers m_a such that $\mu_{ab} = m_b - m_a$. Then the functions $\psi_a = e^{-i(\phi_a + m_a)}$ satisfy both (4.5.4) and (4.5.5).

As to the uniqueness of ψ, or equivalently of the functions ψ_a, it follows from (4.5.4) that any two possible choices differ by multiplication by a global function $g: X \to \mathbb{T}$. The equations (4.5.5) then show that g is constant.

Proof of (4.4.2). We can now prove (4.4.2) very simply. We first construct a circle bundle Y on X with a connection α with curvature ω. For each $\gamma \in \Gamma$ we can pull back Y by the map $\gamma: X \to X$. The resulting bundle $\gamma^* Y$ has a connection α_γ whose curvature is $\gamma^* \omega = \omega$. We now define $\tilde{\Gamma}$ as the group of all pairs (γ, ψ) with $\gamma \in \Gamma$ and $\psi: Y \to \gamma^* Y$ an isomorphism such that $\psi^* \alpha_\gamma = \alpha$. By Proposition (4.5.3) there is a circle of possible choices of ψ for each γ. Ths isomorphism ψ can equally well be regarded as a map $\psi: Y \to Y$ which covers the action of γ on X and satisfies $\psi^* \alpha = \alpha$. In other words $\tilde{\Gamma}$ is simply the group of all fibre-preserving maps $\psi: Y \to Y$ which preserve α and cover an element γ of Γ. Such a map is completely determined by γ and $\psi(y_0)$, where y_0 is an arbitrary base-point in Y. Thus as a manifold $\tilde{\Gamma}$ is the fibre product $\Gamma \times_X Y$, the pull-back of Y by the map $\Gamma \to X$ which takes γ to $\gamma(\pi(y_0))$.

To identify the description of $\tilde{\Gamma}$ just given with the description by triples (γ, p, u) given after the statement of (4.4.2) we associate to (γ, p, u) the unique automorphism of Y which preserves α and maps y_0 to the point obtained by parallel transport of y_0 along p. (We assume

$\pi(y_0) = x_0$.) We leave it to the reader to check that this defines a group isomorphism.

Proposition (4.4.2) proves one half of (4.4.1) (i). The other half, asserting that if a cocycle ω on $L\mathfrak{g}$ corresponds to a group extension then $\omega/2\pi$ is integral, is now fairly obvious, but we shall restate it in the form:

Proposition (4.5.6). *If an extension of Lie groups*

$$\mathbb{T} \to \tilde{\Gamma} \to \Gamma$$

corresponds to the Lie algebra cocycle ω on the Lie algebra of Γ, then $\omega/2\pi$, regarded as a left-invariant 2-form on Γ, represents the Chern class of the bundle $\tilde{\Gamma}$, and is therefore integral.

Proof. This follows at once from the first method of describing a connection. To obtain the cocycle ω we must choose a vector space decomposition of the Lie algebra $\mathrm{Lie}(\tilde{\Gamma})$ of $\tilde{\Gamma}$:

$$\mathrm{Lie}(\tilde{\Gamma}) \cong \mathbb{R} \oplus \mathrm{Lie}(\Gamma).$$

This induces a decomposition of the tangent space at each point of $\tilde{\Gamma}$, i.e. a connection in $\tilde{\Gamma} \to \Gamma$. The splitting map $\mathrm{Lie}(\Gamma) \to \mathrm{Lie}(\tilde{\Gamma})$ can be identified with the horizontal lifting $\xi \mapsto \tilde{\xi}$ of left-invariant vector fields; so by the formula (4.5.1) the curvature ω is given by the same expression

$$\omega(\xi, \eta) = [\tilde{\xi}, \tilde{\eta}] - [\xi, \eta]^{\sim}$$

which defines the Lie algebra cocycle.

The argument we have just given suffices also to complete the two remaining proofs—of (4.4.6) and (4.4.1) (ii)—which were postponed from the preceding section. First, a central extension $A \to E \to \Gamma$ of a simply connected group Γ is trivial if the induced extension of Lie algebras is trivial. For then the principal bundle E has a flat connection, and we can define a map $\Gamma \to E$ which takes γ to the end-point of the horizontal lift of any path in Γ from the identity to γ. The map is a homomorphism because the connection is E-invariant. That completes the proof of (4.4.6).

To complete (4.4.1) (ii) we must show that two extensions E and E' of LG by $\mathbb{T}$ are isomorphic if their Lie algebra cocycles coincide. To do that we form the 'difference' extension $\mathbb{T} \to E'' \to LG$, i.e. we pull E' back to E, and then pass to the quotient by the image of the homomorphism $\mathbb{T} \to E \times_{LG} E'$ which takes u to (u, u^{-1}). The Lie algebra extension corresponding to E'' is trivial, and so E'' itself is trivial, and $E \cong E'$.

4.6 The group extensions when G is semisimple but not simply connected

If G is a semisimple group we can write $G = \tilde{G}/Z$, where $\tilde{G}$ is the simply connected covering group of G, and $Z \cong \pi_1(G)$ is a finite subgroup of the centre of $\tilde{G}$.

We have seen that an integral bilinear form $\langle\ ,\ \rangle$ on $\mathfrak{g}$ gives rise to a unique central extension $\tilde{L}\tilde{G}$ of $L\tilde{G}$. Because $\tilde{G}$ can be identified canonically with a subgroup of $\tilde{L}\tilde{G}$ we can regard Z as a subgroup of $\tilde{L}\tilde{G}$. In fact Z belongs to the centre of $\tilde{L}\tilde{G}$, because its adjoint action on $\tilde{L}\mathfrak{g}$ is trivial. (That follows from (4.3.2).) Thus we have an extension

$$\mathbb{T} \to (\tilde{L}\tilde{G})/Z \to (LG)^0, \qquad (4.6.1)$$

where $(LG)^0 \cong (L\tilde{G})/Z$ is the identity component of LG. But (4.6.1) is usually *not* the restriction of an extension of the whole group LG. To understand this we observe that the form $\langle\ ,\ \rangle$ on $\mathfrak{g}$ induces a pairing

$$c : Z \times Z \to \mathbb{T} \qquad (4.6.2)$$

in the following way. Let $\tilde{T}$ be a maximal torus of $\tilde{G}$. Given $z_1, z_2 \in Z$, choose ζ_1, ζ_2 in the Lie algebra of $\tilde{T}$ so that $\exp(2\pi\zeta_i) = z_i$. Then define

$$c(z_1, z_2) = e^{2\pi i \langle \zeta_1, \zeta_2 \rangle}.$$

The pairing is independent of the choices made.

Lemma (4.6.3). *The extension (4.6.1) is not the restriction of an extension of LG unless the pairing c is trivial.*

Proof. Consider the automorphism A_λ of $L\mathfrak{g}$ induced by conjugating by an element λ of LG which does not belong to the identity component. We can lift A_λ uniquely to an automorphism $\tilde{A}_\lambda$ of $\tilde{L}\mathfrak{g}$: a simple calculation (cf. (4.3.2)) shows that

$$\tilde{A}_\lambda(\xi, a) = (A_\lambda \xi, a - \langle \lambda^{-1}\lambda', \xi \rangle). \qquad (4.6.4)$$

Let us apply this when $\xi \in \mathfrak{t}$ and $\exp(2\pi\xi) \in Z$, and λ is the loop $\theta \mapsto \exp(\theta\zeta)$ in G for some $\zeta \in \mathfrak{t}$ such that $\exp(2\pi\zeta) \in Z = \pi_1(G)$. We find

$$\tilde{A}_\lambda(\xi, 0) = (\xi, -\langle \xi, \zeta \rangle).$$

Because the exponential map of $(\tilde{L}G)^0$ takes $(2\pi\xi, 0)$ to 1 we see that $\tilde{A}_\lambda$ cannot induce an automorphism of $(\tilde{L}G)^0$ unless $\langle \xi, \zeta \rangle \in \mathbb{Z}$. That proves (4.6.3).

In the other direction, however, we can assert

Lemma (4.6.5). *The conjugation action of LG on $\tilde{L}\tilde{G}$ lifts uniquely to an* action on $\tilde{L}\tilde{G}$, and then the action of $Z \cong \pi_1(G)$ on the centre $Z \times \mathbb{T}$ of $\tilde{L}\tilde{G}$ is given by (4.6.2).

Proof. We first notice that the lift, if it exists, will be unique. In fact $\text{Aut}(\tilde{L}\tilde{G})$ can be identified with a subgroup of $\text{Aut}(L\tilde{G})$ because $L\tilde{G}$ is a perfect group (see (3.4.1)) and there are no non-trivial homomorphisms $L\tilde{G} \to \mathbb{T}$.

The extension $L\tilde{G}$ is defined by a 2-form ω on $L\tilde{G}$. The same calculation that gave us (4.6.4) shows that if c_λ denotes conjugation by $\lambda \in LG$ we have

$$c_\lambda^* \omega = \omega - d\beta, \qquad (4.6.6)$$

where β is the left-invariant 1-form given by

$$\beta(\xi) = \langle \lambda^{-1}\lambda', \xi \rangle. \qquad (4.6.7)$$

Let us think of elements of $L\tilde{G}$ as triples (γ, p, u) as in Section 4.4. Then $c_\lambda : L\tilde{G} \to L\tilde{G}$ is covered by the automorphism

$$(\gamma, p, u) \mapsto (c_\lambda \gamma, c_\lambda p, e^{i\beta \cdot p} u), \qquad (4.6.8)$$

where $\beta \cdot p$ denotes the integral of β along p.

We can now describe a class of extensions of LG for semisimple groups G.

Proposition (4.6.9). *For any integral inner product $\langle \, , \, \rangle$ on $\mathfrak{g}$ there is a group $\bar{L}G$ whose identity component is $\tilde{L}\tilde{G}$ and whose group of components is $Z \cong \pi_1(G)$. It is an extension of LG by $\mathbb{T} \times Z$, and the conjugation action of the group of components on $\mathbb{T} \times Z$ is given by (4.6.2).*

The extension $\bar{L}G$ is not determined uniquely by $\langle \, , \, \rangle$, but is unique up to the addition of an arbitrary extension of $\pi_1(G)$ by $\mathbb{T}$.

Remark. An extension of $\pi_1(G)$ by $\mathbb{T}$ clearly gives an extension of LG by $\mathbb{T}$, and this can be added—in the usual sense of addition for group extensions—to $\bar{L}G$ without changing the identity component.

Proof. Let us choose a maximal torus T in G, and let $\Lambda = \check{T} = \text{Hom}(\mathbb{T}; T)$. Thus Λ is a subgroup of LG. The part Λ_0 of Λ which is contained in the identity component of LG can be identified with a subgroup of $L\tilde{G}$. Let $\tilde{\Lambda}_0$ be the extension of Λ_0 by $\mathbb{T}$ induced by $\tilde{L}\tilde{G}$. Suppose that $\tilde{\Lambda}_0$ can be extended to an extension $\tilde{\Lambda}$ of Λ. Then we can form the semidirect product $\tilde{\Lambda} \tilde{\times} \tilde{L}\tilde{G}$, where $\tilde{\Lambda}$ acts on LG through Λ, which acts on $\tilde{L}\tilde{G}$ by (4.6.5). But the antidiagonal map $\tilde{\Lambda}_0 \to \tilde{\Lambda} \tilde{\times} \tilde{L}\tilde{G}$ (i.e. $\lambda \mapsto (\lambda, \lambda^{-1})$) embeds $\tilde{\Lambda}_0$ as a normal subgroup, and we can define $\bar{L}G$ as the quotient group $(\tilde{\Lambda} \tilde{\times} \tilde{L}\tilde{G})/\tilde{\Lambda}_0$.

Returning to the existence of $\tilde{\Lambda}$, we observe that an extension $\tilde{\Lambda}$ of a free abelian group Λ by $\mathbb{T}$ is completely determined by the commutator map $(x, y) \mapsto xyx^{-1}y^{-1}$, which is a skew biadditive map $\Lambda \times \Lambda \to \mathbb{T}$. This can be lifted to a skew biadditive map $\Lambda \times \Lambda \to \mathbb{R}$. Conversely any such map $b : \Lambda \times \Lambda \to \mathbb{R}$ defines an extension of Λ by $\mathbb{T}$ with the cocycle

$$(\lambda, \mu) \mapsto e^{\frac{1}{2}ib(\lambda, \mu)}.$$

Thus $\tilde{\Lambda}$ can be obtained by extending arbitrarily the skew map which defines $\tilde{\Lambda}_0$; the ambiguity is precisely an arbitrary extension of $\Lambda/\Lambda_0 = \pi_1(G)$.

Remark. The extension $\tilde{L}G$ is most appealing when G is simply laced and $\langle\ ,\ \rangle$ is the basic inner product. Then the pairing c of (4.6.2) is nondegenerate, and the centre of $\tilde{L}G$ is exactly $\mathbb{T}$. We shall determine the induced extension of Λ_0 in this case in Section 4.8.

4.7 The basic central extension of LU_n

We shall not bother to discuss the extensions of LG when G is not semisimple except in the particular case $G = U_n$. The extension of LU_n which we shall now describe plays a central role in the theory.

Proposition (4.7.1).
 (i) *There is a canonical extension $\tilde{L}U_n$ of LU_n by $\mathbb{T}$ whose Lie algebra cocycle is given by (4.2.2), where $\langle\ ,\ \rangle$ is the basic inner product on $\mathfrak{u}_n$.*
 (ii) *The subgroup U_n of constant loops is identified canonically with a subgroup of $\tilde{L}U_n$.*
 (iii) *The centre of the identity component of $\tilde{L}U_n$ is $\mathbb{T} \times \mathbb{T}$, where the first $\mathbb{T}$ is the kernel of the extension and the second is the centre of the canonical copy of U_n. Conjugation by a loop of winding number k transforms $\mathbb{T} \times \mathbb{T}$ by*

$$(u, v) \mapsto (uv^{-k}, v).$$

 (iv) *The natural action of $\mathrm{Diff}^+(S^1)$ on $L\mathfrak{u}_n$ comes from a unique action of the double covering of $\mathrm{Diff}^+(S^1)$ on $\tilde{L}U_n$. (But see Remark (4.7.2) below.)*

Remark. The extension $\tilde{L}U_n$ is determined completely by the Lie algebra extension, but only up to non-canonical isomorphism. (For any group Γ the group of automorphisms of Γ which are the identity on Γ^0 and Γ/Γ^0 is $\mathrm{Hom}(\Gamma/\Gamma^0; Z)$, where Z is the centre of Γ^0.) Furthermore there is an infinite dimensional space—namely $\mathrm{Hom}(L\mathfrak{u}_n; \mathbb{R})$—of automorphisms of $\tilde{L}\mathfrak{u}_n$ which induce the identity on $L\mathfrak{u}_n$. These facts make the study of $\tilde{L}U_n$ quite confusing, especially where the action of $\mathrm{Diff}^+(S^1)$ is concerned.

Proof. We shall be brief, as the details are much the same as we have encountered in the preceding sections.

We construct the desired extension of the identity component $(LU_n)^0$ by applying Proposition (4.4.2) to the simply connected space $X = (LU_n)^0/T$, where T is the standard maximal torus of U_n. Then we observe that LU_n is a semidirect product $\mathbb{Z} \,\tilde{\times}\, (LU_n)^0$, where the copy of $\mathbb{Z}$ is generated by any loop λ of winding number 1. We can define $\tilde{L}U_n = \mathbb{Z} \,\tilde{\times}\, (\tilde{L}U_n)^0$ providing we can lift the conjugation action of λ on $(LU_n)^0$ to an automorphism of $(\tilde{L}U_n)^0$. The argument of (4.6.5) shows that the lift is possible, and also proves assertion (iii). The essential point is that we can choose λ to be a loop in the abelian subgroup T, so that conjugation by λ does act on X. The construction depends on the choice of λ, but there is a canonical choice.

To investigate the action of $\text{Diff}^+(S^1)$ on $\tilde{L}U_n$ we introduce the simply connected covering group $\mathcal{D}$ of $\text{Diff}^+(S^1)$. This can be realized as the group of diffeomorphisms $\phi : \mathbb{R} \to \mathbb{R}$ such that $\phi(\theta + 2\pi) = \phi(\theta) + 2\pi$. To prove that the action of $\mathcal{D}$ on LU_n lifts to $\tilde{L}U_n$ it is enough to construct an extension of the semidirect product $\mathcal{D} \,\tilde{\times}\, LU_n$ by $\mathbb{T}$ which restricts to $\tilde{L}U_n$ over LU_n and is trivial over $\mathcal{D}$. That can be done by first constructing the extension over the connected component $\mathcal{D} \,\tilde{\times}\, (LU_n)^0$ by applying our standard procedure to the homogeneous space $Y = \mathcal{D} \,\tilde{\times}\, (LU_n)^0/T$, on which the usual 2-cocycle of $L\mathfrak{u}_n$ defines an invariant integral 2-form. Then one must lift the conjugation action of λ on $\mathcal{D} \,\tilde{\times}\, (LU_n)^0$. The equation (4.6.6) is still valid on the larger space, where now β is the invariant form defined by the linear map

$$\left(f\frac{\mathrm{d}}{\mathrm{d}\theta}, \xi\right) \mapsto \langle \lambda'\lambda^{-1}, \tfrac{1}{2}f\lambda'\lambda^{-1} + \xi\rangle$$

on the Lie algebra of $\mathcal{D} \,\tilde{\times}\, (LU_n)^0$. Thus the argument is as before.

Finally we must calculate the action on $\tilde{L}U_n$ of the central element τ of $\mathcal{D}$ defined by $\tau(\theta) = \theta + 2\pi$. It is enough to calculate the conjugation action of λ on the element $(\tau, p, 1)$ of the extension of $\mathcal{D} \,\tilde{\times}\, (LU_n)^0$, where p is the obvious path in Y from the base-point to τ. If we assume that λ is a homomorphism $\mathbb{T} \to T$ then the path p is left fixed by the conjugation, and so the formula (4.6.8) shows that $(\tau, p, 1)$ is multiplied by the element $e^{i\beta \cdot p}$ of the centre. Because $\langle \lambda'\lambda^{-1}, \lambda'\lambda^{-1}\rangle = 1$ we find that $e^{i\beta \cdot p} = -1$. This means that τ acts on the components of odd winding number in $\tilde{L}U_n$ by multiplication by -1. That completes the proof of (4.7.1).

Remark (4.7.2). The action of $\mathcal{D}$ on $\tilde{L}U_n$ is completely determined by its action on the Lie algebra $\tilde{L}\mathfrak{u}_n$ (for the kernel of $\text{Aut}(\tilde{L}U_n) \to \text{Aut}((\tilde{L}U_n)^0)$ is abelian). But there are automorphisms of $\tilde{L}\mathfrak{u}_n$ which do not extend to $\tilde{L}U_n$, and so the action of $\mathcal{D}$ on $\tilde{L}U_n$ can be changed without affecting its action on $\tilde{L}\mathfrak{u}_n$ up to isomorphism. Thus despite Proposition (4.7.1) (iv) it *is* possible to lift the action of $\text{Diff}^+(S^1)$ from LU_n to $\tilde{L}U_n$. One way is to

4.7 THE BASIC CENTRAL EXTENSION OF $L\mathbb{T}$

make $\phi \in \mathcal{D}$ act on an element $\tilde{\gamma} \in \tilde{L}U_n$ above $\gamma \in LU_n$ by

$$(\phi, \tilde{\gamma}) \mapsto A_\phi(\tilde{\gamma}) \cdot e^{ia(\phi, f)} \tag{4.7.3}$$

where

$$e^{if(\theta)} = \det \gamma(\theta),$$
$$a(\phi, f) = \frac{1}{4\pi} \int_0^{2\pi} (f(\phi^{-1}\theta) - f(\theta)) \, d\theta, \tag{4.7.4}$$

and A_ϕ is the action of $\phi \in \mathcal{D}$ described in Proposition (4.7.1). If ϕ is the translation $\theta \mapsto \theta + \alpha$ then

$$a(\phi, f) = -\tfrac{1}{2}\alpha \Delta_f,$$

where Δ_f is the winding number of $\det \gamma$. This means that the central element $\tau \in \mathcal{D}$ acts trivially.

We shall leave it to the reader to check that this action of $\text{Diff}^+(S^1)$ is unique up to automorphisms of $\tilde{L}U_n$. In Section 6.8 we shall see that the particular choice (4.7.3) arises naturally.

The case $n = 1$

The basic central extension of $LU_1 = L\mathbb{T}$ can be described by an explicit cocycle, which we give here for future reference. We first observe that any element of $L\mathbb{T}$ can be written in the form e^{if}, where f is a smooth function such that

$$f(\theta + 2\pi) = f(\theta) + 2\pi \Delta_f$$

for some integer Δ_f which is the winding number of e^{if}. We shall write $\hat{f}$ for the average of f on the interval $[0, 2\pi]$, i.e.

$$\hat{f} = \frac{1}{2\pi} \int_0^{2\pi} f(\theta) \, d\theta.$$

Proposition (4.7.5). *The basic central extension of $L\mathbb{T}$ is defined by the cocycle c, where*

$$c(e^{if}, e^{ig}) = e^{iS(f,g)}$$

and

$$S(f, g) = \frac{1}{4\pi} \int_0^{2\pi} f(\theta) g'(\theta) \, d\theta + \tfrac{1}{2}\hat{f}\Delta_g + \tfrac{1}{2}\Delta_f(\hat{g} - g(0)).$$

The action on $\tilde{L}\mathbb{T}$ of a diffeomorphism $\phi \in \text{Diff}^+(S^1)$ can be taken to be

$$(e^{if}, u) \mapsto (e^{if\circ\phi^{-1}}, u e^{i(\Delta_f - 1)a(\phi, f)}),$$

where $a(\phi, f)$ is as in (4.7.4), and $\tilde{L}\mathbb{T}$ is identified with $L\mathbb{T} \times \mathbb{T}$ as a set. In particular, the rotation R_α through the angle α acts by

$$(e^{if}, u) \mapsto (R_\alpha e^{if}, u e^{-\tfrac{1}{2}i\alpha\Delta_f(\Delta_f - 1)}).$$

Proof. We leave it to the reader to check that c—which is a cocycle because S is bilinear— does define an extension of $L\mathbb{T}$ with the properties described in (4.7.1).

4.8 The restriction of the extension to LT

If T is a maximal torus of G then the lattice $\Lambda = \text{Hom}(\mathbb{T}; T)$—which we usually denote by $\check{T}$—is a subgroup of LG, and we can consider the restriction to it of an extension $\tilde{L}G$. We know how to describe the resulting extension $\tilde{\Lambda}$ of Λ by $\mathbb{T}$ only when G is a simply laced group, but the result in that case is quite striking, and is the basis for the construction of the basic representation of LG by 'vertex operators' (see Chapter 13). Let us choose representatives ε_λ in $\tilde{\Lambda}$ for the elements $\lambda \in \Lambda$, so that a general element of $\tilde{\Lambda}$ can be written $u\varepsilon_\lambda$, with $u \in \mathbb{T}$.

Proposition (4.8.1). *If G is a simply laced and simply connected group then the representatives ε_λ can be chosen so that the multiplication in $\tilde{\Lambda}$ is given by*

$$\varepsilon_\lambda \cdot \varepsilon_\mu = (-1)^{b(\lambda,\mu)} \varepsilon_{\lambda+\mu},$$

where $b: \Lambda \times \Lambda \to \mathbb{Z}/2$ is any bilinear form such that

$$b(\lambda, \lambda) \equiv \tfrac{1}{2}\langle \lambda, \lambda \rangle \pmod{2},$$

and $\langle \ , \ \rangle$ is the form on $\mathfrak{g}$ which defines $\tilde{L}G$.

Remark. If $\langle \ , \ \rangle$ is the basic form on $\mathfrak{g}$ then this multiplication formula is very reminiscent of the bracket relations (2.5.1) for the generators of $\mathfrak{g}_\mathbb{C}$:

$$[e_\lambda, e_\mu] = (-1)^{b(\lambda,\mu)} e_{\lambda+\mu}.$$

Proof. Because any extension of Λ by $\mathbb{T}$ which is abelian is trivial an extension is completely determined by its commutator map $(\lambda, \mu) \mapsto \varepsilon_\lambda \varepsilon_\mu \varepsilon_\lambda^{-1} \varepsilon_\mu^{-1}$, which is a skew biadditive map

$$\Lambda \times \Lambda \to \mathbb{T}.$$

Thus what we have to show is that $\varepsilon_\lambda \varepsilon_\mu \varepsilon_\lambda^{-1} \varepsilon_\mu^{-1} = (-1)^{\langle \lambda, \mu \rangle}$. If we choose paths p_λ and p_μ in LG from the identity to λ and μ, and use the description of $\tilde{L}G$ given in Section 4.4 then the task is to show that

$$\int_\sigma \omega \equiv \pi \langle \lambda, \mu \rangle \pmod{2\pi}, \tag{4.8.2}$$

where σ is any piece of surface bounded by the two paths $p_\lambda * \lambda . p_\mu$ and $p_\mu * \mu . p_\lambda$ from 1 to $\lambda + \mu$. It is enough to prove the formula when λ and μ are positive coroots, for the coroots span Λ. We also may as well assume that $\langle \ , \ \rangle$ is the basic inner product on $\mathfrak{g}$.

4.8 THE RESTRICTION OF THE EXTENSION TO LT

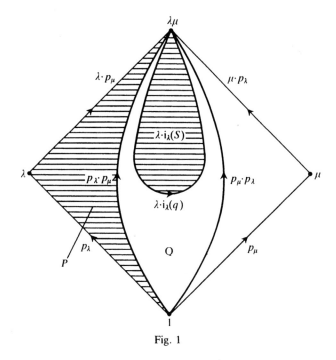

Fig. 1

If λ is a coroot there is a canonical homomorphism $i_\lambda : SU_2 \to G$ whose restriction to the diagonal matrices is λ. Let us define the path p_λ from 1 to λ as $i_\lambda \circ p$, where $p : [0, \pi] \to LSU_2$ is given by

$$p(t)(z) = \begin{pmatrix} \cos t & \sin t \\ -\sin t & \cos t \end{pmatrix} \quad \text{for} \quad 0 \leq t \leq \frac{\pi}{2},$$

and

$$= \begin{pmatrix} -z \cos t & \sin t \\ -\sin t & -z^{-1} \cos t \end{pmatrix} \quad \text{for} \quad \frac{\pi}{2} \leq t \leq \pi.$$

Now consider the piece of surface in LG given by

$$P : \{(s, t) : 0 \leq s \leq t \leq \pi\} \to LG,$$

where $P(s, t) = p_\lambda(t) p_\mu(s)$. Observing that

$$p(t)^{-1} p'(t) = \begin{pmatrix} 0 & 1 \\ -1 & 0 \end{pmatrix} \quad \text{when} \quad 0 \leq t \leq \frac{\pi}{2},$$

and

$$= \begin{pmatrix} 0 & -z^{-1} \\ z & 0 \end{pmatrix} \quad \text{when} \quad \frac{\pi}{2} \leq t \leq \pi,$$

we find that $P^*\omega = 0$. As P is bounded by the paths $p_\lambda * \lambda.p_\mu$ and $p_\lambda \cdot p_\mu$ we conclude that it is enough to integrate ω over a piece of surface bounded by $p_\lambda \cdot p_\mu$ and $p_\mu \cdot p_\lambda$. If $\langle \lambda, \mu \rangle = 0$ then $p_\lambda \cdot p_\mu = p_\mu \cdot p_\lambda$, and so the formula (4.8.2) holds in that case.

We next consider the surface

$$Q = \{(s, t) : 0 \leqslant s \leqslant t \leqslant \pi\} \to LG$$

given by $Q(s, t) = p_\lambda(s)p_\mu(t)p_\lambda(t-s)$. Once again a rather tedious check shows that $Q^*\omega = 0$. The surface Q is bounded by $p_\lambda \cdot p_\mu$, $p_\mu \cdot p_\lambda$, and the path

$$s \mapsto p_\lambda(s)p_\mu(\pi)p_\lambda(\pi - s).$$

But conjugation by

$$i_\mu \begin{pmatrix} u & 0 \\ 0 & u^{-1} \end{pmatrix}$$

normalizes $i_\lambda(SU_2)$, and corresponds to the map

$$\begin{pmatrix} a & b \\ c & d \end{pmatrix} \mapsto \begin{pmatrix} a & bu^{\langle \lambda, \mu \rangle} \\ cu^{-\langle \lambda, \mu \rangle} & d \end{pmatrix}$$

on SU_2. So

$$p_\lambda(s)p_\mu(\pi)p_\lambda(\pi - s) = p_\mu(\pi)i_\lambda(q(s)),$$

where q is some loop in SU_2. The only interesting case is when $\langle \lambda, \mu \rangle = -1$. Then we find

$$q(s) = \begin{pmatrix} z\cos 2s & \sin 2s \\ -\sin 2s & z^{-1}\cos 2s \end{pmatrix} \quad \text{if} \quad 0 \leqslant s \leqslant \frac{\pi}{2},$$

$$= \begin{pmatrix} z\cos 2s & -z\sin 2s \\ z^{-1}\sin 2s & z^{-1}\cos 2s \end{pmatrix} \quad \text{if} \quad \frac{\pi}{2} \leqslant s \leqslant \pi.$$

Because $i_\lambda^*\omega$ is the standard 2-form ω on LSU_2 we are finally reduced to proving that the integral of ω over a piece of surface in LSU_2 bounded by the loop q is $\pm \pi$. A suitable surface is given by

$$S : \{(\xi, \eta, \zeta) : \xi^2 + \eta^2 + \zeta^2 = 1, \eta \geqslant 0\} \to LSU_2,$$

where

$$S(\xi, \eta, \zeta) = \begin{pmatrix} z(\xi + i\eta) & \zeta \\ -\zeta & z^{-1}(\xi - i\eta) \end{pmatrix} \quad \text{for} \quad \zeta \geqslant 0,$$

$$= \begin{pmatrix} z(\xi + i\eta) & -z\zeta \\ z^{-1}\zeta & z^{-1}(\xi - i\eta) \end{pmatrix} \quad \text{for} \quad \zeta \leqslant 0.$$

The calculation of the integral presents no problems.

4.8 THE RESTRICTION OF THE EXTENSION TO LT

We can now write down an explicit cocycle which describes the restriction of the extension $\tilde{L}G$ to LT. An element of LT is of the form $\exp f$, where $f: \mathbb{R} \to \mathfrak{t}$ is a map such that

$$\Delta_f = \frac{1}{2\pi}(f(\theta + 2\pi) - f(\theta))$$

is constant and belongs to the lattice Λ. We shall write $\hat{f} \in \mathfrak{t}$ for the average value of f on the interval $[0, 2\pi]$.

Proposition (4.8.3). *If G is a simply laced and simply connected group then the extension of LT induced by $\tilde{L}G$ is given by the cocycle c, where*

$$c(f, g) = (-1)^{b(\Delta_f, \Delta_g)} e^{iS(f,g)},$$

and

$$S(f, g) = \frac{1}{4\pi} \int_0^{2\pi} \langle f(\theta), g'(\theta) \rangle \, d\theta + \tfrac{1}{2} \langle \hat{f}, \Delta_g \rangle + \tfrac{1}{2} \langle \Delta_f, \hat{g} - g(0) \rangle.$$

The bilinear form b on Λ is as in (4.8.1).

Proof. We first observe that c is well-defined as a map $LT \times LT \to \mathbb{T}$. It is a cocycle because it is bimultiplicative, and so it does define some extension of LT.

Over the identity component of LT the desired extension is completely described by its Lie algebra cocycle—see the discussion of $\tilde{L}U_n$ in Section 4.7. The cocycle c clearly induces the correct cocycle on the Lie algebra. Furthermore c describes the correct extension of the lattice $\Lambda \subset LT$, in view of Proposition (4.8.1), although the coset representative ε_λ of (4.8.1) has now been replaced by $e^{\frac{1}{4}i\pi\langle \lambda, \lambda \rangle}\varepsilon_\lambda$. To complete the proof it is enough to check that c gives the correct adjoint action of Λ on the Lie algebra $\tilde{L}\mathfrak{t}$. The adjoint action was calculated in Proposition (4.3.2), and is easily seen to agree with that given by c.

Remark. The cocycle of (4.8.3) is not invariant under the action of $\text{Diff}^+(S^1)$ on LT, and one cannot choose the coset representatives to make it so. If $\varepsilon(f) \in \tilde{L}T$ is a representative of $f: \mathbb{R} \to \mathfrak{t}$ chosen so that (4.8.3) holds then we have

$$\phi^*\varepsilon(f) = e^{\frac{1}{2}i\langle a(\phi, f), \Delta_f \rangle}\varepsilon(\phi^*f), \tag{4.8.4}$$

where $\phi \in \text{Diff}^+(S^1)$, and

$$a(\phi, f) = \frac{1}{4\pi} \int_0^{2\pi} (f(\phi(\theta)) - f(\theta)) \, d\theta.$$

In particular if ϕ is rotation through the angle α then

$$\phi^*\varepsilon(f) = e^{\frac{1}{2}i\alpha \langle \Delta_f, \Delta_f \rangle}\varepsilon(\phi^*f). \tag{4.8.5}$$

4 CENTRAL EXTENSIONS

Among all the cocycles describing the same extension $\tilde{L}T$ of LT the complicated-looking expression of (4.8.3) was chosen so as to make the formula (4.8.5) as simple as possible.

4.9 The inner product on $\mathbb{R} \oplus \tilde{L}\mathfrak{g}$

A nondegenerate invariant bilinear form $\langle\ ,\ \rangle$ on $\mathfrak{g}$ induces one on $L\mathfrak{g}$ by the formula

$$\langle \xi, \eta \rangle = \frac{1}{2\pi} \int_0^{2\pi} \langle \xi(\theta), \eta(\theta) \rangle\, d\theta. \qquad (4.9.1)$$

There cannot, however, be a nondegenerate invariant form on the extended algebra $\tilde{L}\mathfrak{g}$—at least if $\mathfrak{g}$ is semisimple—because the centre of a Lie algebra is orthogonal to its commutator subalgebra with respect to any invariant form, and $[\tilde{L}\mathfrak{g}, \tilde{L}\mathfrak{g}] = \tilde{L}\mathfrak{g}$. In view of this it is often useful to notice that the Lie algebra of the semidirect product $\mathbb{T} \tilde{\times} \tilde{L}G$—where $\mathbb{T}$ is the group of rigid rotations of S^1, which acts naturally on $\tilde{L}G$—does possess a canonical nondegenerate invariant bilinear form whenever the form $\langle\ ,\ \rangle$ on $\mathfrak{g}$ which defines the extension $\tilde{L}G$ is nondegenerate.

If we identify the Lie algebra $\tilde{L}\mathfrak{g}$ with $L\mathfrak{g} \oplus \mathbb{R}$ as in Section 4.2, and identify the Lie algebra of $\mathbb{T}$ with $\mathbb{R}$ by $a \leftrightarrow a\frac{d}{d\theta}$, then the Lie algebra of $\mathbb{T} \tilde{\times} \tilde{L}G$ is $\mathbb{R} \oplus L\mathfrak{g} \oplus \mathbb{R}$ with the bracket given by

$$[(x_1, \xi_1, y_1), (x_2, \xi_2, y_2)] = (0, [\xi_1, \xi_2] + x_1\xi_2' - x_2\xi_1', \langle \xi_1, \xi_2' \rangle). \qquad (4.9.2)$$

Here $\langle \xi_1, \xi_2' \rangle$ is as in (4.9.1), i.e. it is the cocycle $\omega(\xi_1, \xi_2)$. We define a bilinear form on the algebra by

$$\langle (x_1, \xi_1, y_1), (x_2, \xi_2, y_2) \rangle = \langle \xi_1, \xi_2 \rangle - x_1 y_2 - y_1 x_2. \qquad (4.9.3)$$

It is immediate that this form is invariant under the adjoint action of the Lie algebra on itself, and hence under the adjoint action of the identity component of the group $\tilde{L}G$. This has a very useful computational corollary, which cannot be obtained so simply by other means.

Proposition (4.9.4). *If γ belongs to the identity component of LG then the adjoint action of γ on $\mathbb{R} \oplus \tilde{L}\mathfrak{g}$ is given by*

$$\gamma \cdot (x, \xi, y) = (x,\ \gamma \cdot \xi - x\gamma'\gamma^{-1},\ y - \langle \gamma^{-1}\gamma', \xi \rangle + \tfrac{1}{2}x\langle \gamma^{-1}\gamma', \gamma^{-1}\gamma' \rangle).$$

Proof. We know from (4.3.2) that the formula is true when $x = 0$. We also know a priori that the right-hand side must be of the form $(x, \ldots, \ldots)$. The formula above is then easily seen to be the only possible one which preserves the bilinear form (4.9.3).

In practice we are more often interested in the coadjoint action of γ on the dual space

$$\mathbb{R} \oplus (\tilde{L}\mathfrak{g})^* \cong \mathbb{R} \oplus (L\mathfrak{g})^* \oplus \mathbb{R}.$$

The most important case is when γ is the homomorphism $\theta \mapsto \exp(\theta\eta)$ defined by $\eta \in \check{T} \subset \mathfrak{t}$, which represents an element of the translation part of the affine Weyl group.

Proposition (4.9.5). *The element $\eta \in \check{T} \subset W_{\text{aff}}$ acts on $\mathbb{R} \oplus \mathfrak{t}^* \oplus \mathbb{R}$ by*
$$(n, \lambda, h) \mapsto (n + \langle \lambda, \eta \rangle + \tfrac{1}{2}h\|\eta\|^2, \lambda + h\eta, h),$$
where we have identified $\mathfrak{t}$ and $\mathfrak{t}^*$ by using the inner product.

Note. When γ does not belong to the identity component of LG its action on $\mathbb{R} \oplus \tilde{L}\mathfrak{g}$ involves choices, as we saw in Section 4.7. In that case the action does not necessarily preserve the inner product. Thus for $\tilde{L}U_n$, if we identify $\check{T}$ with $\mathbb{Z}^n$ in the obvious way, the formula corresponding to (4.9.5) is
$$(n, \lambda, h) \mapsto (n + \langle \lambda, \eta \rangle + \tfrac{1}{2}h\Sigma\eta_i(\eta_i - 1), \lambda + h\eta, h), \quad (4.9.6)$$
which is not orthogonal. The proof of (4.9.6) will be given later as Proposition (6.8.7).

4.10 Extensions of Map$(X; G)$

For a general compact manifold X the group Map$(X; G)$ is neither connected nor simply connected. (If $G = U_n$ and $n > \tfrac{1}{2}\dim X$ then the group of components of Map$(X; G)$ is the group $K^{-1}(X)$ of Atiyah and Hirzebruch [3], and its fundamental group is $\tilde{K}^0(X)$.) Let $m(X; G)$ denote the simply connected covering group of Map$(X; G)$, which one should think of as formed from Map$(X; G)$ by killing successively π_0 and π_1.

Proposition (4.10.1).
  (i) *If G is simply connected then there is a canonical central extension $\tilde{m}(X; G)$ of $m(X; G)$ such that $\pi_2(\tilde{m}(X; G)) = 0$.*
  (ii) *The group $\tilde{m}(X; G)$ is the universal central extension of $m(X; G)$, and its Lie algebra is the universal central extension of Map$(X; \mathfrak{g})$ which was described in Section 4.2.*
  (iii) *If G is simple then the kernel of the extension is $\Omega^1(X)/\Omega^1_\mathbb{Z}(X)$, the space of 1-forms on X modulo the 1-forms which have integral periods.*

Proof. We may as well suppose that G is simple. Let A denote the desired kernel $\Omega^1(X)/\Omega^1_\mathbb{Z}(X)$. We observe that the Lie algebra of A is the vector space $\Omega^1(X)/d\Omega^0(X)$ of (4.2.8). Using (4.4.2) it is enough for us to give a closed 2-form ω on $\tilde{m}(X; G)$ with values in $\Omega^1(X)/d\Omega^0(X)$ such that the integral of ω over every 2-cycle in $\tilde{m}(X; G)$ belongs to $\Omega^1_\mathbb{Z}(X)$. The 2-cocycle ω of Map$(X; \mathfrak{g})$ defined (see (4.2.9)) by $\omega(\xi, \eta) = \langle \xi, d\eta \rangle$ provides such a form on $\tilde{m}(X; G)$. The integrality condition holds because for any smooth map $\gamma: S^1 \to X$ the real-valued 2-form $\int_\gamma \omega$ on $\tilde{m}(X; G)$ is the pull-back of an integral form on LG.

The universality of $\tilde{m}(X; G)$, and the vanishing of π_2, are proved exactly as for loop groups.

The Mickelsson–Faddeev extension

Apart from its central extensions there is another—in principle more elementary—extension of $\mathrm{Map}(X; G)$ which has recently aroused interest in quantum field theory. It was introduced by Mickelsson [112], Faddeev [41], and others. (Cf. also Singer [138] and Zumino [157].)

Let us first observe that if Γ is any Lie group and $c \in H^2(\Gamma; \mathbb{Z})$ is any cohomology class which is invariant under left-translation by elements of Γ—this is automatically the case if Γ is connected—then one can find a unique smooth circle bundle Y on Γ whose Chern class is c. Associated to Y there is an extension $\tilde{\Gamma}$ of Γ by the abelian group of smooth maps $\Gamma \to \mathbb{T}$. An element of $\tilde{\Gamma}$ is a smooth $\mathbb{T}$-equivariant map $\tilde{\gamma}: Y \to Y$ which covers the left-translation $\gamma: \Gamma \to \Gamma$ by some $\gamma \in \Gamma$. If ω is a closed 2-form on Γ representing the class c then the Lie algebra extension corresponding to $\tilde{\Gamma}$ is defined by the cocycle

$$(\xi, \eta) \mapsto \omega(\xi, \eta)$$

with values in the vector space of smooth real-valued functions on Γ. (Here elements ξ, η of the Lie algebra of Γ are regarded as left-invariant vector fields on Γ.) The extension of Γ so defined is not a central extension: Γ acts in the natural way on the kernel $\mathrm{Map}(\Gamma; \mathbb{T})$.

When Γ is $\mathrm{Map}(X; G)$ for some n-dimensional manifold X we can obtain an extension of this form by choosing any element of $H^{n+2}(G; \mathbb{Z})$, pulling it back to $X \times \mathrm{Map}(X; G)$, and then integrating over X.

Now suppose that P is any principal Γ-bundle. For any left-invariant $c \in H^2(\Gamma; \mathbb{Z})$ we can find a 2-form ω on P whose restriction to each fibre of P is closed and represents the class c. The preceding discussion can be generalized to give

Proposition (4.10.2). *Suppose that Γ is connected and simply connected. Then there is an extension of Γ by $\mathrm{Map}(P; \mathbb{T})$ naturally associated to c. The extension of Lie algebras is defined by $(\xi, \eta) \mapsto \omega(\xi_P, \eta_P)$, where ξ_P and η_P are the vector fields on P associated to ξ, η in the Lie algebra of Γ.*

Proof. Following the method of Section 4.4, it is enough for us to define a map

$$C: (\text{loops in } \Gamma) \to \mathrm{Map}(P; \mathbb{T})$$

which has the properties (H1) and (H2) of Section 4.4 and is equivariant with respect to Γ. Given a loop ℓ in Γ and a point $p \in P$ we define

4.10 THE MICKELSSON-FADDEEV EXTENSION

$C_p(\ell) \in \mathbb{T}$ as $\exp(i \int_\sigma \omega)$, where σ is any piece of surface in the fibre $\Gamma \cdot p$ whose boundary is the loop $\ell \cdot p$.

Remarks

(i) The hypothesis that Γ is simply connected in (4.10.2) is unnecessary if c is the transgression of an element of $H^3(B\Gamma; \mathbb{Z})$, where $B\Gamma$ is the classifying space of Γ.

(ii) If P is connected and simply connected then the kernel $\mathrm{Map}(P; \mathbb{T})$ has the homotopy type of a circle, and homotopically the extension is simply the circle bundle corresponding to $c \in H^2(\Gamma; \mathbb{Z})$.

The case of interest in quantum field theory is when P is the contractible space of connections in a principal G-bundle on an orientable 3-manifold X, and $\Gamma = \mathrm{Map}(X; G)$. One constructs $c \in H^2(\Gamma; \mathbb{Z})$ by starting from the element of $H^5(G; \mathbb{Z})$ defined by an invariant trilinear form F on $\mathfrak{g}$. If the G-bundle on X is trivial, so that $P = \Omega^1(X; \mathfrak{g})$, then the Lie algebra cocycle associates to $\xi, \eta \in \mathrm{Map}(X; \mathfrak{g})$ the function

$$A \mapsto \int_X F(A, d\xi, d\eta)$$

on P.

4.11 Appendix: The cohomology of LG and its Lie algebra

For a compact group G the cohomology $H^*(G; \mathbb{R})$ with real coefficients can be calculated by de Rham's theorem. If $g \cdot \omega$ denotes the left-translate of a closed form ω on G by $g \in G$ then the averaged form $\int_G g \cdot \omega \, dg$ represents the same cohomology class as ω, for a cohomology class is unchanged by translation. It follows that the cohomology can be calculated from the cochain complex of left-invariant forms on G, i.e. from the cochain complex of the Lie algebra $\mathfrak{g}$. In other words we have

$$H^*(G; \mathbb{R}) \cong H^*(\mathfrak{g}; \mathbb{R}).$$

It is well known that this cohomology is an exterior algebra on ℓ odd-dimensional generators, where ℓ is the rank of G. The generators correspond to the generators of the algebra of invariant polynomial functions on $\mathfrak{g}$ (which themselves form a polynomial algebra on ℓ generators ([20] Chapter 5, Section 5.3)) in the following way. If P is a polynomial of degree k, regarded as a symmetric multilinear function

$$\mathfrak{g} \times \ldots \times \mathfrak{g} \to \mathbb{R},$$

then one can define a skew multilinear function S of $2k - 1$ variables by $S(\xi_1, \ldots, \xi_{2k-1}) =$

$$\sum_\pi (-1)^\pi P([\xi_{\pi_1}, \xi_{\pi_2}], [\xi_{\pi_3}, \xi_{\pi_4}], \ldots, [\xi_{\pi_{2k-3}}, \xi_{\pi_{2k-2}}], \xi_{\pi_{2k-1}}), \quad (4.11.1)$$

where the sum is over all permutations π of $\{1, 2, \ldots, 2k - 1\}$.

If $G = U_n$ then the generators of the ring of invariant polynomials can be taken to be $P_1, P_2, \ldots, P_n$, where $P_k(A) = \text{trace}(A^k)$.

It is a fairly easy result of algebraic topology [16] that the cohomology $H^*(\Omega G; \mathbb{R})$ of the space of based loops on a simply connected group G is a polynomial algebra on the even dimensional classes obtained by transgressing the generators of $H^*(G; \mathbb{R})$, i.e. by pulling them back to $S^1 \times \Omega G$ by the evaluation map, and then integrating over S^1. The class so obtained from (4.11.1) is the $(2k-2)$-form on ΩG whose value at $\gamma \in \Omega G$ on tangent vectors represented by $\xi_1, \ldots, \xi_{2k-2} \in \Omega \mathfrak{g}$ is

$$\frac{1}{2\pi} \int_0^{2\pi} S(\xi_1(\theta), \xi_2(\theta), \ldots, \xi_{2k-2}(\theta), \gamma(\theta)^{-1} \gamma'(\theta)) \, d\theta. \quad (4.11.2)$$

This form is naturally defined on LG. The cohomology $H^*(LG)$ is simply the tensor product $H^*(G) \otimes H^*(\Omega G)$, because $LG \cong G \times \Omega G$ as a space.

The differential form (4.11.2) is evidently not left-invariant, and we have no reason to expect that the cohomology of LG can be represented by left-invariant forms. Nevertheless we have

Proposition (4.11.3). *The $(2k-2)$-form (4.11.2) on LG is cohomologous to a rational multiple of the left-invariant form obtained by making skew the map*

$$(\xi_1, \ldots, \xi_{2k-2}) \mapsto \frac{1}{2\pi} \int_0^{2\pi} P([\xi_1, \xi_2], \ldots, [\xi_{2k-5}, \xi_{2k-4}], \xi_{2k-3}, \xi'_{2k-2}) \, d\theta.$$

Corollary (4.11.4). *The natural map*

$$H^*(L\mathfrak{g}; \mathbb{R}) \to H^*(LG; \mathbb{R})$$

is surjective.

Remarks. Actually the map of (4.11.4) is an isomorphism. We shall prove that in Section 14.6. (Cf. also Kumar [97].) The result should be contrasted with our discovery in Section 4.2 that $H^2(\text{Map}(X; \mathfrak{g}))$ is vastly larger that $H^2(\text{Map}(X; G))$ when $\dim(X) > 1$. Quillen has pointed out to us that the class in $H^{2k-d-1}(\text{Map}(X; G))$ which is obtained by pulling back the class (4.11.1) by the evaluation map $X \times \text{Map}(X; G) \to G$ and integrating it over a cycle of dimension d in X can be represented by a left-invariant form if $k > d$, but usually not otherwise.

Proof of (4.11.3). Let us introduce some more convenient notation, as follows. When we pull back the Maurer–Cartan 1-form $g^{-1} \, dg$ on G (with values in $\mathfrak{g}$) by the evaluation map $S^1 \times LG \to G$ we shall write the resulting form as $\xi + \eta$, where ξ vanishes on tangent vectors in the S^1-direction and η vanishes along LG. (Thus η is $\gamma(\theta)^{-1} \gamma'(\theta) \, d\theta$ at $(\theta, \gamma) \in S^1 \times LG$.) In this notation the forms of (4.11.2) and (4.11.3) are

4.11 THE COHOMOLOGY OF LG AND ITS LIE ALGEBRA

obtained (up to rational multiples) by integrating over S^1 the forms

$$\Theta = P([\xi, \xi], \ldots, [\xi, \xi], \eta)$$

and

$$\Phi = P([\xi, \xi], \ldots, [\xi, \xi], \xi, d'\xi),$$

respectively on $S^1 \times LG$. (We write d' and d" for differentiation of forms in the S^1 and LG directions respectively.)

Because $d(g^{-1} dg) = -\frac{1}{2}[g^{-1} dg, g^{-1} dg]$ on G we find

$$d'\eta = -\tfrac{1}{2}[\eta, \eta],$$
$$d''\xi = -\tfrac{1}{2}[\xi, \xi],$$

and

$$d'\xi + d''\eta = -[\xi, \eta].$$

Now consider the form $\Psi = P([\xi, \xi], \ldots, [\xi, \xi], \xi, \eta)$ on $S^1 \times LG$. We have $d''[\xi, \xi] = 0$, so

$$d''\Psi = -\tfrac{1}{2}P([\xi, \xi], \ldots, [\xi, \xi], \eta) + P([\xi, \xi], \ldots, [\xi, \xi], \xi, d'\xi)$$
$$+ P([\xi, \xi], \ldots, [\xi, \xi], \xi, [\xi, \eta]).$$

Using the invariance of the polynomial P, and the fact that $[[\xi, \xi], \xi] = 0$ because of the Jacobi identity, the third term on the right-hand-side is equal to Θ, so that we have

$$d\Psi = \tfrac{1}{2}\Theta + \Phi.$$

Integrating this relation over S^1 gives the desired result.

5
THE ROOT SYSTEM: KAC–MOODY ALGEBRAS

In general it is a feature of our approach to loop groups that it does not involve any detailed analysis of the structure of the Lie algebras of the groups. These algebras are examples of what are called *Kac–Moody Lie algebras*,† and there is a very extensive literature devoted to their study. (Cf. Kac [86], Macdonald [109], Helgason [72].) In this chapter we have attempted to explain fairly briefly how loop groups fit into that context. The material in Sections 5.1 and 5.2 will be used later in classifying the representations of loop groups, but the Kac–Moody theory proper which is sketched in Section 5.3 will not be referred to again.

5.1 The root system and the affine Weyl group

We have explained in Chapter 2 that the crucial step in studying the Lie algebra $\mathfrak{g}$ of a compact Lie group G is to decompose the complexification $\mathfrak{g}_\mathbb{C}$ under the adjoint action of a maximal torus T of G. One has

$$\mathfrak{g}_\mathbb{C} = \mathfrak{t}_\mathbb{C} \oplus \bigoplus_\alpha \mathfrak{g}_\alpha,$$

where $\mathfrak{t}_\mathbb{C}$ is the complexified Lie algebra of T and $\mathfrak{g}_\alpha$ is the one-dimensional subspace of $\mathfrak{g}_\mathbb{C}$ where T acts by the homomorphism $\alpha: T \to \mathbb{T}$. The homomorphisms α which occur in the decomposition are called the *roots* of G. They form a finite subset of the lattice $\hat{T} = \text{Hom}(T; \mathbb{T})$.

The most obvious decomposition of the complexified algebra $L\mathfrak{g}_\mathbb{C}$ of a loop group is into its Fourier components:

$$L\mathfrak{g}_\mathbb{C} = \bigoplus_{k \in \mathbb{Z}} \mathfrak{g}_\mathbb{C} \cdot z^k.$$

This is the decomposition into eigenspaces of the action of the circle $\mathbb{T}$ which rotates the loops bodily. The rotation action commutes with the adjoint action of *constant* loops, so we can decompose $L\mathfrak{g}_\mathbb{C}$ further according to the action of a maximal torus T of G:

$$L\mathfrak{g}_\mathbb{C} = \bigoplus_{k \in \mathbb{Z}} \mathfrak{t}_\mathbb{C} z^k \oplus \bigoplus_{(k,\alpha)} \mathfrak{g}_\alpha z^k. \tag{5.1.1}$$

---
† The algebras were first studied, almost simultaneously, by Kantor [89], Kac [82], and Moody [117].

5.1 THE ROOT SYSTEM AND THE AFFINE WEYL GROUP

The pieces in this decomposition are indexed by homomorphisms $\mathbb{T} \times T \to \mathbb{T}$, i.e. by elements of $\mathbb{Z} \times \hat{T}$. Once again the homomorphisms (k, α) which occur (with α possibly 0) are called the *roots* of LG.

We can reformulate what has just been said by introducing the group $\mathbb{T} \tilde{\times} LG$, the semidirect product of $\mathbb{T}$ and LG in which $\mathbb{T}$ acts on LG by rotating the loops. The centralizer of $\mathbb{T}$ in $\mathbb{T} \tilde{\times} LG$ is $\mathbb{T} \times G$, and so $\mathbb{T} \times T$ is a maximal abelian subgroup of $\mathbb{T} \tilde{\times} LG$. The complexified Lie algebra of $\mathbb{T} \tilde{\times} LG$ decomposes as

$$(\mathbb{C} \oplus \mathfrak{t}_\mathbb{C}) \oplus \bigoplus_{k \neq 0} \mathfrak{t}_\mathbb{C} \cdot z^k \oplus \bigoplus_{(k,\alpha)} \mathfrak{g}_\alpha z^k,$$

according to the characters of $\mathbb{T} \times T$.

In the finite dimensional case the roots of G are permuted by the Weyl group. This is the group of automorphisms of T which arise from conjugation in G, i.e. $W = N(T)/T$, where $N(T) = \{n \in G : nTn^{-1} = T\}$ is the normalizer of T in G. If n is an element of $N(T)$ then $n \cdot \mathfrak{g}_\alpha = \mathfrak{g}_{n\alpha}$, where $n\alpha : T \to \mathbb{T}$ is given by $n\alpha(t) = \alpha(n^{-1}tn)$.

In exactly the same way the infinite set of roots of LG is permuted by $W_{\text{aff}} = N(\mathbb{T} \times T)/(\mathbb{T} \times T)$, where $N(\mathbb{T} \times T)$ denotes the normalizer in $\mathbb{T} \tilde{\times} LG$. The group W_{aff} is called the *affine Weyl group*, for reasons we shall explain in a moment. Its structure is described by the following proposition. We shall denote the 'coweight' lattice of G by $\check{T}$: it is the lattice of all homomorphisms $\mathbb{T} \to T$. (See Section 2.4.)

Proposition (5.1.2). *W_{aff} is the semidirect product of $\check{T}$ by W, the Weyl group of G.*

Proof. The lattice $\check{T}$ is a subgroup of LG, and obviously centralizes T. On the other hand, if R_u is the operation of rotating by u (i.e. $R_u \in \mathbb{T} \subset \mathbb{T} \tilde{\times} LG$) then for any $f \in LG$ we have

$$f \cdot R_u \cdot f^{-1} = R_u \cdot \phi, \tag{5.1.3}$$

where $\phi(z) = f(uz)f(z)^{-1}$. If f is a homomorphism $\mathbb{T} \to T$ then $\phi(z)$ is the constant $f(u) \in T$, and so $\check{T} \subset N(\mathbb{T} \times T)$. Conversely, if $f \in LG$ belongs to $N(\mathbb{T} \times T)$ then $f(uz)f(z)^{-1}$ must be constant as a function of z for each u, which implies that $z \mapsto f(z)f(1)^{-1}$ is a homomorphism $\mathbb{T} \to T$. Furthermore $f(1)$ must belong to the normalizer N of T in G. It follows that $N(\mathbb{T} \times T)$ is in G, and this proves (5.1.2).

In the finite dimensional theory one usually thinks of the lattice $\hat{T}$, and hence the roots, as lying in the real vector space $\mathfrak{t}^*$, identifying a homomorphism $\alpha : T \to \mathbb{T}$ with the linear map $\dot{\alpha} : \mathfrak{t} \to \mathbb{R}$ such that $\alpha = e^{i\dot{\alpha}}$. One can think of the roots of LG similarly as linear forms on the Lie algebra $\mathbb{R} \times \mathfrak{t}$ of $\mathbb{T} \times T$. But it is more convenient to regard them as

5 THE ROOT SYSTEM: KAC-MOODY ALGEBRAS

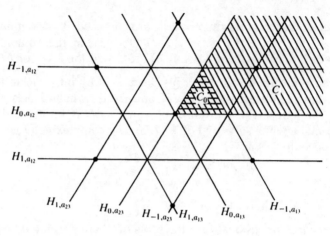

Fig. 2

affine-linear functions on $\mathfrak{t}$, identifying $\mathfrak{t}$ with the affine hyperplane $1 \times \mathfrak{t}$ in $\mathbb{R} \times \mathfrak{t}$. For this reason, and also to distinguish them from the roots of G, the roots of LG are often called *affine roots*. The group W_{aff} acts linearly on $\mathbb{R} \times \mathfrak{t}$: the action of W is obvious, and from (5.1.3) the action of $\lambda \in \check{T}$ is given by

$$\lambda \cdot (x, \xi) = (x, \xi + x\lambda).$$

Thus W_{aff} preserves the hyperplane $1 \times \mathfrak{t}$, and $\lambda \in \check{T}$ acts on it by translation by the vector $\lambda \in \check{T} \subset \mathfrak{t}$.

If $\alpha \neq 0$, the affine root (k, α), regarded as an affine-linear function on $\mathfrak{t}$, is determined up to sign by the affine hyperplane

$$H_{k,\alpha} = \{\xi \in \mathfrak{t} : (\alpha, \xi) = -k\}$$

in $\mathfrak{t}$ where it vanishes. This collection of hyperplanes is called the *diagram* of LG. The picture when $G = SU_3$ is shown in Fig. 2.

The connected components in $\mathfrak{t}$ of the complement of the hyperplanes $H_{k,\alpha}$ are called the *alcoves* of the diagram. Recall that the components of the complement of the $H_{0,\alpha}$ (which form the diagram of G) are called the *chambers*. Each chamber C contains a unique alcove C_0 whose closure contains the origin. If one chooses a chamber C then the roots of G are called positive or negative according to their sign on C. The set

$$\{\xi \in \mathfrak{t} : 0 < \alpha(\xi) < 1 \text{ for all positive roots } \alpha\}$$

is the corresponding alcove C_0. An affine root is called positive or negative according to its sign on C_0. The positive affine roots corresponding to the walls of C_0 are called the *simple* affine roots.

5.1 THE ROOT SYSTEM AND THE AFFINE WEYL GROUP

If G is semisimple then it is well-known that each chamber C is a simplicial cone, bounded by ℓ hyperplanes $H_{0,\alpha_1}, \ldots, H_{0,\alpha_\ell}$ corresponding to the simple roots $\alpha_1, \ldots, \alpha_\ell$ of G. Here ℓ is the dimension of $\mathfrak{t}$, called the *rank* of G. If G is a simple group then it has a highest root $\alpha_{\ell+1}$ (the highest weight of the adjoint representation) which dominates all the other positive roots. In that case the alcove C_0 is an ℓ-dimensional simplex bounded by $H_{0,\alpha_1}, \ldots, H_{0,\alpha_\ell}$ and $H_{1,-\alpha_{\ell+1}}$, and LG has $\ell+1$ simple roots $(0, \alpha_1), (0, \alpha_2), \ldots, (0, \alpha_\ell), (1, -\alpha_{\ell+1})$. In general C_0 is a product of q simplexes, one for each simple factor of G, and LG has $k+q$ simple roots, namely $(0, \alpha_i)$ for $i = 1, \ldots, \ell$, and $(1, -\alpha_i)$ for $i = \ell+1, \ldots, \ell+q$, where $\alpha_{\ell+1}, \ldots, \alpha_{\ell+q}$ are the highest weights occurring in the adjoint representation.

The finite group W is known to be generated by reflections in the hyperplanes $H_{0,\alpha}$, and acts simply transitively on the set of chambers. Corresponding statements hold for W_{aff}.

Proposition (5.1.4). *If G is simply connected, then*
  (i) *W_{aff} is generated by reflections in the hyperplanes $H_{k,\alpha}$, and*
  (ii) *W_{aff} acts simply transitively on the set of alcoves.*

Proof.
  (i) We know that the reflection s_α in $H_{0,\alpha}$ belongs to W. It is given by
$$s_\alpha(\xi) = \xi - \alpha(\xi)h_\alpha,$$
where h_α is the coroot corresponding to α. (See Section 2.4.) Now $\alpha(h_\alpha) = 2$, so the point $-\tfrac{1}{2}kh_\alpha$ belongs to $H_{k,\alpha}$. The reflection $s_{k,\alpha}$ in $H_{k,\alpha}$ is therefore
$$\begin{aligned}s_{k,\alpha}(\xi) &= (\xi + \tfrac{1}{2}kh_\alpha) - \alpha(\xi + \tfrac{1}{2}kh_\alpha)h_\alpha - \tfrac{1}{2}kh_\alpha\\ &= s_\alpha(\xi) - kh_\alpha.\end{aligned} \qquad (5.1.5)$$
But h_α belongs to the lattice $\check{T}$, so $s_{k,\alpha} \in W \tilde{\times} \check{T} = W_{\text{aff}}$.

Conversely, it is enough to show that the translation t_α by $h_\alpha \in \mathfrak{t}$ belongs to the group generated by the $s_{k,\alpha}$, for the coroots h_α generate $\check{T}$. But from (5.1.5) we have
$$t_\alpha = s_{1,-\alpha}s_{0,\alpha}.$$

  (ii) (Cf. Bourbaki [20] Chapter 5, §3.1.) Let A be an arbitrary alcove. We must show that $\gamma A = C_0$ for some $\gamma \in W_{\text{aff}}$. Choose a point a in A. The orbit of a under W_{aff} is a locally finite subset S in $\mathfrak{t}$. We must show that S meets C_0. Choose a point $c \in C_0$ and a point $b \in S$ whose distance from c is minimal. If $b \notin C_0$ then b must be separated from c by at least one wall H of C_0. Reflecting b in H will produce a point of the orbit S closer to c than b is: a contradiction. So W_{aff} acts transitively on the alcoves. Conversely, an element of W_{aff} is completely determined by the alcove to which it takes C_0: that follows at once from the corresponding fact about the action of W on the chambers.

Remark (5.1.6). The proof of (ii) actually shows that W_{aff} is generated by the reflections in the hyperplanes corresponding to the *simple* affine roots.

5.2 Generators and relations

We can now describe the Lie algebra $L\mathfrak{g}_{\mathbb{C}}$, or more precisely its universal central extension, in terms of generators and relations. For a finite dimensional semisimple algebra $\mathfrak{g}$, if one chooses for each root α a non-zero element e_α in the root-space $\mathfrak{g}_\alpha$ (see Section 2.4) then $\mathfrak{g}_{\mathbb{C}}$ is generated by the e_α, and even by $e_j = e_{\alpha_j}$ and $f_j = e_{-\alpha_j}$ for $j = 1, \ldots, \ell$, where the α_j are the simple roots. In fact $[e_j, f_j]$ is a multiple of the coroot $h_j = h_{\alpha_j} \in \mathfrak{t}$, and when the e_j and f_j are normalized so that $[e_j, f_j] = ih_j$ the following is a complete set of relations defining $\mathfrak{g}_{\mathbb{C}}$:

$$\begin{aligned}
&[e_j, f_j] = ih_j \\
&[e_j, f_k] = 0 \quad \text{if} \quad j \neq k \\
&[h_j, e_k] = ia_{jk} e_k \\
&[h_j, f_k] = -ia_{jk} f_k \\
&(\operatorname{ad} e_j)^{1-a_{jk}} e_k = 0 \\
&(\operatorname{ad} f_j)^{1-a_{jk}} f_k = 0.
\end{aligned} \tag{5.2.1}$$

Here 'ad x' means the operation $y \mapsto [x, y]$, and the a_{jk} are integers forming an $\ell \times \ell$ matrix called the *Cartan matrix* of $\mathfrak{g}$. This matrix determines the structure of $\mathfrak{g}$ completely. For a proof of this, see Serre [134].

Now let us turn to loop groups. Let us choose elements e_j, f_j in $L\mathfrak{g}_{\mathbb{C}}$ corresponding to the simple affine roots. In the notation of Section 5.1 we can take e_j and f_j to be the usual elements of $\mathfrak{g}_{\mathbb{C}} \subset L\mathfrak{g}_{\mathbb{C}}$ when $1 \leq j \leq \ell$, and $e_j = ze_{-\alpha_j}$, $f_j = z^{-1} e_{\alpha_j}$ for $\ell < j \leq \ell + q$. (Here q is the number of simple factors in $\mathfrak{g}$.)

Proposition (5.2.2). *If $\mathfrak{g}$ is semisimple then $L_{\text{pol}}\mathfrak{g}_{\mathbb{C}}$ is generated by the elements e_j, f_j corresponding to the simple affine roots.*

Proof. We may as well assume that $\mathfrak{g}$ is simple. Then the e_j and f_j for $j \leq \ell$ generate $\mathfrak{g}_{\mathbb{C}}$. But $e_{\ell+1} = ze_{-\alpha}$, where α is the highest root of $\mathfrak{g}$. Because the adjoint representation is irreducible, all of $z\mathfrak{g}_{\mathbb{C}}$ can be obtained by applying elements of $\mathfrak{g}_{\mathbb{C}}$ to $ze_{-\alpha}$, and so $z\mathfrak{g}_{\mathbb{C}}$ is contained in the algebra generated by the e_j and f_j. But then $z^2 e_{-\alpha}$ is a multiple of $[zh_j, ze_{-\alpha}]$, where $ih_j = [e_j, f_j]$ and $j \leq \ell$ is chosen so that $\alpha(h_j) \neq 0$. This gives us $z^2 \mathfrak{g}_{\mathbb{C}}$. And so on.

5.2 GENERATORS AND RELATIONS

We can now check at once that the set of relations (5.2.1), where j and k run from 1 to $\ell + q$, hold in $L\mathfrak{g}_\mathbb{C}$. The $(\ell + q) \times (\ell + q)$ Cartan matrix is given by

$$a_{jk} = \beta_k(h_{\beta_j})$$

where $\beta_j = \alpha_j$ if $j \leq \ell$ and $\beta_j = -\alpha_j$ if $j > \ell$. Because only ℓ of the β_i are linearly independent, we see that the Cartan matrix has rank ℓ.

Although the relations (5.2.1) hold in $L_{\text{pol}}\mathfrak{g}_\mathbb{C}$ they do not define it. It is a theorem of Gabber and Kac [52]—which we shall not prove in this book—that the relations define the universal central extension $\tilde{L}_{\text{pol}}\mathfrak{g}_\mathbb{C}$ of $L_{\text{pol}}\mathfrak{g}_\mathbb{C}$ by $K_\mathbb{C} \cong \mathbb{C}^q$ which is described by the cocycle ω_K of (4.2.7). We shall content ourselves here with pointing out that the relations (5.2.1) do hold in $\tilde{L}_{\text{pol}}\mathfrak{g}_\mathbb{C}$. To see that, we identify $\tilde{L}_{\text{pol}}\mathfrak{g}_\mathbb{C}$ with $L_{\text{pol}}\mathfrak{g}_\mathbb{C} \oplus K_\mathbb{C}$, and define elements $\tilde{e}_j$, $\tilde{f}_j$, $\tilde{h}_j$ of $\tilde{L}_{\text{pol}}\mathfrak{g}_\mathbb{C}$ by

$$\tilde{e}_j = (e_j, 0)$$
$$\tilde{f}_j = (f_j, 0)$$
$$\tilde{h}_j = (h_j, 0) \quad \text{for} \quad j = 1, \ldots, \ell,$$
$$= (h_j, -\tfrac{1}{2}\langle h_j, h_j \rangle_K) \quad \text{for} \quad j = \ell + 1, \ldots, \ell + q.$$

It is easy to check that the $\tilde{e}_j$, $\tilde{f}_j$, $\tilde{h}_j$ satisfy the relations (5.2.1). Furthermore the elements generate $\tilde{L}_{\text{pol}}\mathfrak{g}_\mathbb{C}$, because the inner products $\langle h_j, h_j \rangle_K$, for $\ell < j \leq \ell + q$, span $K_\mathbb{C}$.

With the preceding formulae in mind it is natural, whenever we are studying a central extension $\tilde{L}G$ of LG defined by a bilinear form $\langle \, , \, \rangle$ on $\mathfrak{g}$, to associate to each affine root $\boldsymbol{\alpha} = (k, \alpha)$ with $\alpha \neq 0$ an *affine coroot* $h_\boldsymbol{\alpha}$ in $\tilde{L}\mathfrak{g}$ defined by

$$h_\boldsymbol{\alpha} = (h_\alpha, -\tfrac{1}{2}k\langle h_\alpha, h_\alpha \rangle). \tag{5.2.3}$$

Taking $h_\boldsymbol{\alpha}$ together with $e_\boldsymbol{\alpha} = (z^k e_\alpha, 0)$ and $e_{-\boldsymbol{\alpha}} = (z^{-k} e_{-\alpha}, 0)$ we then have a copy of the Lie algebra of SU_2 embedded in $\tilde{L}\mathfrak{g}_\mathbb{C}$, and we can exponentiate to obtain a homomorphism

$$i_\boldsymbol{\alpha} : SU_2 \to \tilde{L}G. \tag{5.2.4}$$

The argument of the proof of (3.5.3) clearly implies the following.

Proposition (5.2.5). *If G is simply connected then the $\ell + q$ subgroups $i_\boldsymbol{\alpha}(SU_2)$ corresponding to the simple affine roots generate $\tilde{L}_{\text{pol}}G$.*

Example. The group $L_{\text{pol}}SU_2$ is generated by the subgroup SU_2 of constant loops, and the copy of SU_2 consisting of the elements

$$\begin{pmatrix} a & bz \\ -\bar{b}z^{-1} & \bar{a} \end{pmatrix}$$

with $|a|^2 + |b|^2 = 1$. The latter subgroup is the transform of the former by the outer automorphism of LSU_2 corresponding to the non-trivial element of the centre of SU_2. (Cf. (3.4.4).)

5.3 Kac–Moody Lie algebras

It is well known that the Cartan matrix $A = (a_{ij})$ of any semisimple Lie algebra $\mathfrak{g}_\mathbb{C}$ satisfies the following two conditions (see [20] Chapter 6, where a_{ij} is written $n(\alpha_i, \alpha_j)$).

(C1) $a_{ij} \in \mathbb{Z}$ for all i, j; $a_{ii} = 2$ for all i; $a_{ij} \leq 0$ if $i \neq j$; $a_{ij} = 0$ whenever $a_{ji} = 0$.

(C2) A is positive definite, in the sense that all the principal minors of A are >0.

Conversely, given any $\ell \times \ell$ matrix A satisfying (C1), the relations (5.2.1) define an abstract complex Lie algebra $\mathfrak{g}'(A)$, which is essentially the *Kac–Moody Lie algebra* defined by the Cartan matrix A. If A satisfies (C2) as well, then $\mathfrak{g}'(A)$ will be finite dimensional and semisimple, but if (C2) does not hold, $\mathfrak{g}'(A)$ will be infinite dimensional. It is natural to ask what can be said about $\mathfrak{g}'(A)$ in this infinite dimensional case. It turns out that, after modifying $\mathfrak{g}'(A)$ slightly in a way which corresponds to passing from LG to the semidirect product $\tilde{\mathbb{T}} \ltimes LG$, one obtains an algebra to which much of the finite dimensional structure theory can be carried over. As it is not our purpose in this book to give a systematic exposition of Kac–Moody Lie algebras, we shall only describe the beginning of the theory, referring the reader to the excellent survey article of Macdonald [109] or the book of Kac [86] for further details.

The first thing to notice is that the elements h_i are linearly independent and generate a maximal abelian subalgebra $\mathfrak{h}'$ of $\mathfrak{g}'(A)$. The analogy with the finite dimensional case now suggests that one should define the simple roots $\alpha_j' \in \mathfrak{h}'^*$ of $\mathfrak{g}'(A)$ by the formula $\alpha_j'(h_i) = a_{ij}$. Unfortunately, if A is not invertible, the resulting simple roots will be linearly dependent. This problem can be overcome by passing to a semidirect product $\mathfrak{g}(A) = \mathfrak{d} \oplus \mathfrak{g}'(A)$, where $\mathfrak{d}$ is a space of derivations of $\mathfrak{g}'(A)$ defined as follows. Let $\mathfrak{d}'$ be spanned by the derivations d_i, $i = 1, \ldots, \ell$, defined by $d_i(e_j) = \delta_{ij} e_j$, and $d_i(f_j) = -\delta_{ij} f_j$. Then $\operatorname{ad}(\mathfrak{h}')$ is a subspace of $\mathfrak{d}'$, since $\operatorname{ad}(h_i) = \sum a_{ij} d_j$; let $\mathfrak{d}$ be a complementary subspace, so that $\dim \mathfrak{d} = \operatorname{corank} A$. Then $\mathfrak{g}(A) = \mathfrak{d} \oplus \mathfrak{g}'(A)$ is the Kac–Moody Lie algebra with Cartan matrix A; it is independent of the choice of $\mathfrak{d}$ up to isomorphism.

With this modification, $\mathfrak{h} = \mathfrak{d} \oplus \mathfrak{h}'$ is a maximal abelian subalgebra of $\mathfrak{g}(A)$, and the simple roots α_i can be defined so that $[h, e_i] = \alpha_i(h) e_i$ for

5.3 KAC-MOODY LIE ALGEBRAS

all $h \in \mathfrak{h}$. Then $\alpha_j(h_i) = a_{ij}$ and the α_i are linearly independent. The remaining roots are defined in the obvious way.

To proceed further we need a complex-valued invariant symmetric bilinear form on $\mathfrak{g}(A)$. Unfortunately $\mathfrak{g}(A)$ may not have such a form; in fact a necessary and sufficient condition for this is that the Cartan matrix is *symmetrizable*, which means that there is an invertible diagonal matrix D such that DA is symmetric. When it exists, the form is non-degenerate, and it is unique up to a constant factor if A is indecomposable, i.e. cannot be written as a non-trivial direct sum of two other matrices. (If A is decomposable then $\mathfrak{g}(A)$ is the direct product of the corresponding subalgebras.)

If there is an invariant form $\langle \ , \ \rangle$ on $\mathfrak{g}(A)$ then its restriction to $\mathfrak{h}$ is non-degenerate. Relative to the corresponding symmetric bilinear form on $\mathfrak{h}^*$ one has $\langle \alpha_i, \alpha_i \rangle > 0$ for all i. In general, $\langle \alpha, \alpha \rangle$ is real for each root α, but is not always positive unless $\mathfrak{g}(A)$ is finite dimensional; the roots α for which $\langle \alpha, \alpha \rangle > 0$ are called real, the others imaginary.

The Weyl group W of $\mathfrak{g}(A)$ can now be defined as the group of isometries of $\mathfrak{h}$ generated by the reflections in the planes $H_{\alpha_i} = \ker \alpha_i$. Obviously W takes real roots to real roots; in fact, the real roots are precisely the W-orbit of the simple roots.

We must now explain where loop algebras fit into this picture. The Cartan matrix of a Lie algebra $\tilde{L}\mathfrak{g}_\mathbb{C}$ satisfies (C1) and

(C2') $\det A = 0$ and all the proper principal minors of A are >0.

(see [86]). Conversely, the Kac–Moody Lie algebras corresponding to Cartan matrices satisfying (C1) and (C2') are usually called *affine Lie algebras*. Thus $\tilde{L}\mathfrak{g}_\mathbb{C}$ is an affine Lie algebra, or, more precisely, the semidirect product $\mathbb{C}^q \oplus \tilde{L}\mathfrak{g}_\mathbb{C}$, where q is the number of simple factors in $\mathfrak{g}_\mathbb{C}$. (If $\mathfrak{g}_\mathbb{C} = \mathfrak{g}_{1,\mathbb{C}} \oplus \ldots \oplus \mathfrak{g}_{q,\mathbb{C}}$ then $\mathbb{C}^q \oplus \tilde{L}\mathfrak{g}_\mathbb{C}$ means the product of the algebras $\mathbb{C} \oplus \tilde{L}\mathfrak{g}_{i,\mathbb{C}}$, where the factor $\mathbb{C}$ is generated by the obvious derivation $-i\dfrac{d}{d\theta} = z\dfrac{d}{dz}$ of $\tilde{L}\mathfrak{g}_{i,\mathbb{C}}$.)

Only about half of the affine Lie algebras arise in this way. The remainder are the Lie algebras of the twisted loop groups introduced in Section 3.7. In the algebraic context one chooses an outer automorphism α of $\mathfrak{g}_\mathbb{C}$ of finite order k, so that $k = 1$, 2 or 3, and replaces $L\mathfrak{g}_\mathbb{C}$ by its subalgebra $L_{(\alpha)}\mathfrak{g}_\mathbb{C}$ consisting of the loops $f \in L\mathfrak{g}_\mathbb{C}$ which are equivariant:

$$f(\varepsilon^{-1}z) = \alpha(f(z)), \qquad (5.3.1)$$

where ε is a primitive k^{th} root of unity.

All affine Lie algebras admit an invariant symmetric bilinear form, and their root systems are well understood (see Helgason [72] Chapter X §5).

5 THE ROOT SYSTEM: KAC–MOODY ALGEBRAS

We have already discussed the root system of $\tilde{L}\mathfrak{g}_{\mathbf{C}}$ in the preceding section, and the twisted algebras present little extra difficulty.

If $G_{\mathbf{C}}$ is a simply connected Lie group with Lie algebra $\mathfrak{g}_{\mathbf{C}}$, then any automorphism of $\mathfrak{g}_{\mathbf{C}}$ lifts uniquely to an automorphism of $G_{\mathbf{C}}$ and one can define the twisted loop group $L_{(\alpha)}G_{\mathbf{C}}$ by the same formula (5.3.1), where f is now interpreted as an element of $LG_{\mathbf{C}}$. Thus every affine Lie algebra comes from a Lie group. The extent to which this is true for an arbitrary Kac–Moody Lie algebra is an open question. Certainly no concrete realizations of the groups are known. (See Tits [145], [146] for the current state of affairs.)

6
LOOP GROUPS AS GROUPS OF OPERATORS IN HILBERT SPACE

In this chapter we shall study the natural embedding of the loop group of $GL_n(\mathbb{C})$ in the restricted general linear group of Hilbert space. This embedding will play a fundamental role throughout the rest of the book.

6.1 Loops as multiplication operators

Let $H^{(n)}$ denote the Hilbert space $L^2(S^1; \mathbb{C}^n)$ of square-summable $\mathbb{C}^n$-valued functions on the circle. The group $L_{\text{cts}}GL_n(\mathbb{C})$ of continuous maps $S^1 \to GL_n(\mathbb{C})$ acts on $H^{(n)}$ by multiplication operators: if γ is a matrix-valued function on the circle, we denote the corresponding multiplication operator by M_γ.

The norm $\|M_\gamma\|$ of the operator M_γ is $\|\gamma\|_\infty$, the supremum of $\|\gamma(\theta)\|$ for $\theta \in S^1$. It follows that $\gamma \mapsto M_\gamma$ embeds the Banach Lie group $L_{\text{cts}}GL_n(\mathbb{C})$ as a closed subgroup (with the induced topology) of the Banach Lie group $GL(H^{(n)})$ of all invertible bounded operators in $H^{(n)}$, with the operator-norm topology. We recall that $GL(H^{(n)})$ is an open subset of the Banach algebra $\mathscr{B}(H^{(n)})$ of all bounded operators in $H^{(n)}$ [34].

The operators M_γ all commute with M_z, the operation of multiplication by the scalar-valued function $z = e^{i\theta}$ on the circle. Indeed $L_{\text{cts}}GL_n(\mathbb{C})$ is not far from being the commutant of M_z in $GL(H^{(n)})$.

Theorem (6.1.1). *The commutant of M_z in $GL(H^{(n)})$ is the group $L_{\text{meas}}GL_n(\mathbb{C})$ of bounded measurable maps $S^1 \to GL_n(\mathbb{C})$.*

To say here that γ is bounded means that both $\|\gamma(\theta)\|$ and $\|\gamma(\theta)^{-1}\|$ are bounded outside a set of measure zero.

Proof. If $A \in GL(H^{(n)})$ commutes with M_z, let $\phi_i = A\varepsilon_i \in H^{(n)}$, where ε_i is the ith basis vector of $\mathbb{C}^n$, identified with the corresponding constant function in $H^{(n)}$. Thinking of the ϕ_i as taking values which are n-component column vectors, we put $\phi_1, \phi_2, \ldots, \phi_n$ side by side to form a measurable matrix-valued function ϕ. Then $A = M_\phi$. For we can approximate any $f \in H^{(n)}$ by elements of the form $\sum p_i(z)\varepsilon_i$ where each p_i is a polynomial in z and z^{-1}; and because A commutes with M_z we have

$$A\left(\sum p_i(z)\varepsilon_i\right) = \sum p_i(z)\phi_i = M_\phi\left(\sum p_i(z)\varepsilon_i\right).$$

6 LOOP GROUPS AS GROUPS OF OPERATORS

It follows that $A = M_\phi$, and that $\|\phi(\theta)\|$ and $\|\phi(\theta)^{-1}\|$ are essentially bounded.

6.2 The restricted general linear group of Hilbert space

To obtain more refined results we must introduce a restricted general linear group. (This group was first studied by Shale [136].) It is defined for a Hilbert space H which is equipped with a *polarization*, i.e. a decomposition $H = H_+ \oplus H_-$ as the orthogonal sum of two closed subspaces. The decomposition can be conveniently given by the unitary operator $J: H \to H$ which is $+1$ on H_+ and -1 on H_-. The restricted general linear group consists of the operators which are fairly close to preserving the decomposition $H = H_+ \oplus H_-$.

Definition (6.2.1). *$GL_{\text{res}}(H)$ is the subgroup of $GL(H)$ consisting of operators A such that the commutator $[J, A]$ is a Hilbert–Schmidt operator.*

We recall (cf. [125]) that an operator $T: H_1 \to H_2$ is Hilbert–Schmidt if for some (and hence every) complete orthonormal sequence $\{e_i\}$ in H_1 the sequence $\sum \|Te_i\|^2$ converges. The Hilbert–Schmidt norm $\|T\|_2$ is then $(\sum \|Te_i\|^2)^{\frac{1}{2}}$. The Hilbert–Schmidt operators in H form a two-sided ideal $\mathcal{I}_2(H)$ in $\mathcal{B}(H)$—it follows from this that $GL_{\text{res}}(H)$ really is a group—and they are themselves a Hilbert space under the norm $\| \ \|_2$.

The definition of $GL_{\text{res}}(H)$ can be reformulated as follows. If an element A of $GL(H)$ is written as a 2×2 matrix

$$A = \begin{pmatrix} a & b \\ c & d \end{pmatrix} \qquad (6.2.2)$$

with respect to the decomposition $H = H_+ \oplus H_-$, then A belongs to $GL_{\text{res}}(H)$ if and only if b and c are Hilbert–Schmidt operators.

For yet another formulation, we introduce the Banach algebra $\mathcal{B}_J(H)$ of all bounded operators $A: H \to H$ such that $[J, A]$ is Hilbert–Schmidt. The norm $\| \ \|_J$ is defined by

$$\|A\|_J = \|A\| + \|[J, A]\|_2.$$

The group $GL_{\text{res}}(H)$ is the group of units of $\mathcal{B}_J(H)$: we give it the topology defined by $\| \ \|_J$, and it is then a complex Banach Lie group.

We shall also define the restricted unitary group.

Definition (6.2.3). *$U_{\text{res}}(H)$ is the subgroup of $GL_{\text{res}}(H)$ consisting of unitary operators.*

$U_{\text{res}}(H)$ is a real Banach Lie group. The standard polar decomposition (cf. [125]) of operators in Hilbert space shows that $GL_{\text{res}}(H)$ is the

6.2 RESTRICTED GENERAL LINEAR GROUP

topological product of $U_{\text{res}}(H)$ and the contractible space of positive definite elements.† $GL_{\text{res}}(H)$ is the complexification of $U_{\text{res}}(H)$.

If the operator A of (6.2.2) belongs to $GL_{\text{res}} = GL_{\text{res}}(H)$ then its components a and d are Fredholm operators, i.e. they have finite dimensional kernels (null spaces) and cokernels. For an operator is Fredholm if it is invertible modulo compact operators, and the invertibility of A implies that a and d are invertible modulo Hilbert–Schmidt operators, which are compact. (An account of Fredholm operators from a topologist's point of view can be found in the appendix to Atiyah [3]. In Douglas [40] there is a detailed discussion from the point of view of operator theory.) A Fredholm operator a has an integer invariant $\chi(a)$, its *index*, defined by

$$\chi(a) = \dim \ker(a) - \dim \operatorname{coker}(a).$$

This is invariant under continuous deformation, and divides up the space of Fredholm operators into its connected components. It follows that the group GL_{res} is disconnected into pieces characterized by the integer $\chi(a)$. (Notice that $\chi(a) = -\chi(d)$, because $a \oplus d$ can be deformed linearly through Fredholm operators to the invertible operator A.) In fact two elements A_1 and A_2 are in the same connected component if $\chi(a_1) = \chi(a_2)$: that follows from the following much more precise result, which shows that GL_{res} has the homotopy type of the space which topologists call $\mathbb{Z} \times BU$.

Proposition (6.2.4). *The map $A \mapsto a$ from $GL_{\text{res}}(H)$ to the space $\operatorname{Fred}(H_+)$ of Fredholm operators in H_+ is a homotopy equivalence.*

Proof. Consider the map

$$\begin{pmatrix} a & b \\ c & d \end{pmatrix} \mapsto \begin{pmatrix} a \\ c \end{pmatrix}$$

which assigns to an element of GL_{res} its first column. This is a map onto an open subset $\mathscr{F}$ of $\operatorname{Fred}(H_+) \times \mathscr{I}_2(H_+; H_-)$, where $\mathscr{I}_2(H_+; H_-)$ is the space of Hilbert–Schmidt operators $H_+ \to H_-$. On the other hand $\mathscr{F}$ is also the homogeneous space $GL_{\text{res}}/\mathscr{B}$, where $\mathscr{B}$ is the subgroup of all elements of the form

$$\begin{pmatrix} 1 & b \\ 0 & d \end{pmatrix}.$$

This subgroup is contractible, for it is the semidirect product of $GL(H_-)$ and the vector space $\mathscr{I}_2(H_-; H_+)$. (We recall—see Kuiper [96]—that the general linear group of Hilbert space is contractible.) So $GL_{\text{res}} \to \mathscr{F}$ is a homotopy equivalence.

† One must check that taking the positive square-root of positive elements of $\mathscr{B}_J(H)$ is well-defined and continuous.

But the projection $\mathscr{F} \to \text{Fred}(H_+)$ is also a homotopy equivalence, for the inverse image of an element a is the contractible open set of $\mathscr{I}_2(H_+; H_-)$ consisting of all operators c such that $c \,|\, \ker(a)$ is injective. (A map with contractible fibres is a homotopy equivalence providing certain local conditions are satisfied. (See Dold [39].) These hold automatically for a projection from an open set of a Banach space.)

6.3 The map $LGL_n(\mathbb{C}) \to GL_{\text{res}}(H^{(n)})$

Returning to loop groups, we have seen that the continuous loops in $GL_n(\mathbb{C})$ can be regarded as a subgroup of $GL(H^{(n)})$. The smooth loops are contained in $GL_{\text{res}}(H^{(n)})$. In defining the restricted group we shall always decompose $H^{(n)} = L^2(S^1; \mathbb{C}^n)$ as $H_+^{(n)} \oplus H_-^{(n)}$, where

$$H_+^{(n)} = \{\text{functions whose negative Fourier coefficients vanish}\}$$
$$= \left\{f \in H^{(n)} : f(\theta) = \sum_{k \geq 0} f_k e^{ik\theta} \text{ with } f_k \in \mathbb{C}^n\right\},$$
$$= \{f \in H^{(n)} : f \text{ is the boundary value of a}$$
$$\text{function holomorphic in } |z| < 1\},$$

and

$$H_-^{(n)} = (H_+^{(n)})^\perp$$
$$= \left\{f \in H^{(n)} : f(\theta) = \sum_{k < 0} f_k e^{ik\theta}\right\}.$$

In other words, we have decomposed $H^{(n)}$ essentially into the positive and negative eigenspaces of the infinitesimal rotation operator $-i\,d/d\theta$.

Proposition (6.3.1). *If $\gamma : S^1 \to GL_n(\mathbb{C})$ is continuously differentiable, then the multiplication operator M_γ belongs to $GL_{\text{res}}(H^{(n)})$.*

We shall give two proofs of this, as both are instructive.

First Proof. Let us write γ as a Fourier series

$$\gamma(\theta) = \sum_{k \in \mathbb{Z}} \gamma_k e^{ik\theta},$$

where the γ_k are $n \times n$ matrices. With respect to the obvious orthonormal basis of $H^{(n)}$ the operator M_γ is represented by a $\mathbb{Z} \times \mathbb{Z}$ matrix (M_{pq}) whose entries are $n \times n$ matrices. In fact $M_{pq} = \gamma_{p-q}$. We must show that the $(H_+^{(n)} \to H_-^{(n)})$ and $(H_-^{(n)} \to H_+^{(n)})$ components of M_γ are Hilbert-Schmidt, in other words that

$$\sum_{p \geq 0, q < 0} \|M_{pq}\|^2 < \infty \quad \text{and} \quad \sum_{p < 0, q \geq 0} \|M_{pq}\|^2 < \infty.$$

6.3 THE MAP $LGL_n(\mathbb{C}) \to GL_{\text{res}}(H^{(n)})$

This is equivalent to

$$\sum_{k \in \mathbb{Z}} (|k| + 1) \|\gamma_k\|^2 < \infty,$$

which is certainly true if γ is differentiable, as the square of the L^2 norm of the derivative γ' is $\sum k^2 \|\gamma_k\|^2$.

Second Proof. The operator J defining the decomposition $H_+^{(n)} \oplus H_-^{(n)}$ is a singular integral operator

$$(Jf)(\theta) = \frac{1}{2\pi} PV \int_0^{2\pi} K(\theta, \phi) f(\phi) \, d\phi$$

whose kernel K is given by

$$K(\theta, \phi) = \sum_{k \geq 0} e^{ik(\theta - \phi)} - \sum_{k < 0} e^{ik(\theta - \phi)}$$
$$= 1 + i \cot \tfrac{1}{2}(\theta - \phi). \qquad (6.3.2)$$

Here PV denotes the 'principal value' of the integral, i.e.

$$\lim_{\varepsilon \to 0} \left(\int_0^{\theta - \varepsilon} + \int_{\theta + \varepsilon}^{2\pi} \right).$$

(J is the analogue for the circle of the *Hilbert transform* $\tilde{J}$ of functions on the line (cf. [158] Vol. 2, p. 243) defined by

$$(\tilde{J}f)(x) = \frac{1}{2\pi i} PV \int_{-\infty}^{\infty} \frac{f(y)}{x - y} \, dy. \;)$$

The commutator $[M_\gamma, J]$ is therefore described by the kernel $K(\theta, \phi)(\gamma(\theta) - \gamma(\phi))$, and is Hilbert–Schmidt when the kernel is square-summable, i.e. when

$$\int_0^{2\pi} \int_0^{2\pi} \frac{\|\gamma(\theta) - \gamma(\phi)\|^2}{\sin^2 \tfrac{1}{2}(\theta - \phi)} \, d\theta \, d\phi < \infty.$$

That is certainly true if γ is continuously differentiable, for then the integrand is a continuous function on $S^1 \times S^1$.

The loop group $LGL_n(\mathbb{C})$ is of course not a topological subgroup of $GL_{\text{res}}(H)$: it has a much finer topology than its image. In fact the topology on $LGL_n(\mathbb{C})$ induced by $GL_{\text{res}}(H)$ can be described in the following way.

If $\gamma = \sum \gamma_k z^k$ is a matrix-valued function on S^1 then the Hilbert–Schmidt norm of the commutator $[M_\gamma, J]$ is

$$\|\gamma\|_{2, \frac{1}{2}} = \left\{ \sum |k| \, \|\gamma_k\|^2 \right\}^{\frac{1}{2}}.$$

84   6 LOOP GROUPS AS GROUPS OF OPERATORS

This is commonly known as the Sobolev norm corresponding to '$\frac{1}{2}$-differentiable' functions [144]. Let A denote the Banach algebra of measurable matrix-valued functions γ on S^1 such that

$$|\!|\!|\gamma|\!|\!| = \|\gamma\|_\infty + \|\gamma\|_{2,\frac{1}{2}} < \infty.$$

(Here $\|\gamma\|_\infty$ denotes the L^∞ norm.) The group $GL_n(A)$ will be denoted by $L_{\frac{1}{2}}GL_n(\mathbb{C})$. It is a Banach Lie group. Clearly we have

Proposition (6.3.3). *$L_{\frac{1}{2}}GL_n(\mathbb{C})$ is the commutant of the multiplication operator M_z in $GL_{\mathrm{res}}(H^{(n)})$.*

The group $L_{\frac{1}{2}}GL_n(\mathbb{C})$ is of interest to us because it is the largest loop group to which much of the theory of this book applies: in particular it is the largest group for which the crucial central extension can be constructed and the basic irreducible representation defined. On the other hand it is hard to describe its elements explicitly. It contains all loops of class $C^{(1)}$, but it neither contains nor is contained in the group of continuous loops, and the smooth loops are *not* dense in it. We shall now give some examples to illustrate these facts.

Examples.
(i) Piecewise smooth loops belong to $L_{\frac{1}{2}}$ if and only if they are continuous: the typical discontinuous example is

$$\sum_{k>0} \frac{\sin k\theta}{k} = \tfrac{1}{2}(\pi - \theta) \quad \text{for} \quad 0 < \theta < 2\pi.$$

(ii) The function

$$f = \sum_{k>1} \frac{\cos k\theta}{k \log k}$$

satisfies $\|f\|_{2,\frac{1}{2}} < \infty$ but is not bounded near $\theta = 0$ (because $\sum 1/k \log k = \infty$: see Zygmund [158] Chapter 5 §1), and hence not continuous.

(iii) The function

$$g = \sum_{k>1} \frac{\sin k\theta}{k(\log k)^{\frac{1}{2}}}$$

is continuous, but $\|g\|_{2,\frac{1}{2}} = \infty$.

(iv) The function e^{if}, where f is as in (ii), belongs to $L_{\frac{1}{2}}\mathbb{T}$, but is not continuous. To see that e^{if} belongs to $L_{\frac{1}{2}}\mathbb{T}$ we begin with e^a, where

$$a = -\sum_{k>1} \frac{\sin k\theta}{k \log k}.$$

This function a is bounded and continuous and belongs to the Banach algebra A, so e^a belongs to $L_{\frac{1}{2}}\mathbb{C}^\times$. But e^a has the unique factorization $e^a = e^{if} \cdot e^h$, where $h(z) = -i \sum_{k>1} z^k/k$ is the boundary value of a function holomorphic for $|z| < 1$. We shall see later—from Proposition (8.3.5)—that this implies that e^{if} belongs to $L_{\frac{1}{2}}\mathbb{T}$.

6.4 Bott periodicity

The inclusion $LGL_n(\mathbb{C}) \to GL_{\text{res}}(H^{(n)})$ is essentially the map known to algebraic topologists as the inverse of the 'Bott periodicity' map. Bott's theorem asserts that its restriction to the subgroup $\Omega GL_n(\mathbb{C})$ of 'based' loops, i.e. those such that $\gamma(0) = 1$, induces an isomorphism of homotopy groups π_i for $i < 2n - 1$. Because $\pi_i \Omega GL_n(\mathbb{C}) \cong \pi_{i+1} GL_n(\mathbb{C})$, while, as we shall see in Section 6.6,

$$\pi_i GL_{\text{res}} \cong \pi_{i-1} GL_n(\mathbb{C}) \qquad (6.4.1)$$

when $i < 2n + 1$, this means that

$$\pi_i GL_n(\mathbb{C}) \cong \pi_{i+2} GL_n(\mathbb{C})$$

when $i < 2n - 2$. We shall return to this subject in Section 8.8. Meanwhile let us notice the interesting fact that the inverse Bott map (ordinarily defined only up to homotopy) has been realized as a homomorphism of infinite dimensional Lie groups.

6.5 The isomorphism $H^{(n)} \cong H$ and the embedding $L\mathbb{T} \to LU_n$

Although it seems at first sight a very unnatural thing to do, we shall find it surprisingly useful to identify the Hilbert space $H^{(n)} = L^2(S^1; \mathbb{C}^n)$ with the standard Hilbert space $H = H^{(1)} = L^2(S^1; \mathbb{C})$ by means of the obvious lexicographic correspondence between their orthonormal bases: if $\{\varepsilon_i : 1 \leq i \leq n\}$ is the standard basis of $\mathbb{C}^n$, we make $\varepsilon_i z^k \in H^{(n)}$ correspond to $z^{nk+i-1} \in H$. More invariantly, given a vector-valued function with components $(f_1, f_2, \ldots, f_n)$, we associate to it the scalar-valued function $\tilde{f}$ given by

$$\tilde{f}(\zeta) = f_1(\zeta^n) + \zeta f_2(\zeta^n) + \ldots + \zeta^{n-1} f_n(\zeta^n). \qquad (6.5.1a)$$

Conversely, given $\tilde{f} \in H$, we obtain $(f_i) \in H^{(n)}$ by

$$f_{i+1}(z) = \frac{1}{n} \sum_\zeta \zeta^{-i} \tilde{f}(\zeta), \qquad (6.5.1b)$$

where ζ runs through the nth roots of z.

The isomorphism $H^{(n)} \cong H$ is an isometry. It makes continuous functions correspond to continuous ones, and also preserves all other reasonable classes of functions, for example: smooth, real analytic, rational, polynomial. Furthermore the decomposition $H^{(n)} = H^{(n)}_+ \oplus H^{(n)}_-$ corresponds to the decomposition $H = H_+ \oplus H_-$.

The multiplication operator M_z on $H^{(n)}$ corresponds to M_{z^n} on H.

Identifying the commutant of M_{z^n} on H with $L_{\text{meas}}GL_n(\mathbb{C})$ by (6.1.1), and noticing that it must contain the commutant of M_z on H, we have an inclusion

$$L_{\text{meas}}\mathbb{C}^\times \subset L_{\text{meas}}GL_n(\mathbb{C}),$$

inducing

$$L\mathbb{C}^\times \subset LGL_n(\mathbb{C}),$$
$$L\mathbb{T} \subset LU_n,$$

and so on. The last inclusion has already been described in Proposition (3.6.4).

6.6 The central extension of $GL_{\text{res}}(H)$

We shall now define a central extension of GL_{res} by the multiplicative group $\mathbb{C}^\times$ of non-zero complex numbers. The motivation for the definition will become clear in Chapter 7.

We begin by recalling a few facts about traces and determinants for operators in Hilbert space. For proofs and further details we refer the reader to Simon [137].

(i) An operator $T: H_1 \to H_2$ (where H_1 and H_2 are Hilbert spaces) is of *trace class* if it is of the form

$$Tv = \sum \lambda_k \langle u_k, v \rangle w_k,$$

where $\{u_k\}$ and $\{w_k\}$ are orthonormal families in H_1 and H_2, and $\sum |\lambda_k| < \infty$. The *trace norm* $\|T\|_1$ of T is then $\sum |\lambda_k|$, and the *trace* of T, defined if $H_1 = H_2$, is given by

$$\text{tr}(T) = \sum \lambda_k \langle u_k, w_k \rangle.$$

(ii) The operators of trace class in $\mathscr{B}(H)$ form a two-sided ideal $\mathscr{I}_1(H)$ contained in the ideal $\mathscr{I}_2(H)$ of Hilbert–Schmidt operators. The product of two Hilbert–Schmidt operators is of trace class.

(iii) An operator $A: H \to H$ has a determinant (by definition) if and only if $A - 1$ is of trace class. If A has a determinant it is invertible if and only if $\det(A) \neq 0$. If A_1 and A_2 have determinants, then so does $A_1 A_2$, and $\det(A_1 A_2) = \det(A_1)\det(A_2)$.

To obtain the central extension of GL_{res} we begin by constructing an extension

$$\mathscr{T} \to \mathscr{E} \to GL_{\text{res}}^0 \tag{6.6.1}$$

6.6 THE CENTRAL EXTENSION OF $GL_{res}(H)$

of the identity component of GL_{res} by the group $\mathcal{T}$ of all invertible operators $q : H_+ \to H_+$ which have a determinant. (The topology of $\mathcal{T}$ is defined by using the trace-norm as a metric.) The extension $\mathscr{E}$ is of interest in its own right. We shall see that it is a *contractible* Banach Lie group. The exact sequence of homotopy groups associated to the fibration (6.6.1) shows that

$$\pi_i(GL_{res}^0) \cong \pi_{i-1}(\mathcal{T}).$$

The last group is well-known to coincide with $\pi_{i-1}(GL_n(\mathbb{C}))$ when $i - 1 < 2n$—see, for example, Palais [121]—and this gives us the isomorphism (6.4.1) already mentioned.

The definition of $\mathscr{E}$ is very simple. The identity component of GL_{res} consists of the operators

$$A = \begin{pmatrix} a & b \\ c & d \end{pmatrix}$$

such that the Fredholm operator a has index zero. Because a has index zero one can add to it an operator t of finite rank so that $q = a + t$ is invertible. We define $\mathscr{E}$ as a subgroup of $GL_{res} \times GL(H_+)$:

$$\mathscr{E} = \{(A, q) \in GL_{res} \times GL(H_+) : a - q \text{ is of trace class}\}.$$

We give it, however, not the subgroup topology, but that induced by its embedding

$$(A, q) \mapsto (A, a - q)$$

as an open set of $GL_{res} \times \mathcal{I}_1(H_+)$. It is then a Banach Lie group. The motivation for the definition of $\mathscr{E}$ will appear, as we have said, in Chapter Seven, but it is at any rate clear that $\mathscr{E}$ is an extension of GL_{res}^0 by $\mathcal{T}$.

Proposition (6.6.2). *The group $\mathscr{E}$ is contractible.*

Proof. Consider the diagram

$$\begin{array}{ccc} \mathscr{E} & \to & GL(H_+) \times \mathcal{I}_1(H) \\ \downarrow & & \downarrow \\ GL_{res}^0 & \to & \mathrm{Fred}^0(H_+), \end{array}$$

where the upper horizontal map is $(A, q) \mapsto (q, aq^{-1} - 1)$,
 the lower horizontal map is $A \mapsto a$,
 the right-hand vertical map is $(q, t) \mapsto (1 + t)q$, and
 $\mathrm{Fred}^0(H_+)$ denotes the Fredholm operators of index 0 in H_+.
Both vertical maps are fibrations with the group $\mathcal{T}$ as fibre, and the diagram is cartesian (i.e. the fibration on the left is the pull-back of that on the right). We know from (6.2.4) that the lower horizontal map is a homotopy equivalence; it follows that the upper one is too. But $GL(H_+) \times \mathcal{I}_1(H)$ is contractible.

We can use the determinant homomorphism $\det: \mathcal{T} \to \mathbb{C}^\times$ to obtain from $\mathscr{E}$ an extension—obviously a central extension—of GL_{res}^0 by $\mathbb{C}^\times$. This extension is simply $\mathscr{E}/\mathcal{T}_1$, where $\mathcal{T}_1$ is the kernel of det. We wish, however, to have an extension of all of GL_{res}, not just of its identity component. Now GL_{res} is the semidirect product of its identity component by $\mathbb{Z}$, where we can take for $\mathbb{Z}$ the subgroup generated by any element in the ± 1 components, for example a shift operator $\sigma: H \to H$ which embeds H_+ in itself with codimension one. The automorphism $A \mapsto \sigma A \sigma^{-1}$ of GL_{res}^0 is covered by the endomorphism $\tilde{\sigma}$ of $\mathscr{E}$ defined by

$$(A, q) \mapsto (\sigma A \sigma^{-1}, q_\sigma),$$

where

$$q_\sigma = \begin{cases} \sigma q \sigma^{-1} & \text{on} \quad \sigma(H_+) \\ 1 & \text{on} \quad H_+ \ominus \sigma(H_+). \end{cases}$$

The endomorphism $\tilde{\sigma}$ induces $q \mapsto q_\sigma$ on the normal subgroup $\mathcal{T}$ of $\mathscr{E}$, and is not an automorphism. Indeed, though we shall not give the proof here, there is *no* automorphism of $\mathscr{E}$ which covers $A \mapsto \sigma A \sigma^{-1}$, and so there is no extension of GL_{res} by $\mathcal{T}$ which restricts to $\mathscr{E}$. On the other hand, because $\det(q_\sigma) = \det(q)$, the endomorphism of $\mathscr{E}$ does induce an automorphism of $\mathscr{E}/\mathcal{T}_1$, and we can form the semidirect product $\mathbb{Z} \tilde{\times} (\mathscr{E}/\mathcal{T}_1)$ to obtain a central extension $GL_{\mathrm{res}}^\sim$ of GL_{res} by $\mathbb{C}^\times$. This is the extension we have been seeking.

Remark (6.6.3). It is easy to check that if σ is *any* element of GL_{res} such that $\sigma(H_+) \subset H_+$, and $\tilde{\sigma} \in GL_{\mathrm{res}}^\sim$ is a representative of σ, then the above formula

$$\tilde{\sigma} \cdot (A, q) \cdot \tilde{\sigma}^{-1} = (\sigma A \sigma^{-1}, q_\sigma)$$

always holds in $GL_{\mathrm{res}}^\sim$.

There is no continuous cross-section of $GL_{\mathrm{res}}^\sim \to GL_{\mathrm{res}}$ (indeed it follows from (6.6.2) that its Chern class is the universal first Chern class in $H^2(\mathbb{Z} \times BU)$), and so the extension cannot be described by a continuous cocycle. But there is a cross-section of $\mathscr{E} \to GL_{\mathrm{res}}$ defined in the subset U of GL_{res} where a is invertible: it is given by $A \mapsto \tilde{A} = (A, a)$. In that region, which is a dense open subset of the identity component of GL_{res}, we have

Proposition (6.6.4). *If $A_1 A_2 = A_3$ in GL_{res}, and A_1, A_2, A_3 all belong to U, then*

$$\tilde{A}_1 \tilde{A}_2 = c(A_1, A_2) \tilde{A}_3$$

in $GL_{\mathrm{res}}^\sim$, where

$$c(A_1, A_2) = \det(a_1 a_2 a_3^{-1}).$$

Notice that the operators a_1, a_2, a_3 do not themselves have determinants, but only the combination $a_1 a_2 a_3^{-1}$.

The main practical utility of Proposition (6.6.4) is that it enables us to read off the Lie algebra cocycle of the extension $\widetilde{GL}_{\text{res}}$. If an extension of a Lie group Γ is defined by a smooth cocycle $c: \Gamma \times \Gamma \to K$ then the corresponding Lie algebra cocycle is

$$(\xi, \eta) \mapsto D_1 D_2 c(\xi, \eta) - D_1 D_2 c(\eta, \xi),$$

where $D_1 D_2 c$ denotes the mixed second partial derivative of c at the identity. The Lie algebra of GL_{res} consists of all bounded operators A such that $[J, A]$ is Hilbert–Schmidt. As usual we shall write them

$$A = \begin{pmatrix} a & b \\ c & d \end{pmatrix},$$

where b and c are Hilbert–Schmidt. We find

Proposition (6.6.5). *The Lie algebra cocycle corresponding to the extension $\widetilde{GL}_{\text{res}}$ is given by*

$$\begin{aligned}(A_1, A_2) &\mapsto \operatorname{trace}([a_1, a_2] - a_3) \\ &= \operatorname{trace}(c_1 b_2 - b_1 c_2) \\ &= \tfrac{1}{4}\operatorname{trace}(J[J, A_1][J, A_2]),\end{aligned} \quad (6.6.6)$$

where $A_3 = [A_1, A_2]$.

To conclude this section let us notice that it is natural to define $\widetilde{U}_{\text{res}}$ as the intersection of $\widetilde{GL}_{\text{res}}$ with $U(H) \times U(H_+)$. Then $\widetilde{U}_{\text{res}}$ is an extension of $U_{\text{res}} = U_{\text{res}}(H)$ by $\mathbb{T}$, and its complexification is GL_{res}.

6.7 The central extension of $LGL_n(\mathbb{C})$

We can use the homomorphism $M: LGL_n(\mathbb{C}) \to GL_{\text{res}}$ to pull back the extension $\widetilde{GL}_{\text{res}}$. We obtain a central extension $\tilde{L}_{\mathbb{C}}$ of $LGL_n(\mathbb{C})$ by $\mathbb{C}^{\times}$. It is a complex Lie group, because the homomorphism M is holomorphic. The subgroup LU_n of $LGL_n(\mathbb{C})$ maps into the unitary group U_{res}, so the extension of U_{res} by $\mathbb{T}$ pulls back to an extension $\tilde{L}$ of LU_n by $\mathbb{T}$; and $\tilde{L}_{\mathbb{C}}$ is obviously a complexification of $\tilde{L}$.

Proposition (6.7.1). *The extension $\tilde{L}$ of LU_n induced by $\widetilde{U}_{\text{res}}$ is the basic extension constructed in Section 4.7.*

Proof. We saw in Section 4.7 that the basic extension of LU_n is characterized by its Lie algebra cocycle. We shall calculate the Lie algebra cocycle of $\tilde{L}$.

Let ξ and η be elements of $L\mathfrak{u}_n$, and A_1, A_2, A_3 be the operators on

$H^{(n)}$ corresponding to ξ, η, and $[\xi, \eta]$. In view of (6.6.5) we must show that
$$\text{trace}\{[a_1, a_2] - a_3\} = i\omega(\xi, \eta), \tag{6.7.2}$$
where
$$\omega(\xi, \eta) = \frac{1}{2\pi} \int_0^{2\pi} \langle \xi(\theta), \eta'(\theta) \rangle \, d\theta,$$
and $\langle \ , \ \rangle$ is the standard inner product on $\mathfrak{u}_n$ given by $\langle X, Y \rangle = -\text{trace}(XY)$.

By linearity we can suppose that $\xi = Xz^k$ and $\eta = Yz^m$, with $X, Y \in \mathfrak{gl}_n(\mathbb{C})$. If $k + m \neq 0$ then the left-hand-side of (6.7.2) is zero because the matrices of $[a_1, a_2]$ and a_3 have no diagonal entries; and $\omega(\xi, \eta) = 0$ also. If $m = -k$, on the other hand, then both operators $[a_1, a_2]$ and a_3 preserve each of the subspaces $\mathbb{C}^n \cdot z^q$ of which $H_+^{(n)}$ is the sum (for $q \geq 0$). If $q \geq k$ then $[a_1, a_2]$ and a_3 coincide on $\mathbb{C}^n \cdot z^q$. If $q < k$ then $[a_1, a_2]$ acts as $-YX$ on $\mathbb{C}^n \cdot z^q$, while a_3 acts as $[X, Y]$. The left-hand-side of (6.7.2) is therefore $-k \cdot \text{trace}(XY)$, i.e. $k\langle X, Y \rangle$. The right-hand-side is the same.

Remark. Although it is not needed for the proof of (6.7.1) it is instructive to use the remark (6.6.3) to calculate explicitly the effect of conjugation by a loop γ of winding number m on the centre of the identity component $\tilde{L}^0$ of $\tilde{L}$. The centre of $\tilde{L}^0$ is canonically $\mathbb{T} \times \mathbb{T}$, where the first $\mathbb{T}$ is the scalar matrices in $U_n \subset LU_n$, and the second $\mathbb{T}$ belongs to the extension. An element u of the first $\mathbb{T}$ can be represented by $(u, u) \in \mathscr{E} \subset GL_{\text{res}} \times GL(H_+)$. We can assume that M_γ maps H_+ into itself with codimension m. Then the automorphism of $\mathscr{E}$ corresponding to M_γ takes (u, u) to (u, v), where $v : H_+ \to H_+$ is multiplication by u on $M_\gamma H_+$, and is the identity on the m-dimensional space $H_+ \ominus M_\gamma H_+$. So $\det(vu^{-1}) = u^{-m}$, as we want.

We can now prove that *all* of the extensions
$$\mathbb{T} \to \tilde{L}G \to LG$$
considered in Chapter 4 possess complexifications
$$\mathbb{C}^\times \to \tilde{L}G_\mathbb{C} \to LG_\mathbb{C}.$$
It is enough to consider the case where G is simply connected and simple. If we choose a unitary representation ρ of G on $\mathbb{C}^n$, then by pulling back the extension $\tilde{L}_\mathbb{C}$ of $LGL_n(\mathbb{C})$ found above we get an extension $\tilde{L}_\rho G_\mathbb{C}$ which corresponds to the trace form $\langle \ , \ \rangle_\rho$ of ρ on $\mathfrak{g}$:
$$\langle \xi, \eta \rangle_\rho = -\text{trace}(\rho(\xi)\rho(\eta)).$$
Now any integral invariant form on $\mathfrak{g}$ is an integral multiple of the basic form $\langle \ , \ \rangle$ (see Section 4.4). The extension $\tilde{L}G_\mathbb{C}$ corresponding to $\langle \ , \ \rangle$

6.8 Embedding Diff$^+(S^1)$ in $U_{\text{res}}(H)$

The group of diffeomorphisms of the circle acts on $H^{(n)} = L^2(S^1; \mathbb{C}^n)$. In fact one can make it act in more than one way. We shall choose to make the action *unitary*, i.e. to regard the elements of $H^{(n)}$ as $\frac{1}{2}$-densities on S^1. Thus a diffeomorphism $f: S^1 \to S^1$ acts on functions $\xi: S^1 \to \mathbb{C}^n$ by $\xi \mapsto f \cdot \xi$, where

$$(f \cdot \xi)(\theta) = \xi(g(\theta)) \cdot |g'(\theta)|^{\frac{1}{2}}, \qquad (6.8.1)$$

and g is the inverse of f.

Proposition (6.8.2). $\quad \text{Diff}^+(S^1) \subset U_{\text{res}}(H^{(n)})$.

Proof. This can be proved by either of the methods of (6.3.1). It is done by the first method in [131], so here we shall sketch the alternative argument. We represent J by the kernel K of (6.3.2). From (6.8.1) the kernel representing the action of f is $\delta(g(\theta) - \phi)g'(\theta)^{\frac{1}{2}}$, where δ is the Dirac δ-function. The kernel of the commutator $[f, J]$ is therefore

$$\int_0^{2\pi} \{\delta(g(\theta) - \psi)g'(\theta)^{\frac{1}{2}} K(\psi, \phi) - K(\theta, \psi)\delta(g(\psi) - \phi)g'(\psi)^{\frac{1}{2}}\} \, d\psi.$$

This reduces to

$$g'(\theta)^{\frac{1}{2}} K(g(\theta), \phi) - K(\theta, f(\phi)) f'(\phi)^{\frac{1}{2}}. \qquad (6.8.3)$$

From (6.3.2) we see that K is a smooth function of both its variables except on the diagonal, where

$$K(\theta, \phi) = 2i/(\theta - \phi) + (\text{smooth function}).$$

Inserting this in (6.8.3), we find that the kernel of $[f, J]$ is continuous (indeed smooth) everywhere, and hence that $[f, J]$ is Hilbert–Schmidt.

There is, however, an important difference between $LGL_n(\mathbb{C})$ and $\text{Diff}^+(S^1)$ in relation to GL_{res}. The former maps smoothly into GL_{res}, whereas the inclusion of $\text{Diff}^+(S^1)$ is not even continuous. Indeed the norm topology of $GL(H^{(n)})$—and hence â fortiori that of GL_{res}—induces the discrete topology on $\text{Diff}^+(S^1)$. (To see that, observe that for any diffeomorphism f except the identity there is a unit vector $\xi \in H^{(n)} = L^2(S^1; \mathbb{C}^n)$ such that $\langle \xi, f^*\xi \rangle = 0$: take ξ to have support in a small neighbourhood of a point of S^1 not fixed by f.) If one attempts to calculate formally the homomorphism of Lie algebras induced by

Diff$^+(S^1) \to GL_{\text{res}}$ then one finds that vector fields on S^1 correspond to *unbounded* operators in $H^{(n)}$: consider, for example, $d/d\theta$.

Despite this, the central extension of Diff$^+(S^1)$—considered as an abstract group—induced by $\widetilde{GL}_{\text{res}}$ is indeed a Lie group, and its Lie algebra cocycle can be calculated exactly as was done in (6.7.1) above, ignoring the unboundedness of the operators. The formal calculation is carried out in [131]; the reason for its validity is that the *composite* map

$$\text{Diff}^+(S^1) \to U_{\text{res}} \to U_{\text{res}}/(U_+ \times U_-), \tag{6.8.4}$$

where $U_+ \times U_- = U(H_+) \times U(H_-)$ is the commutant of J in U_{res}, is smooth, and the 2-cocycle of $\mathfrak{u}_{\text{res}}$, regarded as an invariant form on U_{res}, actually comes from $U_{\text{res}}/(U_+ \times U_-)$. To see that (6.8.4) is smooth we observe first that the map $A \mapsto [A, J]A^{-1}$ defines a smooth immersion from a neighbourhood of the base-point in $U_{\text{res}}/(U_+ \times U_-)$ into the space of Hilbert–Schmidt operators, and then that when f is a diffeomorphism the commutator $[f, J]$ is represented by a smooth kernel which depends smoothly on f.

We shall record the result here.

Proposition (6.8.5). *The central extension of* Diff$^+(S^1)$ *induced by* $\widetilde{GL}_{\text{res}}(H^{(n)})$ *is trivial over* $SL_2(\mathbb{R})$, *and has the Lie algebra cocycle*

$$(\xi, \eta) \mapsto \frac{n}{12\pi} \int_0^{2\pi} (\xi'''(\theta) + \xi'(\theta))\eta(\theta)\, d\theta$$

for $\xi = \xi(\theta)\, d/d\theta$, $\eta = \eta(\theta)\, d/d\theta$ *in* Vect(S^1).

Because Diff$^+(S^1)$ is contained in U_{res} it acts on the extension $\widetilde{U}_{\text{res}}$ by conjugation, and hence on the subgroup $\check{L}U_n$ which covers $LU_n \subset GL_{\text{res}}$. This way of seeing that Diff$^+(S^1)$ acts on $\check{L}U_n$ is simpler and more natural than the one given in Section 4.7.

Proposition (6.8.6). *The action of* Diff$^+(S^1)$ *on* $\check{L}U_n$ *induced by the embedding in* $\widetilde{GL}_{\text{res}}(H^{(n)})$ *is given by the formula* (4.7.3).

Proof. From the discussion in Section 4.7 we know that it is enough to check the action on the Lie algebra $\check{L}\mathfrak{u}_n$, and therefore even enough to find the action of $\xi \in \text{Vect}(S^1)$ on $\tilde{\eta} \in \check{L}\mathfrak{u}_n$. In notation corresponding to that of (6.7.2) we must show that

$$\text{trace}\{[a_1, a_2] - a_3\} = -\frac{1}{4\pi} \int_0^{2\pi} \xi(\theta) \text{trace } \eta'(\theta)\, d\theta.$$

This is proved by a calculation exactly like the proof of (6.7.2): when the action of the vector field $\xi = \sum \xi_k e^{ik\theta}\, d/d\theta$ is written as a $\mathbb{Z} \times \mathbb{Z}$ matrix whose entries are $n \times n$ blocks, the $(p, q)^{\text{th}}$ entry is $-\frac{1}{2}i(p+q)\xi_{p-q}\mathbf{1}_n$, where $\mathbf{1}_n$ is the $n \times n$ identity matrix.

We can also prove the following explicit formula which will be useful later. (It agrees with (4.7.5) when $n = 1$.)

Proposition (6.8.7). *If $\gamma: S^1 \to U_n$ is the homomorphism $\theta \mapsto \exp(\theta\xi)$, and $\tilde{\gamma}$ is a representative of γ in $\tilde{L}U_n$, then the action on $\tilde{\gamma}$ of the rotation R_α through the angle α is given by*

$$R_\alpha \tilde{\gamma} = e^{-\frac{1}{2}i\alpha(\|\xi\|^2 - m)} \cdot \tilde{\gamma} \cdot \gamma(\alpha)^{-1},$$

where m is the winding number of γ.

Proof. It is enough to consider the case when $\gamma H_+ \subset H_+$, and then we can use Remark (6.6.3) to calculate $\tilde{\gamma} R_\alpha \tilde{\gamma}^{-1}$ in $\widetilde{GL}_{\text{res}}$. One finds

$$\tilde{\gamma} R_\alpha \tilde{\gamma}^{-1} = R_\alpha \cdot \gamma(\alpha) \cdot u^{-1},$$

where u is the determinant of the action of R_α on $H_+ \ominus \gamma H_+$. To calculate u we can assume that γ is the diagonal loop $z \mapsto \text{diag}(z^{k_1}, \ldots, z^{k_n})$. Then

$$u = \exp - i\alpha \sum \tfrac{1}{2} k_j(k_j - 1) = \exp - \tfrac{1}{2}i\alpha(\|\xi\|^2 - m).$$

6.9 Other polarizations of H: replacing the circle by the line, and the introduction of 'mass'

In two-dimensional quantum field theory one is interested primarily in functions defined on the line $\mathbb{R}$, which represents physical space, rather than on the circle. One can identify $\mathbb{R} \cup \infty$ with S^1 by stereographic projection, i.e.

$$e^{i\theta} \in S^1 \leftrightarrow 2 \tan \tfrac{1}{2}\theta \in \mathbb{R}, \tag{6.9.1}$$

where we choose $\theta \in (-\pi, \pi]$. The Hilbert space $H = L^2(S^1; \mathbb{C})$ is then isometrically isomorphic to $H^\mathbb{R} = L^2(\mathbb{R}; \mathbb{C})$ by the correspondence

$$\phi \in H \leftrightarrow \tilde{\phi} \in H^\mathbb{R},$$

where

$$\tilde{\phi}(2 \tan \tfrac{1}{2}\theta) = \phi(\theta) \cos \tfrac{1}{2}\theta. \tag{6.9.2}$$

The natural polarization of $H^\mathbb{R}$ is given by the positive and negative eigenspaces of $-id/dx$ on $\mathbb{R}$, i.e. $H^\mathbb{R} = H^\mathbb{R}_+ \oplus H^\mathbb{R}_-$, where $H^\mathbb{R}_+$ consists of the functions f whose Fourier transform $\hat{f}$, given by

$$\hat{f}(\xi) = \frac{1}{\sqrt{2\pi}} \int_\mathbb{R} f(x) e^{-ix\xi} \, dx,$$

vanishes for $\xi < 0$. Under the isomorphism $H \cong H^\mathbb{R}$ of (6.9.2) we find

$$H^\mathbb{R}_+ \leftrightarrow e^{\frac{1}{2}i\theta} H_+,$$

94   6 LOOP GROUPS AS GROUPS OF OPERATORS

where $e^{\frac{1}{2}i\theta}$ is an L^∞ function on the circle with a jump discontinuity at $\theta = \pm\pi$. The Fourier series of $e^{\frac{1}{2}i\theta}$ is $\sum a_k e^{ik\theta}$, with

$$a_k = 2 \cdot (-1)^{k+1}/(2k-1)\pi.$$

Because $\sum |ka_k^2|$ does not converge, the subspace $H_+^{\mathbb{R}}$ does not belong to $\mathrm{Gr}(H)$. Conjugation by $e^{\frac{1}{2}i\theta}$ therefore does not define an automorphism of $GL_{\mathrm{res}}(H)$, and the groups $GL_{\mathrm{res}}(H)$ and $GL_{\mathrm{res}}(H_{\mathbb{R}})$ defined with respect to the natural polarizations are not mapped to each other by the isomorphism $H \cong H^{\mathbb{R}}$ of (6.9.2), even though they are isomorphic groups.

The group LT acts on $H^{\mathbb{R}}$ by the correspondence (6.9.1), and similarly LU_n acts on $H^{\mathbb{R},(n)} = L^2(\mathbb{R}; \mathbb{C}^n)$. But the embedding $LU_n \to GL_{\mathrm{res}}(H^{\mathbb{R},(n)})$ so obtained is not interestingly different from the standard $LU_n \to GL_{\mathrm{res}}(H^{(n)})$, for conjugation by $e^{\frac{1}{2}i\theta}$ induces the identity on the image of LU_n.

More interesting is to polarize the space $H^\Delta = L^2(\mathbb{R}; \mathbb{C}^2)$ according to the positive and negative parts of the spectrum of the operator

$$D_m = \begin{pmatrix} -i & 0 \\ 0 & i \end{pmatrix} \frac{\mathrm{d}}{\mathrm{d}x} + \begin{pmatrix} 0 & m \\ m & 0 \end{pmatrix}$$

where m is some positive number. One should think of H^Δ as the space of solutions ψ of the Dirac equation with mass m

$$\left\{ \begin{pmatrix} 0 & 1 \\ 1 & 0 \end{pmatrix} \frac{\partial}{\partial t} + \begin{pmatrix} 0 & 1 \\ -1 & 0 \end{pmatrix} \frac{\partial}{\partial x} \right\} \psi = im\psi \qquad (6.9.3)$$

for functions $\psi: \mathbb{R}^2 \to \mathbb{C}^2$; the decomposition $H^\Delta = H_+^\Delta \oplus H_-^\Delta$ is then the decomposition according to the spectrum of the 'energy' operator $-i(\partial/\partial t)$. As usual we can form the tensor product $H^{\Delta,(n)} = H^\Delta \otimes \mathbb{C}^n$, and the loop group LU_n acts on this by multiplication operators. (We are again using the identification (6.9.1).)

Proposition (6.9.4). *The action of LU_n on $H^{\Delta,(n)}$ induces an embedding $i_m : LU_n \to GL_{\mathrm{res}}(H^{\Delta,(n)})$.*

Proof. If we replace functions on $\mathbb{R}$ by their Fourier transforms then D_m becomes the multiplication operator on $H^\Delta \otimes \mathbb{C}^n$ by the matrix-valued function

$$\begin{pmatrix} \xi & m \\ m & -\xi \end{pmatrix} \otimes 1.$$

The operator J corresponding to the polarization is therefore multiplication by the function $J(\xi) \otimes 1$, where

$$J(\xi) = \frac{1}{E(\xi)} \begin{pmatrix} \xi & m \\ m & -\xi \end{pmatrix} \otimes 1,$$

6.9 OTHER POLARIZATIONS OF H

and $E(\xi) = +\sqrt{(\xi^2 + m^2)}$. (Notice that $J(\xi)^2 = 1$.) If $\gamma: \mathbb{R} \to U_n$ is an element of LU_n it is enough for us to show that the commutator $[M_{\gamma - \gamma(\infty)}, J]$ is Hilbert–Schmidt. Let $\hat{\gamma}$ be the Fourier transform of $\gamma - \gamma(\infty)$. The commutator is represented by the kernel (see the second proof of Proposition (6.3.1))

$$(J(\xi) - J(\eta)) \otimes \hat{\gamma}(\xi - \eta). \tag{6.9.5}$$

Now the trace of the matrix $(J(\xi) - J(\eta))^2$ is

$$a(\xi, \eta) = \frac{4(E(\xi)E(\eta) - \xi\eta - m^2)}{E(\xi)E(\eta)},$$

and it is easy to show that

$$\int_{\mathbb{R}} a(\eta + \zeta, \eta) \, d\eta \le C |\zeta|, \tag{6.9.6}$$

where C is some constant which depends only on m. The Hilbert–Schmidt norm of the commutator (6.9.5) is therefore dominated by

$$\int_{\mathbb{R}} |\zeta| |\hat{\gamma}(\zeta)|^2 \, d\zeta < \infty,$$

as we want.

The embedding i_m gives us nothing new when $m = 0$, for then H^Δ breaks into two independent copies of $H^\mathbb{R}$ on which we have used the standard polarization and its opposite. It follows that the central extension of GL_{res} restricts trivially to LU_n. This remains true in general.

Proposition (6.9.7). *The central extension of $GL_{\text{res}}(H^{\Delta,(n)})$ is trivial over $i_m(LU_n)$.*

Proof. We use the formula (6.6.6) to calculate the Lie algebra cocycle. Suppose that two elements of $L\mathfrak{u}_n$ are represented by matrix-valued functions f and g on $\mathbb{R}$. We can suppose that f and g are square-summable, for the value of the cocycle is not changed by replacing them by $f - f(\infty)$ and $g - g(\infty)$. Then the cocycle is

$$\frac{1}{4} \iint \text{trace}\{J(\xi)(J(\xi) - J(\eta))\hat{f}(\xi - \eta)(J(\eta) - J(\xi))\hat{g}(\eta - \xi)\} \, d\xi \, d\eta$$

$$= \frac{1}{4} \iint \text{trace}\{J(\xi)(J(\xi) - J(\eta))^2\} \langle \hat{f}(\xi - \eta), \hat{g}(\eta - \xi) \rangle \, d\xi \, d\eta.$$

But this vanishes, because

$$J(\xi)(J(\xi) - J(\eta))^2 = 2J(\xi) - J(\eta) - J(\xi)J(\eta)J(\xi)^{-1},$$

which has trace 0.

Before mentioning our final variant in this direction we should point out that the main interest of the Dirac polarization of H^Δ is not for constructing representations of LU_n. More important is that it leads to a new representation of the so-called 'canonical commutation relations'. We shall prove the basic result here, but we shall postpone discussing its significance till Chapter 10.

Let M_f denote the operator on $H^\Delta = L^2(\mathbb{R}; \mathbb{C}^2)$ given by multiplication by

$$\begin{pmatrix} f & 0 \\ 0 & f \end{pmatrix}$$

where $f: \mathbb{R} \to \mathbb{C}$ is a smooth function which is constant outside a finite interval (but perhaps with different values at the two ends of the line). The commutator $[J, M_f]$ is Hilbert–Schmidt. (We proved this above when $f(+\infty) = f(-\infty)$, but the argument works in general.) Thus the operators M_f form an abelian subalgebra $\mathscr{F}$ of the Lie algebra $\mathfrak{gl}_{\text{res}}(H^\Delta)$, and the central extension of $\mathfrak{gl}_{\text{res}}$ is trivial when restricted to $\mathscr{F}$.

Now let N_f denote the operator on H^Δ given by

$$\begin{pmatrix} f & 0 \\ 0 & -f \end{pmatrix},$$

where $f: \mathbb{R} \to \mathbb{C}$ is smooth with compact support. Again the commutator $[J, N_f]$ is Hilbert–Schmidt: its kernel is

$$(J(\xi)A - AJ(\eta))\hat{f}(\xi - \eta),$$

where

$$A = \begin{pmatrix} 1 & 0 \\ 0 & -1 \end{pmatrix},$$

and the calculation is essentially as before, except that because $J(\xi)A - AJ(\eta)$ does not vanish when $\xi = \eta$ we need f to vanish at $\pm\infty$. (The estimate (6.9.6) is replaced by

$$\int_\mathbb{R} \text{trace}(J(\eta + \zeta)A - AJ(\eta))^2 \, d\eta \leq C_1 + C_2 |\zeta|.)$$

The operators N_f form an abelian subalgebra $\mathscr{F}'$ of $\mathfrak{gl}_{\text{res}}(H^\Delta)$, and the central extension restricts trivially to it.

Let us now think of $\mathscr{F}$ and $\mathscr{F}'$ as subalgebras of the extension $\widetilde{\mathfrak{gl}}_{\text{res}}$, and if f is a smooth function with compact support let us introduce the notation

$$\Phi(f) = 2^{-\frac{1}{2}} M_F$$
$$\dot{\Phi}(f) = 2^{-\frac{1}{2}} N_f$$

6.9 OTHER POLARIZATIONS OF H

where $F(x) = \int_{-\infty}^{x} f(y)\,dy$. The motivation for this notation is that we have

$$-i\dot{\Phi}(f) = [D_m, \Phi(f)] \qquad (6.9.8)$$

as operators on H^Δ. The remarkable result is

Proposition (6.9.9). *In $\mathfrak{gl}_{res}^\sim$ the operators $\Phi(f)$ and $\dot{\Phi}(f)$ satisfy the 'canonical commutation relations', i.e.*

$$[\Phi(f), \Phi(g)] = [\dot{\Phi}(f), \dot{\Phi}(g)] = 0,$$

and

$$[\dot{\Phi}(f), \Phi(g)] = -i \int_{\mathbb{R}} f(x)g(x)\,dx.$$

Proof. The formula (6.6.6) for the cocycle tells us that $[\dot{\Phi}(f), \Phi(g)]$ is given by

$$\frac{1}{8} \iint \mathrm{trace}\{J(\xi)(J(\xi)A - AJ(\eta))(J(\eta) - J(\xi))\hat{f}(\xi - \eta)\hat{G}(\eta - \xi)\}\,d\xi\,d\eta.$$

But

$$\mathrm{trace}\{J(\xi)(J(\xi)A - AJ(\eta))(J(\eta) - J(\xi))\} = 2\,\mathrm{trace}\,A(J(\eta) - J(\xi))$$

$$= 4\left[\frac{\eta}{E(\eta)} - \frac{\xi}{E(\xi)}\right]$$

$$= 4b(\xi, \eta), \text{ say.}$$

Because $\xi/E(\xi) \to \pm 1$ as $\xi \to \pm\infty$ it is clear that

$$\int_{\mathbb{R}} b(\eta + \zeta, \eta)\,d\eta = -2\zeta,$$

so the commutator is

$$\int_{\mathbb{R}} \zeta \hat{f}(\zeta) \hat{G}(-\zeta)\,d\zeta = -i \int_{\mathbb{R}} f(x)g(x)\,dx.$$

We conclude this section with one last subgroup of $U_{res}(H^\Delta)$. It is the group $\Lambda\mathbb{T}$ of smooth maps $\gamma: \mathbb{R} \to \mathbb{T}$ which have compact support—i.e. $\gamma(x) = 1$ when $|x|$ is large. We make it act on H^Δ by associating to γ the multiplication operator

$$\begin{pmatrix} \gamma & 0 \\ 0 & 1 \end{pmatrix}.$$

It follows from the preceding discussion that this belongs to $U_{res}(H^\Delta)$, and that the induced central extension of $\Lambda\mathbb{T}$ is the basic one (i.e. the one coming from the inclusion $\Lambda\mathbb{T} \subset L\mathbb{T}$).

More generally, we can embed ΛU_n in $U_{res}(H^{\Delta,(n)})$ in a precisely analogous way. The interest of these embeddings is that Carey and Ruijsenaars [24] have shown that when $m > 0$ the standard representation of U_{res} restricts to give a type III factor representation of ΛU_n. (Cf. Section 10.7.)

6.10 Generalizations to other groups of maps

In the definition (6.2.1) of GL_{res} the ideal $\mathscr{I}_2 = \mathscr{I}_2(H)$ of Hilbert–Schmidt operators can be replaced by any other symmetrically normed two-sided ideal $\mathscr{I}$ (cf. Simon [137]). Let us denote the resulting group by $GL_\mathscr{I}$. The biggest such $\mathscr{I}$ is the ideal $\mathscr{K}$ of compact operators; and for any $p \geq 1$ there is the ideal $\mathscr{I}_p$ consisting of operators T such that $(T^*T)^{p/2} \in \mathscr{I}_1$. All the groups $GL_\mathscr{I}$ have properties very similar to GL_{res}. Their homotopy type is independent of $\mathscr{I}$ (see Palais [121]), and there is an extension

$$\mathscr{T}_{\mathscr{I}^2} \to \mathscr{E}_\mathscr{I} \to GL_\mathscr{I},$$

where $\mathscr{E}_\mathscr{I}$ is contractible, and $\mathscr{T}_{\mathscr{I}^2}$ is the group of invertible operators belonging to $1 + \mathscr{I}^2$. But only if $\mathscr{I}^2 \subset \mathscr{I}_1$, i.e. if $\mathscr{I} \subset \mathscr{I}_2$, is there a determinant homomorphism $\mathscr{T}_{\mathscr{I}^2} \to \mathbb{C}^\times$ enabling one to construct a central extension by $\mathbb{C}^\times$. In other words, unless $\mathscr{I} \subset \mathscr{I}_2$ the basic 2-dimensional cohomology class of the space $GL_\mathscr{I}$ cannot be represented by a left-invariant differential form.

Now let us consider how far the theory of this chapter can be generalized from the loop group LGL_n to the group $\text{Map}(X; GL_n)$, where X is a compact smooth manifold.

It is easy to find embeddings

$$\text{Map}(X; GL_n) \to GL_\mathscr{K},$$

and their classification is an interesting question in algebraic topology. If X is of odd dimension $d = 2m - 1$ and is a 'spin manifold'—i.e. it is orientable and satisfies the additional mild global condition that its second Stiefel-Whitney class vanishes—then there is a complex vector bundle E on X called the bundle of 'spinors'. The fibres of E have dimension 2^{m-1}. There is also a self-adjoint first order differential operator D, the Dirac operator, which acts on the space of sections of E. If H is the space of L^2 sections of E then we can write $H = H_+ \oplus H_-$, where H_+ (resp. H_-) is spanned by the eigenfunctions of D with positive (resp. negative) eigenvalues. The group $\text{Map}(X; \mathbb{C}^\times)$ acts by multiplication operators on H, and similarly $\text{Map}(X; GL_n)$ acts on $H \otimes \mathbb{C}^n$, the space of sections of $E \otimes \mathbb{C}^n$. This action defines an embedding of $\text{Map}(X; GL_n)$ in $GL_\mathscr{K}(H \otimes \mathbb{C}^n)$.

More generally, an embedding

$$i_{(E,J)}: \text{Map}_{\text{cts}}(X; GL_n) \to GL_\mathscr{K}(H \otimes \mathbb{C}^n) \qquad (6.10.1)$$

6.10 GENERALIZATIONS TO OTHER GROUPS OF MAPS

is defined by any pair (E, J), where E is a vector bundle on X and J is a self-adjoint operator in the Hilbert space H of L^2 sections of E such that

(i) $J^2 = 1$, and
(ii) $[J, M_f]$ is compact for every continuous function f on X.

If $n > \frac{1}{2} \dim X$ then the group of connected components of $\text{Map}(X; GL_n)$ is, almost by definition, the generalized cohomology group $K^{-1}(X)$ of Atiyah and Hirzebruch [3]. This group is closely related to the classical cohomology

$$H^{\text{odd}}(X; \mathbb{Z}) = \bigoplus_{k \text{ odd}} H^k(X; \mathbb{Z}),$$

and becomes isomorphic to it when tensored with the rationals. Passing to connected components in (6.10.1) gives a homomorphism

$$\text{ind}_{(E,J)} : K^{-1}(X) \to \mathbb{Z}. \qquad (6.10.2)$$

With a little more care it is not hard to show that (E, J) defines an element $\sigma_{(E,J)}$ of the generalized homology group $K_{-1}(X)$, and that $\text{ind}_{(E,J)}$ is the natural pairing with $\sigma_{(E,J)}$. Furthermore, every element of $K_{-1}(X)$ arises in this way.

Among the embeddings $i_{(E,J)}$ the one defined by the Dirac operator is basic: its class $\sigma_{(E,J)}$ is the fundamental class of the manifold X, and the corresponding map (6.10.2) is the 'Gysin' map in K-theory. If X is a sphere, then $i_{(E,J)}$ is the Bott periodicity map.

The preceding statements are a rapid summary of the easy part of an extensive theory which has been developed by Atiyah [4], Kasparov [90], and Connes [32]. A slightly different way of looking at the same material is found in the work of Brown, Douglas, and Fillmore [23], who prove that elements of $K_{-1}(X)$ can be identified with isomorphism classes of algebra extensions

$$\mathcal{K} \to A \to C(X),$$

where $C(X)$ is the algebra of continuous complex-valued functions on X. Such an extension of algebras clearly defines a group extension

$$\mathcal{T}_{\mathcal{K}} \to GL_n(A) \to GL_n(C(X)) = \text{Map}_{\text{cts}}(X; GL_n).$$

This is the extension got by pulling back $\mathscr{E}_{\mathcal{K}}$ by $i_{(E,J)}$.

But from the point of view of the present book the group $GL_{\mathcal{K}}$ is not of very much use, because nothing is known about its representations. To embed $\text{Map}(X; GL_n)$ in $GL_{\mathcal{I}_2}$ we must use pairs (E, J) such that $[J, M_f]$ is Hilbert–Schmidt when f is smooth. In the language of Brown, Douglas, and Fillmore we must study extensions

$$\mathcal{I}_1 \to A \to \mathcal{D}(X)$$

of the algebra of smooth functions by the ideal of trace-class operators.

These extensions were studied by Helton and Howe [73]. They correspond to elements of $H_1(X; \mathbb{Z})$, which is a canonical subgroup of $K_{-1}(X)$. The corresponding extensions are the 'uninteresting' ones which we found in Chapter 4.

The best way to think of the situation is probably in terms of pseudo-differential operators [144]. In practice J will be given by a pseudo-differential operator of order zero. The commutator $[J, M_f]$, when f is a smooth function, will then be an operator of order -1. On a manifold of dimension d such an operator belongs to the ideal $\mathcal{I}_r$ if $r > d$. It will thus not normally be Hilbert–Schmidt if $d > 1$.

Example. Let us consider the polarization corresponding to the Dirac operator on a torus X of odd dimension $d = 2m - 1$, i.e. $X = \mathbb{R}^d/2\pi\mathbb{Z}^d$. The spin bundle on X is a trivial bundle whose fibre $\Delta \cong \mathbb{C}^N$ (where $N = 2^{m-1}$) is an irreducible module for the Clifford algebra C_d generated by elements $e_1, \ldots, e_d$ such that $e_i^2 = 1$ and $e_i e_j = -e_j e_i$ when $i \neq j$. The Dirac operator on the space H of maps $X \to \Delta$ is

$$D = -i \sum e_j \frac{\partial}{\partial \theta_j}.$$

If we expand the functions in Fourier series, so that H is identified with $\ell^2(\mathbb{Z}^d; \Delta)$ then D becomes the multiplication operator

$$\{f_p\} \mapsto \{pf_p\}.$$

(Here $p \in \mathbb{Z}^d$, and $pf_p \in \Delta$ is got by acting with $p \in \mathbb{R}^d \subset C_d$ on $f_p \in \Delta$.) The corresponding polarization operator J is multiplication by $p/\|p\|$. The commutator $[J, M_f]$, where M_f is multiplication by the scalar-valued function $f = \sum f_p e^{i\langle p, \theta \rangle}$, is represented by the kernel

$$(p, q) \mapsto f_{p-q} \cdot \{p/\|p\| - q/\|q\|\} \tag{6.10.3}$$

on $\mathbb{Z}^d \times \mathbb{Z}^d$. Now $p/\|p\| - q/\|q\|$ is a self-adjoint operator on Δ whose square is

$$2\left(1 - \frac{\langle p, q \rangle}{\|p\| \|q\|}\right) = 4 \sin^2 \frac{\phi}{2},$$

where ϕ is the angle between p and q. If $p - q$ is held fixed then $4 \sin^2(\phi/2)$ decays like $\|p\|^{-2}$ as $p \to \infty$. The kernel (6.10.3) is therefore square-summable only if $\dim(X) = 1$. In general it belongs to the Schatten class $\mathcal{I}_r$ when $r > \dim(X)$.

7
THE GRASSMANNIAN OF HILBERT SPACE AND THE DETERMINANT LINE BUNDLE

Because we are studying loop groups by regarding them as groups of operators in Hilbert space we shall need to have a rather detailed knowledge of the structure of the Grassmannian of Hilbert space. This chapter is devoted to that subject. The most important part is the construction of the determinant line bundle in Section 7.7, and the reader interested in that can omit everything between Sections 7.1 and 7.7 except for the definition of an 'admissible basis' in Section 7.5.

7.1 The definition of Gr(H)

Suppose that H is a separable Hilbert space with a given polarization $H = H_+ \oplus H_-$: we assume that H_+ and H_- are infinite dimensional orthogonal closed subspaces. We shall study the Grassmannian of closed subspaces of H which are 'comparable' in size with H_+. Before giving the formal definition of this class of subspaces, let us explain that they are a completion of the class of subspaces W which are *commensurable* with H_+, i.e. those such that $W \cap H_+$ has finite codimension in both W and H_+. They may, however, have zero intersection with H_+: for example the graph W_T of every Hilbert–Schmidt operator $T: H_+ \to H_-$ is included, but W_T is commensurable with H_+ only if T is of finite rank.

Definition (7.1.1). Gr(H) *is the set of all closed subspaces W of H such that*
  (i) *the orthogonal projection* $\mathrm{pr}_+ : W \to H_+$ *is a Fredholm operator,*
and
  (ii) *the orthogonal projection* $\mathrm{pr}_- : W \to H_-$ *is a Hilbert–Schmidt operator.*

Fredholm and Hilbert–Schmidt operators have been discussed already in Section 6.2. We recall that a bounded operator is Fredholm if its kernel and cokernel are finite dimensional.

Another way of stating the definition (7.1.1) is: W belongs to Gr(H) if it is the image of an operator $w : H_+ \to H$ such that $\mathrm{pr}_+ \circ w$ is Fredholm and $\mathrm{pr}_- \circ w$ is Hilbert–Schmidt. As the sum of a Fredholm operator and a Hilbert–Schmidt operator is Fredholm, we see that if W belongs to Gr(H) then so does the graph of every Hilbert-Schmidt operator $W \to W^\perp$. These graphs form the subset U_W of Gr(H) consisting of all W' for which the orthogonal projection $W' \to W$ is an isomorphism: it is in one-to-one correspondence with the Hilbert space $\mathcal{I}_2(W; W^\perp)$ of Hilbert–Schmidt operators $W \to W^\perp$. In fact

Proposition (7.1.2). Gr(H) *is a Hilbert manifold modelled on* $\mathcal{I}_2(H_+; H_-)$.

Before proving this we need one further observation. The group $GL_{\text{res}}(H)$ introduced in Section 6.2 acts on the set Gr(H). We have

Proposition (7.1.3). *The subgroup* $U_{\text{res}}(H)$ *of* $GL_{\text{res}}(H)$ *acts transitively on* Gr(H), *and the stabilizer of* H_+ *is* $U(H_+) \times U(H_-)$.

Proof of (7.1.3). Suppose $W \in \text{Gr}(H)$; we shall find $A \in U_{\text{res}}(H)$ such that $A(H_+) = W$. Let $w: H_+ \to H$ be an isometry with image W, and $w^\perp: H_- \to H$ an isometry with image $W^\perp$. Then

$$w \oplus w^\perp : H_+ \oplus H_- \to H_+ \oplus H_-$$

is a unitary transformation A such that $A(H_+) = W$. We write it

$$A = \begin{pmatrix} w_+ & w_+^\perp \\ w_- & w_-^\perp \end{pmatrix}.$$

Because W belongs to Gr(H) we know that w_+ is Fredholm and w_- is Hilbert–Schmidt. But because A is unitary it follows that $w_+^\perp$ is Hilbert–Schmidt also (for $w_+^* w_+^\perp + w_-^* w_-^\perp = 0$), and so A belongs to $U_{\text{res}}(H)$.

The assertion about the stabilizer of H_+ is obvious.

Proof of (7.1.2). Suppose that U_{W_0} and U_{W_1} are the subsets of Gr(H) described above corresponding to the Hilbert spaces $I_0 = \mathcal{I}_2(W_0; W_0^\perp)$ and $I_1 = \mathcal{I}_2(W_1; W_1^\perp)$. Let $U_{W_0} \cap U_{W_1}$ correspond to I_{01} in I_0 and I_{10} in I_1. We must show that I_{01} and I_{10} are open sets, and that the 'change of coordinates' $I_{01} \to I_{10}$ is smooth.

Let the matrix of the identity transformation

$$W_0 \oplus W_0^\perp \to W_1 \oplus W_1^\perp$$

be

$$\begin{pmatrix} a & b \\ c & d \end{pmatrix} \qquad (7.1.4)$$

(i.e. a is a map $W_0 \to W_1$, etc.) From the proof of (7.1.3) we know that a and d are Fredholm, and b and c are Hilbert–Schmidt. Suppose that $W \in \text{Gr}(H)$ is simultaneously the graph of $T_0: W_0 \to W_0^\perp$ and $T_1: W_1 \to W_1^\perp$. Then the operators

$$\begin{pmatrix} a & b \\ c & d \end{pmatrix} \begin{pmatrix} 1 \\ T_0 \end{pmatrix} \quad \text{and} \quad \begin{pmatrix} 1 \\ T_1 \end{pmatrix} q$$

from W_0 to $W_1 \oplus W_1^\perp$ must coincide, where q is some isomorphism

7.1 THE DEFINITION OF Gr(H)

$W_0 \to W_1$. We conclude that

$$T_1 = (c + dT_0)(a + bT_0)^{-1}. \tag{7.1.5}$$

Thus T_1 is a holomorphic function of T_0 in the open set

$$I_{01} = \{T_0 \in I_0 : a + bT_0 \text{ is invertible}\}.$$

For a subspace W of H which is commensurable with H_+ it is natural to define the *virtual dimension* of W relative to H_+ as

$$\dim(W/W \cap H_+) - \dim(H_+/W \cap H_+).$$

The generalization of this for an arbitrary $W \in \mathrm{Gr}(H)$ is the *index* of the perpendicular projection $\mathrm{pr}_+ : W \to H_+$, i.e.

$$\mathrm{virt.dim}\, W = \dim(\ker \mathrm{pr}_+) - \dim(\mathrm{coker}\, \mathrm{pr}_+).$$

Equivalently,

$$\mathrm{virt.dim}\, W = \dim(W \cap H_-) - \dim(W^\perp \cap H_+).$$

The virtual dimension separates $\mathrm{Gr}(H)$ into disconnected pieces. In fact the subspaces with a given virtual dimension form a connected set; we shall see presently, for example, that the spaces of virtual dimension zero are the closure of the coordinate patch consisting of the graphs of all Hilbert-Schmidt operators $H_+ \to H_-$. Notice also that if

$$A = \begin{pmatrix} a & b \\ c & d \end{pmatrix}$$

belongs to $GL_{\mathrm{res}}(H)$, then

$$\mathrm{virt.dim}\, A(W) = \mathrm{virt.dim}\, W + \chi(a),$$

where $\chi(a)$ is the index of the Fredholm operator a.

To proceed further we shall introduce an orthonormal basis in H. That amounts to identifying H with the space $L^2(S^1; \mathbb{C})$ with its natural basis $\{z^k\}_{k \in \mathbb{Z}}$. (As usual $z = e^{i\theta}$.) We then have a collection of special points $\{H_S\}$ in $\mathrm{Gr}(H)$: H_S is just the closed subspace spanned by z^s for $s \in S$, where S is a subset of $\mathbb{Z}$ which has finite difference from the positive integers $\mathbb{N}$ (i.e. S is bounded below, and contains all sufficiently large integers). We shall write $\mathscr{S}$ for the collection of such sets S. Notice that

$$\mathrm{virt.dim}\, H = \mathrm{card}(S - \mathbb{N}) - \mathrm{card}(\mathbb{N} - S).$$

We shall call this number the *virtual cardinal* of S.

Proposition (7.1.6). *For any $W \in \mathrm{Gr}(H)$ there is a set $S \in \mathscr{S}$ such that the orthogonal projection $W \to H_S$ is an isomorphism. In other words the sets $\{U_S\}_{S \in \mathscr{S}}$, where $U_S = U_{H_S}$, form an open covering of $\mathrm{Gr}(H)$.*

Proof. Because the projection $W \to H_+$ has finite dimensional kernel one can find $S_0 \in \mathscr{S}$ such that the projection $W \to H_{S_0}$ is injective. If it is not also surjective there is some $s \in S_0$ such that z^s is not in its range. Then the projection $W \to H_{S_1}$, where $S_1 = S_0 - \{s\}$, is still injective. Repeating this finitely many times gives us the desired S.

We now have quite explicit coordinate charts on $\mathrm{Gr}(H)$, indexed by $\mathscr{S}$. A point of U_S is the graph of a Hilbert–Schmidt operator $H_S \to H_S^\perp$, and is represented by an $\tilde{S} \times S$ matrix, where $\tilde{S} = \mathbb{Z} - S$. The transitions between the charts are given by (7.1.5) where the matrix (7.1.4) is a permutation matrix; in particular, the components b and c have only finitely many non-zero entries.

7.2 Some dense submanifolds of Gr(H)

We shall describe in terms of the coordinate charts just introduced four important dense submanifolds of $\mathrm{Gr}(H)$. The reason for being interested in them will appear later.

(i) $\mathrm{Gr}_0(H)$ consists of all subspaces W such that $z^k H_+ \subset W \subset z^{-k} H_+$ for some k. Such subspaces can be identified with subspaces of $H_{-k,k} = z^{-k} H_+/z^k H_+$, and so $\mathrm{Gr}_0(H)$ is the union of the finite dimensional classical Grassmannians $\mathrm{Gr}(H_{-k,k})$. In terms of the coordinate charts, $\mathrm{Gr}_0(H)$ consists of the graphs of operators $H_S \to H_S^\perp$ with only finitely many non-zero matrix entries: these are dense in $\mathscr{I}_2(H_S; H_S^\perp)$.

(ii) $\mathrm{Gr}_1(H)$ consists of all subspaces W which are commensurable with H_+. These are the graphs of all operators $H_S \to H_S^\perp$ of *finite rank*.

(iii) $\mathrm{Gr}_\omega(H)$ consists of the graphs of all operators $T: H_S \to H_S^\perp$ whose matrix entries T_{pq} (for $p \in \tilde{S}$, $q \in S$) are such that $r^{p-q} T_{pq}$ is bounded for some r with $0 < r < 1$.

(iv) $\mathrm{Gr}_\infty(H)$ consists of graphs of all operators $T: H_S \to H_S^\perp$ whose entries T_{pq} are *rapidly decreasing*, i.e. such that $|p-q|^m T_{pq}$ is bounded for each m.

Without entering fully into the motivation for introducing these subspaces, let us notice that if W belongs to $\mathrm{Gr}_\infty(H)$ then it has a dense subspace consisting of smooth functions: that is so because the finite linear combinations of the smooth functions

$$w_q = z^q + \sum_p T_{pq} z^p$$

are dense in W. Similarly, if W belongs to $\mathrm{Gr}_\omega(H)$ then real-analytic functions are dense in W, and if W belongs to $\mathrm{Gr}_0(H)$ then trigonometric

7.2 SOME DENSE SUBMANIFOLDS OF Gr(H)

polynomials are dense in it. These conditions do not, however, characterize Gr_0, Gr_ω and Gr_∞. Thus the graph W_T of $T: H_+ \to H_-$, where

$$Tz^k = \frac{1}{k} z^{-k},$$

does not belong to any of them, though the trigonometric polynomials are obviously dense in it. (At the other extreme, it is not hard to show that a generic $W \in \text{Gr}(H)$ contains no non-zero smooth function at all.)

The subspace $\text{Gr}_\infty(H)$ can be described in the following way.

Proposition (7.2.1). $\text{Gr}_\infty(H)$ *consists precisely of the subspaces $W \in \text{Gr}(H)$ for which the images of both orthogonal projections*

$$\text{pr}_-: W \to H_- \quad \text{and} \quad \text{pr}_+: W^\perp \to H_+$$

consist of smooth functions.

Proof. That the images do consist of smooth functions if W belongs to Gr_∞ is immediate from the definition. (Notice that if W is the graph of $T: H_S \to H_S^\perp$ then $W^\perp$ is the graph of $-T^*: H_S^\perp \to H_S$.) Conversely, if W is the graph of $T: H_S \to H_S^\perp$ and the image of pr_- consists of smooth functions then so does the image of T. Thus T defines a map from H_S to the space of smooth functions on the circle. By the closed graph theorem this must be continuous, and it can be thought of as a smooth map from the circle into the dual of H_S. Its smoothness is equivalent to the condition that

$$|p|^m \left\{ \sum_q |T_{pq}|^2 \right\}^{\frac{1}{2}}$$

is bounded as $p \to \infty$ for each m. Similarly the smoothness of the image of $\text{pr}_+: W^\perp \to H_+$ is equivalent to the boundedness of

$$|q|^m \left\{ \sum_p |T_{pq}|^2 \right\}^{\frac{1}{2}}$$

as $q \to \infty$ for each m; and the two conditions together imply that W belongs to Gr_∞.

An exactly analogous description can be given of $\text{Gr}_0(H)$ and $\text{Gr}_\omega(H)$: for the former the proof is trivial.

The four subspaces can be considered as manifolds in their own right. The most important for us will be $\text{Gr}_\infty(H)$, which we shall refer to as the *smooth* Grassmannian. The description above shows that it is a manifold modelled on the metrizable nuclear space (cf. [60]) of matrices $T = \{T_{pq}: p < 0, q \geq 0\}$ whose topology is defined by the sequence of seminorms ρ_m, where

$$\rho_m(T) = \sup_{p,q} |p-q|^m |T_{pq}|.$$

One can also describe $\mathrm{Gr}_\infty(H)$ directly in terms of the space C^∞ of smooth functions on the circle (with its usual topology): it consists of all closed subspaces W of C^∞ such that the projection $W \to C_+^\infty$ is Fredholm and the projection $W \to C_-^\infty$ is compact. (Here $C_\pm^\infty = C^\infty \cap H_\pm$.) We shall, however, omit the justification of this.

We conclude this section with a very simple application of the existence of the dense submanifold $\mathrm{Gr}_0 = \mathrm{Gr}_0(H)$.

Proposition (7.2.2). *Every holomorphic function $f: \mathrm{Gr}(H) \to \mathbb{C}$ is constant on each connected component.*

Proof. It is enough to show that f is locally constant on Gr_0. But Gr_0 is the union of the finite dimensional Grassmannians $\mathrm{Gr}(H_{-n,n})$. As these are compact algebraic varieties every holomorphic function on them is locally constant.

7.3 The stratification of Gr(H)

A generic element W of $\mathrm{Gr}(H)$, if it has virtual dimension zero, is transversal to H_-, i.e. $W \cap H_- = 0$ and $W + H_- = H$. These generic elements form a dense open subset in their connected component. The other elements W in the same component meet H_- non trivially: it follows from the discussion below that those such that $\dim(W \cap H_-) \geq k$ form a closed subset of codimension k^2. The most obvious stratification of $\mathrm{Gr}(H)$ would be by the dimension of the intersection $W \cap H_-$, which is necessarily finite. We shall need, however, a finer stratification, which records the dimension of $W \cap z^m H_-$ for every m.

Let us say that an element f of $H = L^2(S^1; \mathbb{C})$ is of *finite order s* if it is of the form

$$\sum_{k=-\infty}^{s} f_k z^k \qquad (7.3.1)$$

with $f_s \neq 0$. In other words, f is the boundary value of a function f which is holomorphic in the hemisphere $|z| > 1$ except for a pole of order s at $z = \infty$. For any $W \in \mathrm{Gr}(H)$, let W^{fin} denote the set of elements of finite order in W. Because elements of finite order are dense in any H_S, and because the projection $W \to H_S$ is an isomorphism for suitable S, we have

Proposition (7.3.2). W^{fin} *is dense in W.*

The elements of W of order $\leq m$ form the finite dimensional space

7.3 THE STRATIFICATION OF Gr(H)

$W_m = W \cap z^{m+1}H_-$. For given W we define

$$S_W = \{s \in \mathbb{Z} : W \text{ contains an element of order } s\}.$$

The set S_W belongs to $\mathscr{S}$, and its virtual cardinal is the virtual dimension d of W, for the number of elements of S_W which are $\leq m$ is dim W_m, which is $m + 1 + d$ providing m is large enough for the projection $W \to z^{m+1}H_+$ to be surjective.

For each $s \in S_W$ let w_s be an element of W of the form (7.3.1) with $f_s = 1$. Evidently $\{w_s\}$ is a basis of W^{fin} in the algebraic sense, and the projection $W \to H_{S_W}$ is an isomorphism. We can choose w_s uniquely so that it projects to z^s; we shall call this the *canonical basis* of W. (This choice is precisely the process of choosing a basis for a subspace in 'reduced echelon form', familiar in elementary linear algebra.)

For given $S \in \mathscr{S}$ we shall call the set

$$\Sigma_S = \{W \in \text{Gr}(H) : S_W = S\}$$

the *stratum* of Gr(H) corresponding to S. In other words, Σ_S consists of all W such that $\dim(W_m) = d_m(S)$ for all m, where $d_m(S)$ is the number of elements of S which are $\leq m$.

An indexing set S of virtual cardinal d can be written canonically

$$S = \{s_{-d}, s_{-d+1}, s_{-d+2}, \ldots\},$$

with $s_{-d} < s_{-d+1} < s_{-d+2} < \ldots$ and $s_k = k$ for large k. We shall order the sets of the same virtual cardinal by defining

$$S \leq S' \Leftrightarrow s_k \geq s'_k \text{ for all } k$$
$$\Leftrightarrow d_m(S) \leq d_m(S') \text{ for all } m.$$

We shall also define the *length* $\ell(S)$ of S by

$$\ell(S) = \sum_{k \geq 0} (k - s_k).$$

Then $S < S'$ implies $\ell(S) < \ell(S')$.

Finally, it will be convenient to introduce the 'strictly lower triangular' subgroup $\mathcal{N}_-$ of GL_{res}, consisting of all elements A such that $A(z^k H_-) = z^k H_-$ and $(A - 1)(z^k H_-) \subset z^{k-1} H_-$ for all k.

The stratification is described by

Proposition (7.3.3).
 (i) *The stratum Σ_S is a contractible closed submanifold of the open set U_S, of codimension $\ell(S)$.*
 (ii) *Σ_S is the orbit of H_S under $\mathcal{N}_-$.*
 (iii) *If $W \in U_S$ then $S \geq S_W$.*
 (iv) *The closure of Σ_S is the union of the strata $\Sigma_{S'}$ with $S' \geq S$.*

108   7 THE GRASSMANNIAN OF HILBERT SPACE

Proof.
(i) We have already shown that Σ_S is contained in U_S. Now if $W \in U_S$ then $W \to H_S$ is an isomorphism, and so W has a unique basis $\{w_s\}$ which projects to $\{z^s\}$. Because of the uniqueness, W belongs to Σ_S if and only if w_s has order s for each s. If W is described as the graph of $T: H_S \to H_{\tilde{S}}^\perp$, so that $w_s = z^s + Tz^s$, then W belongs to Σ_S precisely when the matrix elements T_{pq} vanish when $p > q$. The number of pairs (p, q) in $\tilde{S} \times S$ such that $p > q$ is the length $\ell(S)$. Thus Σ_S corresponds to a sub-Hilbert-space of codimension $\ell(S)$.

(ii) Suppose that $W \in \Sigma_S$ is the graph of $T: H_S \to H_{\tilde{S}}^\perp$. Let $\mathrm{pr}_S: H \to H_S$ be the projection. Then $A = 1 + T \circ \mathrm{pr}_S$ belongs to $\mathcal{N}_-$, and $A(H_S) = W$.

(iii) Perpendicular projection on to H_S can only lower the order of an element, and so if $W \to H_S$ is an isomorphism then H_S must have at least as many linearly independent elements of order $\leq m$ as W does, i.e.

$$\mathrm{card}\{s \in S : s \leq m\} \geq \mathrm{card}\{s \in S_W : s \leq m\}$$

for each m. This is equivalent to the assertion $S \geq S_W$.

(iv) It follows from (iii) that the closure of Σ_S is contained in the union of the $\Sigma_{S'}$ with $S' \geq S$. But if $S' > S$ let W_t be the subspace spanned by

$$(1-t)z^{s_k} + tz^{s'_k}$$

for $k \geq -d$. If $0 \leq t < 1$ then W_t belongs to Σ_S; if $t = 1$ then $W_t = H_{S'} \in \Sigma_{S'}$. This proves that the closure of Σ_S meets $\Sigma_{S'}$. The closure must then contain $\Sigma_{S'}$ because $\Sigma_{S'}$ is a single orbit of the group $\mathcal{N}_-$.

7.4 The cellular decomposition of $\mathrm{Gr}_0(H)$

The Grassmannian of a finite dimensional vector space has a classical decomposition into Schubert cells. (Cf. [68] or [116].) Our Grassmannian $\mathrm{Gr}_0(H)$ is the union of the finite dimensional Grassmannians $\mathrm{Gr}(H_{-n,n})$, and it too can be decomposed into Schubert cells. This decomposition is dual to the stratification of $\mathrm{Gr}(H)$ described in the last section in the following sense:

(i) the same set $\mathcal{S}$ indexes the cells $\{C_S\}$ and the strata $\{\Sigma_S\}$;
(ii) the dimension of C_S is the codimension of Σ_S;
(iii) C_S meets Σ_S transversally in a single point, and meets no other stratum of the same codimension.

To describe C_S we begin by defining the *co-order* of a polynomial element

$$f = \sum_{k=-N}^{N} f_k z^k$$

of H as the smallest k such that $f_k \neq 0$. Then for $W \in \text{Gr}_0(H)$ the set
$$S^W = \{s \in \mathbb{Z} : W \text{ contains an element of co-order } s\}$$
belongs to $\mathscr{S}$, and for $S \in \mathscr{S}$ we define
$$C_S = \{W \in \text{Gr}_0(H) : S^W = S\}.$$

Proposition (7.4.1).
(i) C_S *is a closed submanifold of the open set U_S of* $\text{Gr}(H)$ *and is diffeomorphic to* $\mathbb{C}^{\ell(S)}$.
(ii) C_S *is the orbit of H_S under the 'strictly upper triangular' subgroup $\mathcal{N}_+$ of* GL_{res}.
(iii) *If $W \in \text{Gr}_0(H)$ belongs to U_S then $S \leq S^W$.*
(iv) *The closure of C_S is the union of the $C_{S'}$ with $S' \leq S$.*
(v) C_S *intersects $\Sigma_{S'}$ if and only if $S \geq S'$, and C_S intersects Σ_S transversally in the single point H_S.*

The strictly upper triangular subgroup $\mathcal{N}_+$ consists of all A such that $A(z^k H_+) = z^k H_+$ and $(A-1)(z^k H_+) \subset z^{k+1} H_+$ for all k.

Proof. This is precisely analogous to (7.3.3). The essential observation is that C_S consists of the graphs of all operators $T : H_S \to H_S^\perp$ whose matrix elements T_{pq} vanish unless $p > q$.

7.5 The Plücker embedding

Points of a finite dimensional Grassmannian are traditionally described by Plücker coordinates. We can do exactly the same with $\text{Gr}(H)$.

We have pointed out in Section 7.3 that any $W \in \text{Gr}(H)$ has a canonical basis. We shall find it useful, however, to introduce a class of 'admissible bases' for W. Suppose that W has virtual dimension d.

Definition (7.5.1). *A sequence $\{w_k\}_{k \geq -d}$ in W is called an* admissible basis *for W if*
(i) *the linear map $w : z^{-d} H_+ \to W$ which takes z^k to w_k is a continuous isomorphism, and*
(ii) *the composite $\text{pr} \circ w$, where $\text{pr} : W \to z^{-d} H_+$ is the orthogonal projection, is an operator with a determinant.*

Remarks.
(i) We recall that an operator with a determinant is one which differs from the identity by an operator of trace class. (Cf. Section 6.6.)
(ii) We shall usually not distinguish between the basis $\{w_k\}$ and the corresponding linear map w.
(iii) The canonical basis for W is admissible: for it, the composite $\text{pr} \circ w$ differs from the identity by an operator of finite rank.

It is clear from the definitions that any two admissible bases for the same space W are related to each other by a matrix which has a determinant. Furthermore, if w is an admissible basis for W, and $S \in \mathcal{S}$ is a set of virtual cardinal d, and $\text{pr}_S : W \to H_S$ is the projection, then $\text{pr}_S \circ w$ is also an operator with a determinant. We define the *Plücker coordinate* $\pi_S(w)$ of the basis w as the determinant $\det(\text{pr}_S \circ w)$. If $S \in \mathcal{S}$ does not have virtual cardinal d we define $\pi_S(w) = 0$. If w' is another admissible basis for w then

$$\pi_S(w') = \Delta_{ww'} \cdot \pi_S(w),$$

where $\Delta_{ww'}$ is the determinant of the matrix relating w' and w; so if one thinks of $\{\pi_S\}_{S \in \mathcal{S}}$ as *projective* coordinates then they depend only on W.

Proposition (7.5.2). *The Plücker coordinates $\{\pi_S\}_{S \in \mathcal{S}}$ define a holomorphic embedding*

$$\pi : \text{Gr}(H) \to P(\mathcal{H})$$

into the projective space of the Hilbert space $\mathcal{H} = \ell^2(\mathcal{S})$.

Remark. In Section 7.7 we shall give a more invariant description of $\mathcal{H}$.

Proof. We must first show that for an admissible basis w we have

$$\sum_{S \in \mathcal{S}} |\pi_S(w)|^2 < \infty.$$

In fact we shall prove

$$\sum_S |\pi_S(w)|^2 = \det(w^*w). \quad (7.5.3)$$

(The right-hand-side is defined, for if we write $w : z^{-d}H_+ \to H$ as $w_+ \oplus w_-$ with respect to $H = z^{-d}H_+ \oplus z^{-d}H_-$ then $w^*w = w_+^*w_+ + w_-^*w_-$, which has a determinant because w_+ has a determinant and w_- is Hilbert–Schmidt.)

It is enough to prove (7.5.3) for any one admissible basis w for each subspace W. And by continuity it is enough to prove it when W belongs to $\text{Gr}_0(H)$. So we may assume that w_+ differs from the identity matrix in only finitely many entries, and that w_- has only finitely many non-zero entries. In that case (7.5.3) reduces to the following assertion:

If P and Q are $n \times m$ and $m \times n$ matrices, with $n \leq m$, then

$$\det(PQ) = \sum_S \det(P_S)\det(Q_S),$$

where S runs through the n element subsets of $\{1, 2, \ldots, m\}$, and P_S, Q_S are the corresponding $n \times n$ submatrices of P and Q. (This assertion simply expresses the functoriality of the nth exterior power: $\Lambda^n(P \circ Q) = (\Lambda^n P) \circ (\Lambda^n Q)$.)

To prove that π is an embedding, let us consider first the case of a subspace W which is the graph of an operator $T: H_+ \to H_-$. The canonical basis $\{w_k\}_{k \geq 0}$ for W is given by

$$w_q = z^q + \sum_{p<0} T_{pq} z^p.$$

Suppose that $S \in \mathscr{S}$ has virtual cardinal 0, and write $A = S - \mathbb{N}$, $B = \mathbb{N} - S$. These are two finite sets of the same size. A moment's reflection reveals that $\pi_S(w)$ is the determinant of the finite submatrix of (T_{pq}) formed from the rows A and columns B. In particular, each entry T_{pq} occurs among the Plücker coordinates. So π is certainly an embedding in the coordinate patch $U_\mathbb{N}$.

The other coordinate patches can be treated in exactly the same way. We remark finally that W belongs to the patch U_S if and only if $\pi_S(w) \neq 0$.

The stratification of $\mathrm{Gr}(H)$, and also the three dense subspaces Gr_0, Gr_ω and Gr_∞, can be described very simply in terms of Plücker coordinates, as follows.

Proposition (7.5.4).

(i) $W \in U_S \Leftrightarrow \pi_S(W) \neq 0$,
(ii) $W \in \Sigma_S \Leftrightarrow \pi_S(W) \neq 0$ and $\pi_{S'}(W) = 0$ when $S' < S$,
(iii) $W \in C_S \Leftrightarrow \pi_S(W) \neq 0$ and $\pi_{S'}(W) = 0$ unless $S' \leq S$,
(iv) $W \in \mathrm{Gr}_0 \Leftrightarrow \pi_S(W) = 0$ except for finitely many S,
(v) $W \in \mathrm{Gr}_\omega \Leftrightarrow r^{-\ell(S)} \pi_S(W)$ is bounded for $S \in \mathscr{S}$, for some $r < 1$,
(vi) $W \in \mathrm{Gr}_\infty \Leftrightarrow \ell(S)^m \pi_S(W)$ is bounded for $S \in \mathscr{S}$, for each m.

All of these assertions are obvious except perhaps for the last two, whose validity will become clear in the next section.

7.6 The $\mathbb{C}^\times_{\leq 1}$-action

The circle $\mathbb{T}$ acts unitarily on $H = L^2(S^1; \mathbb{C})$ by rotating S^1, and the action preserves the polarization $H = H_+ \oplus H_-$. This means that $\mathbb{T}$ acts on $\mathrm{Gr}(H)$. It is easy to see that the fixed points are precisely the subspaces H_S for $S \in \mathscr{S}$. We shall write $R_u: \mathrm{Gr}(H) \to \mathrm{Gr}(H)$ for the action of $u \in \mathbb{T}$.

The map $\mathbb{T} \times \mathrm{Gr}(H) \to \mathrm{Gr}(H)$ describing the action is continuous, but not differentiable. In the coordinate chart $U_S \cong \mathscr{I}_2(H_S; H_S^\perp)$ the action of R_u on $T: H_S \to H_S^\perp$ multiplies the matrix element T_{pq} by u^{q-p}. From this we find

Proposition (7.6.1). *The $\mathbb{T}$-orbit of a point $W \in \mathrm{Gr}(H)$ is smooth (i.e. the map $u \mapsto R_u W$ is smooth) if and only if W belongs to $\mathrm{Gr}_\infty(H)$. The orbit is*

real-analytic if and only if W belongs to $\mathrm{Gr}_\omega(H)$. Furthermore, $\mathbb{T}$ acts smoothly on the manifold $\mathrm{Gr}_\infty(H)$ with its own C^∞ topology.

The description in terms of coordinate charts shows that the action of $\mathbb{T}$ extends to an action

$$\mathbb{C}^\times_{\leq 1} \times \mathrm{Gr}(H) \to \mathrm{Gr}(H) \qquad (7.6.2)$$

of the semigroup $\mathbb{C}^\times_{\leq 1}$ of non-zero complex numbers of modulus ≤ 1. The map (7.6.2) is holomorphic on the open set $\mathbb{C}^\times_{<1} \times \mathrm{Gr}(H)$. If $|u| < 1$ then R_u maps $\mathrm{Gr}(H)$ into $\mathrm{Gr}_\omega(H)$. On the submanifold $\mathrm{Gr}_0(H)$ the action of $\mathbb{T}$ extends to a holomorphic action of the whole group $\mathbb{C}^\times$.

The action of the semigroup $\mathbb{C}^\times_{\leq 1}$ is very closely connected with the stratification of $\mathrm{Gr}(H)$.

Proposition (7.6.3).
(i) *Σ_S consists precisely of the points $W \in \mathrm{Gr}(H)$ such that $R_u W$ tends to H_S as $u \to 0$.*
(ii) *C_S consists precisely of the points $W \in \mathrm{Gr}_0(H)$ such that $R_u W$ tends to H_S as $u \to \infty$.*

If we restrict u to real values then the situation described in Proposition (7.6.3) is very reminiscent of Morse theory. If the trajectories $u \mapsto R_u W$ were the gradient flow of a function F on $\mathrm{Gr}(H)$ then the H_S would be the critical points of F, and Σ_S and C_S would be the *stable* and *unstable* manifolds of H_S in the sense of Morse theory [142]. This picture is essentially valid; the only qualification is that the function F is defined only on the smooth Grassmannian Gr_∞, where the trajectories are smooth. We shall find the function F in Section 7.8.

Proposition (7.6.3) follows at once from the behaviour of the Plücker coordinates with respect to the $\mathbb{T}$-action.

Proposition (7.6.4). *We have*

$$\pi_S(R_u W) = \lambda u^{\ell(S)} \pi_S(W),$$

where λ is non-zero and independent of S.
In other words the Plücker embedding

$$\pi: \mathrm{Gr}(H) \to P(\mathcal{H})$$

is equivariant with respect to $\mathbb{C}^\times_{\leq 1}$ when R_u acts on $\mathcal{H} = \ell^2(\mathcal{S})$ by

$$(R_u \xi)_S = u^{\ell(S)} \xi_S.$$

Remark. This proposition will be superseded in the next section by the more precise result (7.7.5).

Proof. If $w: z^{-d} H_+ \to W$ is an admissible basis for W then we can take $R_u \circ w \circ R_u^{-1}$ as a basis for $R_u W$. By continuity it is enough to prove the result for a dense set of ws. So for the component of virtual dimension zero we can suppose

$$w_q = z^q + \sum_p T_{pq} z^p.$$

If $S - \mathbb{N} = A = \{a_1, \ldots, a_k\}$, and $\mathbb{N} - S = B = \{b_1, \ldots, b_k\}$, then $\ell(S) = \Sigma (b_i - a_i)$. The Plücker coordinate $\pi_S(w)$ is the determinant of the submatrix of T formed from the rows A and columns B. Conjugation by R_u multiplies this determinant by $z^{-\Sigma a_i + \Sigma b_i} = z^{\ell(S)}$. The other connected components can be treated similarly.

Remark. It is worth pointing out that if $\mathcal{H}_\infty$ and $\mathcal{H}_\omega$ are the smooth and real-analytic vectors in $\mathcal{H}$ in the sense of representation theory [153], i.e. the vectors whose $\mathbb{T}$-orbits are smooth or real-analytic, then

$$\mathrm{Gr}_\infty(H) = \pi^{-1} P(\mathcal{H}_\infty), \text{ and}$$
$$\mathrm{Gr}_\omega(H) = \pi^{-1} P(\mathcal{H}_\omega).$$

7.7 The determinant bundle

In this section we shall construct a holomorphic line bundle Det on the Grassmannian $\mathrm{Gr}(H)$. Its fibre $\mathrm{Det}(W)$ at $W \in \mathrm{Gr}(H)$ is to be thought of as the 'top exterior power' of W. We can make sense of this by using the concept of an 'admissible basis', introduced in Section 7.5. An element of $\mathrm{Det}(W)$ is represented by definition by a formal expression

$$\lambda w_{-d} \wedge w_{-d+1} \wedge w_{-d+2} \wedge \ldots, \tag{7.7.1}$$

where $\lambda \in \mathbb{C}$ and $w = \{w_k\}$ is an admissible basis of W. We shall denote the expression (7.7.1) simply by $[\lambda, w]$. If w' is another admissible basis of W, then $[\lambda, w]$ is identified with $[\lambda \det(t), w']$, where $t = (t_{ij})$ is the matrix relating w and w':

$$w_i = \sum_j t_{ij} w'_j.$$

$\mathrm{Det}(W)$ is clearly a one-dimensional complex vector space, and the union of the $\mathrm{Det}(W)$ for $W \in \mathrm{Gr}(H)$ is the line bundle Det. We must, however, explain how Det is a complex manifold, and why the bundle is locally trivial.

For each indexing set $S \in \mathcal{S}$ we have the open set U_S of $\mathrm{Gr}(H)$, identified with the graphs of Hilbert–Schmidt operators $T: H_S \to H_S^\perp$. The graph W_T of T has the admissible basis $\{w_i\}$, where

$$w_i = z^q + \sum_{p \notin S} T_{pq} z^p \tag{7.7.2}$$

with $q = s_i$ and $S = \{s_{-d}, s_{-d+1}, \ldots\}$. We identify the part of Det above

U_S with $\mathbb{C} \times U_S$ by

$$(\lambda, W_T) \in \mathbb{C} \times U_S \leftrightarrow [\lambda, w] \in \text{Det},$$

where w is given by (7.7.2). The transitions between these local trivializations are as follows. Suppose that W_T belongs to $U_S \cap U_{S'}$, and $W_T = W_{T'}$, where $T': H_{S'} \to H_{S'}^\perp$. We know from (7.1.5) that

$$T' = (c + dT)(a + bT)^{-1},$$

where

$$\begin{pmatrix} a & b \\ c & d \end{pmatrix}$$

is the matrix of the permutation relating S to S'. Then

$$(\lambda, W_T) \in \mathbb{C} \times U_S \leftrightarrow (\lambda', W_{T'}) \in \mathbb{C} \times U_{S'},$$

where

$$\lambda' = \lambda \det(a + bT).$$

This is a holomorphic function of (λ, T), as we require. (To be quite concrete, $\det(a + bT)$ is simply the finite dimensional determinant formed from the rows A and columns B of T, where $A = S' - S$ and $B = S - S'$.)

The Grassmannian $\text{Gr}(H)$ is a homogeneous space under the action of the restricted general linear group $GL_{\text{res}}(H)$. It would be natural to expect the action of GL_{res} to lift to an action on the line bundle Det. This, however, is not quite the case, for if w is an admissible basis for $W \in \text{Gr}(H)$, and $A \in GL_{\text{res}}$, then Aw is not in general an admissible basis of $A(W)$. The extension $\widetilde{GL}_{\text{res}}$ of GL_{res} by $\mathbb{C}^\times$ described in Chapter 6 was constructed precisely to deal with this situation.

Theorem (7.7.3). *The action of GL_{res} on $\text{Gr}(H)$ is covered by an action of $\widetilde{GL}_{\text{res}}$ on the line bundle* Det.

Proof. Let us first consider the connected component Gr^0, consisting of spaces W of virtual dimension 0. An admissible basis for such a W is an isomorphism $w: H_+ \to W$, which we can write as a $\mathbb{Z} \times \mathbb{N}$ matrix

$$w = \begin{pmatrix} w_+ \\ w_- \end{pmatrix}$$

such that $w_+: H_+ \to H_+$ has a determinant. Recall that the subgroup $\mathscr{E}$ of $GL_{\text{res},0} \times GL(H_+)$ is defined as the set of pairs (A, q) such that aq^{-1} has a determinant, where

$$A = \begin{pmatrix} a & b \\ c & d \end{pmatrix}.$$

7.7 THE DETERMINANT BUNDLE

We define an action of $\mathscr{E}$ on the set of admissible bases by

$$(A, q) \cdot w = Awq^{-1}.$$

This is well-defined because $(Awq^{-1})_+ = aw_+q^{-1} + bw_-q^{-1}$ has a determinant. Then $\mathscr{E}$ acts on Det by

$$(A, q) \cdot [\lambda, w] = [\lambda, (A, q)w].$$

The subgroup $\mathscr{T}_1$ of $\mathscr{E}$, which consists of pairs $(1, q)$ with $\det(q) = 1$, acts trivially on Det, and so we have defined an action of $\mathscr{E}/\mathscr{T}_1$. This is the identity component $GL^\sim_{\text{res},0}$ of $GL^\sim_{\text{res}}$.

To make $GL^\sim_{\text{res},0}$ act on the part of Det over Gr^d, the set of subspaces W of virtual dimension d, recall that we defined an automorphism $\tilde{\sigma}$ of $GL^\sim_{\text{res},0}$ which covered the automorphism $A \mapsto \sigma A \sigma^{-1}$ of $GL_{\text{res},0}$. Here $\sigma: H \to H$ is the shift map, given by multiplication by z. We define the action of $\tilde{A} \in GL^\sim_{\text{res},0}$ on Det $|\, Gr^d$ as the action of $\sigma^{-d} \circ \tilde{\sigma}^d(\tilde{A}) \circ \sigma^d$, where

$$\sigma: \text{Det} \to \text{Det}$$

is defined by $\sigma \cdot [\lambda, w] = [\lambda, \sigma w]$. As $GL^\sim_{\text{res}}$ is the semidirect product of $GL^\sim_{\text{res},0}$ and the cyclic subgroup generated by σ, we now have an action of $GL^\sim_{\text{res}}$ on Det.

Remarks.

(i) The group extension $GL^\sim_{\text{res},0}$ can be constructed directly from the line bundle Det. For $GL^\sim_{\text{res},0}$ is the group of all holomorphic automorphisms of Det $|\, Gr^0$ which cover the actions of elements of $GL_{\text{res},0}$ on Gr^0. (If $\tilde{A}_0$ and $\tilde{A}_1$ are automorphisms of Det $|\, Gr^0$ which cover the same map on Gr^0, then $\tilde{A}_0^{-1}\tilde{A}_1$ must be the operation of multiplication by a non-vanishing holomorphic function on Gr^0. But any such function is constant (see Proposition (7.2.2)).

(ii) The line bundle Det has a natural hermitian metric for which

$$\|[\lambda, w]\|^2 = |\lambda|^2 \det(w^*w).$$

This is preserved by the action of $U^\sim_{\text{res}}$. The unit circle bundle in Det can therefore be identified with $U^\sim_{\text{res}}/U(H_+) \times U(H_-)$, and its Chern class is represented by the invariant form defined by (6.6.5).

Let us return to the Plücker embedding defined in Section 7.5. Each Plücker coordinate π_S can be regarded as a holomorphic section of the line bundle Det* dual to Det. For a holomorphic section of Det* is a holomorphic function Det $\to \mathbb{C}$ which is linear on each fibre. The coordinate π_S defines such a function by

$$[\lambda, w] \mapsto \lambda \pi_S(w).$$

The Hilbert space $\mathcal{H}$ of Proposition (7.5.2) is therefore contained in the dual of the space of all holomorphic sections of Det*. We shall see in

Chapter 10 that it is a dense subspace of the dual. Meanwhile, let us notice simply that the embedding $\pi: \mathrm{Gr}(H) \to P(\mathcal{H})$ arises from a holomorphic map

$$\pi: \mathrm{Det} \to \mathcal{H} \qquad (7.7.4)$$

which is linear on each fibre. The line bundle Det is thus the pull-back of the tautological line bundle on $P(\mathcal{H})$. (Cf. Section 2.9.)

The map $\pi: \mathrm{Det} \to \mathcal{H}$ is norm-preserving, as we see from the formula (7.5.3).

Proposition (7.7.5). *The map $\pi: \mathrm{Det} \to \mathcal{H}$ is equivariant with respect to $\mathbb{C}^{\times}_{\leq 1}$ when $R_u \in \mathbb{C}^{\times}_{\leq 1}$ acts on $\{\xi_S\} \in \mathcal{H}$ by*

$$(R_u \xi)_S = u^{\ell^*(S)} \xi_S.$$

Here $\ell^*(S) = \ell(S) + \tfrac{1}{2} d(d+1)$, where $d = \mathrm{card}(S)$.

Proof. We combine the proof of (7.6.4) with the fact that the action of R_u on $[\lambda, w] \in \mathrm{Det}$, where $w: z^{-d} H_+ \to H$ is an admissible basis, is given by

$$R_u [\lambda, w] = [\lambda u^{\tfrac{1}{2} d(d+1)}, R_u w R_u^{-1}].$$

More general determinant bundles

The determinant bundle can actually be defined on a larger space than $\mathrm{Gr}(H)$. Let $\mathrm{Gr}_{\mathrm{cpt}}(H)$ denote the set of closed subspaces W of H such that the projection $W \to H_+$ is Fredholm and the projection $W \to H_-$ is *compact*. Then our construction applies without change to define a holomorphic line bundle Det on $\mathrm{Gr}_{\mathrm{cpt}}(H)$. The crucial difference, however, is that the bundle on $\mathrm{Gr}_{\mathrm{cpt}}(H)$ is not homogeneous: it is acted on only by the subgroup of $\widetilde{GL}_{\mathrm{res}}(H)$ consisting of elements whose off-diagonal blocks are of trace class.

The line bundle Det on $\mathrm{Gr}_{\mathrm{cpt}}(H)$ is essentially the same thing as the determinant bundle defined by Quillen [124] on the space $\mathrm{Fred}(H_+)$ of Fredholm operators in H_+. The fibre of Quillen's bundle at $T: H_+ \to H_+$ is

$$\det(\ker T)^* \otimes \det(\mathrm{coker}\, T).$$

The relation between the two bundles is the following. Let $\mathscr{B}$ denote the space of injective maps $w: H_+ \to H$ such that $w(H_+)$ belongs to $\mathrm{Gr}_{\mathrm{cpt}}(H)$. Then we have holomorphic maps

$$\mathrm{Gr}_{\mathrm{cpt}}(H) \leftarrow \mathscr{B} \to \mathrm{Fred}(H_+).$$

Both of these maps have contractible fibres. The determinant bundles on $\mathrm{Gr}_{\mathrm{cpt}}(H)$ and $\mathrm{Fred}(H_+)$ pull back to the same bundle on $\mathscr{B}$, and the bundle on $\mathrm{Fred}(H_+)$ is the quotient of the one on $\mathscr{B}$ by the obvious free action of $GL(H_+)$.

7.8 Gr(H) as a Kähler manifold and a symplectic manifold

Because the group U_{res} acts transitively on Gr(H) we can define a hermitian metric on Gr(H) by giving a hermitian form on its tangent space at the base-point H_+ which is invariant under the action of the isotropy group $U(H_+) \times U(H_-)$. The tangent space at H_+ is the space $\mathscr{I}_2(H_+; H_-)$ of Hilbert–Schmidt operators $H_+ \to H_-$ (on which the isotropy group acts by left- and right-composition), and the unique invariant inner product is

$$(X, Y) \mapsto 2\,\text{trace}(X^*Y),$$

up to a scalar multiple. This inner product defines a *Kähler* structure on Gr(H). Indeed its imaginary part

$$\omega(X, Y) = -i\,\text{trace}(X^*Y - Y^*X) \tag{7.8.1}$$

is the closed 2-form which we have already encountered in Proposition (6.6.5) as the form on the Lie algebra $\mathfrak{u}_{\text{res}}$ which defines the central extension $\tilde{\mathfrak{u}}_{\text{res}}$. (We recall that an invariant differential form on the homogeneous space $U_{\text{res}}/(U_+ \times U_-)$ is the same thing as a skew form ω on $\mathfrak{u}_{\text{res}}$ which is invariant under the adjoint action of $U_+ \times U_-$ and in addition satisfies $\omega(\xi, \eta) = 0$ when ξ or η belongs to $\mathfrak{u}_+ \oplus \mathfrak{u}_-$.) To see that the form (7.8.1) coincides with that of (6.6.5) we map $\mathscr{I}_2(H_+; H_-)$ into $\mathfrak{u}_{\text{res}}$ by

$$X \mapsto \begin{pmatrix} 0 & -X^* \\ X & 0 \end{pmatrix}.$$

We saw at the end of the last section (see Remark (ii), p. 115) that the form ω represents the Chern class of the line bundle Det on Gr(H). An equivalent statement is that the Kähler structure of Gr(H) is induced from the standard structure on the projective space $P(\mathscr{H})$ by the Plücker embedding.

On a simply connected symplectic manifold X—even if it is infinite dimensional—any vector field ξ which preserves the 2-form ω arises from a so-called *Hamiltonian function* $F: X \to \mathbb{R}$, in the sense that the gradient dF is the 1-form $\omega(\xi, \ )$ on X. On Gr(H) the vector field defined by any element of the Lie algebra $\mathfrak{u}_{\text{res}}$ preserves the form ω, and we can ask for the corresponding function.

Proposition (7.8.2). *The Hamiltonian function* $F: \text{Gr}(H) \to \mathbb{R}$ *which defines the flow on* Gr(H) *corresponding to* $\xi \in \mathfrak{u}_{\text{res}}$ *is given by*

$$F(W) = -i\,\text{trace}\,\xi(J_W - J).$$

Here J and J_W are the operators of square 1 which define the decompositions $H = H_+ \oplus H_-$ and $H = W \oplus W^\perp$. We leave it to the reader to check that the operator $\xi(J_W - J)$ is necessarily of trace class.

Proof. The gradient of F at W along the tangent vector corresponding to $\eta \in \mathfrak{u}_{\text{res}}$ is

$$dF(W; \eta) = -i \text{ trace } \xi[\eta, J_W].$$

Suppose that $W = gH_+$, with $g \in U_{\text{res}}$. Then $J_W = gJg^{-1}$, and the value of the invariant form ω at W on the tangent vectors defined by ξ, η is

$$\begin{aligned}\omega(W; \xi, \eta) &= \omega(g^{-1}\xi g, g^{-1}\eta g) \\ &= -i \text{ trace } g^{-1}\xi g[g^{-1}\eta g, J] \\ &= -i \text{ trace } \xi[\eta, J_W] \\ &= dF(W; \eta).\end{aligned} \qquad (7.8.3)$$

(Here (7.8.3) is obtained from (6.6.5) by noticing that trace $(c_1 b_2 - b_1 c_2)$ = trace $A_1[A_2, J]$.)

We cannot apply Proposition (7.8.2) directly to the rotation action of $\mathbb{T}$ on $\text{Gr}(H)$, for we saw in Section 7.6 that the action was smooth only on the submanifold $\text{Gr}_\infty(H)$. This corresponds to the fact that the infinitesimal generator $-d/d\theta$ is an unbounded operator on H and does not belong to the Lie algebra $\mathfrak{u}_{\text{res}}$. Nevertheless (7.8.2) does hold for the rotation flow on $\text{Gr}_\infty(H)$. We shall call the corresponding Hamiltonian function the *energy* $\mathscr{E}: \text{Gr}_\infty(H) \to \mathbb{R}$. Thus

$$\mathscr{E}(W) = \text{trace}\left(i\frac{d}{d\theta}\right)(J_W - J). \qquad (7.8.4)$$

The critical points of $\mathscr{E}$ are the stationary points of the rotation action, i.e. the points H_S for $S \in \mathscr{S}$. Let us notice that

$$\mathscr{E}(H_S) = \ell^*(S) = \ell(S) + \tfrac{1}{2}d(d+1),$$

where $d = \text{card}(S)$ (Cf. (7.7.5).) More generally, we have

Proposition (7.8.5).

$$\mathscr{E}(W) = \sum_S \ell^*(S) |\pi_S(W)|^2$$

$$= \left\langle \Omega_W, i\frac{d}{d\theta} \cdot \Omega_W \right\rangle,$$

where $\{\pi_S(W)\}$ are the Plücker coordinates of W, normalized so that $\Sigma |\pi_S(W)|^2 = 1$, and Ω_W is the corresponding unit vector in $\mathscr{H}$.

Thus $\mathscr{E}$ takes only positive values.

In the language of quantum mechanics we can regard $\text{Gr}(H)$ as the space of states of a classical system, and $P(\mathscr{H})$ as the corresponding

7.8 Gr(H) AS A SYMPLECTIC MANIFOLD

quantum state space. Then Ω_W represents the quantum state corresponding to W, and (7.8.5) asserts that the classical energy $\mathscr{E}(W)$ is the expected value of the quantum energy operator $i(d/d\theta)$ in the state Ω_W. The result follows from the fact that $\text{Gr}(H)$ has the Kähler structure induced from $P(\mathscr{H})$. For in general if T is any skew-adjoint operator in $\mathscr{H}$ then the Hamiltonian function corresponding to the flow on $P(\mathscr{H})$ induced by T is

$$\xi \mapsto \langle \xi, iT\xi \rangle.$$

It can be shown fairly easily that the Morse decomposition of $\text{Gr}(H)$ into the ascending and descending stable manifolds of the stationary points of the gradient flow of $\mathscr{E}$ is precisely the stratification and cell decomposition which we found in Sections 7.3 and 7.4. We shall not pursue this discussion any further—but see Section 8.9.

8
THE FUNDAMENTAL HOMOGENEOUS SPACE

8.1 Introduction: the factorization theorems

The most important results proved in this chapter are three factorization theorems. We shall state them here for the loop group of the general linear group $GL_n(\mathbb{C})$, but this can be replaced by $LG_\mathbb{C}$ for any compact G. The first involves the subgroup $L^+GL_n(\mathbb{C})$ of $LGL_n(\mathbb{C})$ consisting of loops γ which are the boundary values of holomorphic maps

$$\gamma : \{z \in \mathbb{C} : |z| < 1\} \to GL_n(\mathbb{C}).$$

Theorem (8.1.1). *Any loop $\gamma \in LGL_n(\mathbb{C})$ can be factorized uniquely*

$$\gamma = \gamma_u \cdot \gamma_+,$$

with $\gamma_u \in \Omega U_n$ and $\gamma_+ \in L^+GL_n(\mathbb{C})$. In fact the product map

$$\Omega U_n \times L^+GL_n(\mathbb{C}) \to LGL_n(\mathbb{C})$$

is a diffeomorphism.

Here ΩU_n denotes the base-point-preserving loops in LU_n, i.e. those such that $\gamma(1) = 1$. Proposition (8.1.1) will be proved in Section 8.3.

The second theorem, which is due to Birkhoff [11, 12], involves also the subgroup $L^-GL_n(\mathbb{C})$ consisting of loops $\gamma \in LGL_n(\mathbb{C})$ which are the boundary values of holomorphic maps

$$\gamma : \{z \in \mathbb{C} \cup \infty : |z| > 1\} \to GL_n(\mathbb{C}).$$

Theorem (8.1.2). *Any loop $\gamma \in LGL_n(\mathbb{C})$ can be factorized*

$$\gamma = \gamma_- \cdot \lambda \cdot \gamma_+,$$

where $\gamma_- \in L^-GL_n(\mathbb{C})$, $\gamma_+ \in L^+GL_n(\mathbb{C})$, and $\lambda \in \check{T}$ is a loop which is a homomorphism from S^1 into the diagonal matrices in $GL_n(\mathbb{C})$, i.e. λ is of the form

$$z^{\mathbf{a}} = \begin{pmatrix} z^{a_1} & & & \\ & z^{a_2} & & \\ & & \ddots & \\ & & & z^{a_n} \end{pmatrix}.$$

8.1 INTRODUCTION: THE FACTORIZATION THEOREMS

The factor λ is uniquely determined by γ up to conjugation in $GL_n(\mathbb{C})$, i.e. up to the order of $\{a_1, \ldots, a_n\}$. Loops for which $\lambda = 1$ form a dense open subset of the identity component of $LGL_n(\mathbb{C})$, and the multiplication map

$$L_1^- \times L^+ \to LGL_n(\mathbb{C}),$$

where $L_1^- = \{\gamma_- \in L^- : \gamma_-(\infty) = 1\}$, is a diffeomorphism on to this subset.

We shall describe two important applications of Birkhoff's theorem in the next section. The theorem will be proved in Section 8.4.

Both theorems (8.1.1) and (8.1.2) have exact analogues for the groups of real-analytic, rational, and polynomial loops, but are false for continuous loops. The third theorem, however, applies only to the group of polynomial loops. We shall refer to it as the *Bruhat factorization*. (Cf. [79].)

Theorem (8.1.3). *Any polynomial loop $\gamma \in L_{\text{pol}}GL_n(\mathbb{C})$ can be factorized*

$$\gamma = \gamma_+^{(1)} \cdot \lambda \cdot \gamma_+^{(2)},$$

where $\gamma_+^{(1)}$ and $\gamma_+^{(2)}$ both belong to L_{pol}^+, and λ is a homomorphism from S^1 into the diagonal matrices.

The three theorems are precise analogues of the following three well-known facts about $GL_n(\mathbb{C})$.

 (i) Any $A \in GL_n(\mathbb{C})$ is the product of a unitary matrix and an upper triangular matrix.

 (ii) Any $A \in GL_n(\mathbb{C})$ can be factorized

$$A = P\pi Q,$$

where P is lower triangular, Q is upper triangular, and π is a permutation matrix. Furthermore π is determined uniquely by A, and $\pi = 1$ for a dense open subset of $GL_n(\mathbb{C})$—in fact for all A whose leading principal minors do not vanish.

 (iii) The same statement as (ii), but with P and Q both upper triangular, and π anti-diagonal for a dense open subset.

Of course (ii) and (iii) are trivially equivalent; they are called the 'Bruhat decomposition' of $GL_n(\mathbb{C})$.

The theorems for loop groups are proved in exactly the same way as the finite dimensional results. The unitary–upper–triangular factorization (i) is simply the 'Gram–Schmidt process' for replacing an arbitrary basis of $\mathbb{C}^n$—the columns of A—by an orthonormal basis. More geometrically, it is the assertion that any flag in $\mathbb{C}^n$ (see Section 2.8) contains an orthonormal basis, i.e. that U_n acts transitively on the flag manifold $GL_n(\mathbb{C})/B$. (B denotes the subgroup of upper-triangular matrices.) The

8 THE FUNDAMENTAL HOMOGENEOUS SPACE

Bruhat decomposition, likewise, expresses the decomposition of the flag manifold into its Schubert cells.

For a loop group LG the space that plays the role of the flag manifold is the complex homogeneous space $X = LG_{\mathbb{C}}/L^+G_{\mathbb{C}}$. Theorem (8.1.1) is the assertion that LG acts transitively on X, so that $X \cong LG/G$. We shall call X the *fundamental homogeneous space* of LG. The analogy between it and the flag manifold is far-reaching. In particular:

(i) it is a complex projective algebraic variety;
(ii) it has a canonical stratification and cell decomposition;
(iii) irreducible representations of LG can be constructed as spaces of holomorphic sections of line bundles on it.

In Section 8.3 we shall show that when G is U_n there is a beautiful description of the space LG/G as a kind of Grassmannian; from this we shall derive the factorization theorems. A similar description is possible for the other classical groups, and a slightly more complicated one for a general compact Lie group.

The idea of the Grassmannian model comes from 'scattering theory' in the sense of Lax and Phillips [99]. We shall say a little about that point of view in Section 8.12, as an appendix to this chapter. From a completely different point of view the Grassmannian model is an expression of the Bott periodicity theorem; Bott's theorem has been mentioned in Section 6.4, but we shall return to it in Section 8.8.

The Grassmannian model reduces the study of LG/G to linear algebra. It is also interesting, however, to think of LG/G as a Kähler manifold, and to study the Morse theory of the energy function on it. We have discussed that approach in Section 8.9. Another completely different point of view, described in Section 8.10, is to regard LG/G directly as a space of holomorphic vector bundles on the Riemann sphere.

The space LG/G is not the only complex homogeneous space of LG. Indeed $L^+G_{\mathbb{C}}$ should be thought of as a maximal parabolic subgroup of $LG_{\mathbb{C}}$ in the sense of algebraic groups, so $LG_{\mathbb{C}}/L^+G_{\mathbb{C}}$ is more accurately to be compared with a Grassmannian $GL_n(\mathbb{C})/P$, where P is a group of echelon matrices

$$\begin{pmatrix} * & * \\ 0 & * \end{pmatrix},$$

than with the flag manifold $GL_n(\mathbb{C})/B$. The difference is not, however, very important: the subgroup B^+ of $L^+GL_n(\mathbb{C})$ consisting of loops γ such that $\gamma(0)$ is upper triangular is a minimal parabolic subgroup, and we shall see in Section 8.7 that $LGL_n(\mathbb{C})/B^+$ can be regarded as a space of 'periodic flags' in Hilbert space. Much more interesting, however, is the existence of a quite different complex homogeneous space for $LGL_n(\mathbb{C})$ which is associated to a Riemann surface. This is described in Section 8.11.

8.2 APPLICATIONS OF THE BIRKHOFF FACTORIZATION

We end this section with a technical remark. The Lie group LG is the semidirect product of the subgroup G of constant loops and the normal subgroup ΩG of loops γ such that $\gamma(1) = 1$. (G acts on ΩG by conjugation.) In particular $LG = G \times \Omega G$ as a manifold; and the homogeneous space LG/G can be identified with ΩG. We shall often make this identification without comment, and shall think of ΩG as a homogeneous space of LG. The action of $\gamma \in LG$ on ΩG is therefore $\omega \mapsto \tilde{\omega}$, where

$$\tilde{\omega}(z) = \gamma(z)\omega(z)\gamma(1)^{-1},$$

and the rotation R_α of S^1 through the angle α acts on ΩG by

$$(R_\alpha \omega)(\theta) = \omega(\theta - \alpha)\omega(-\alpha)^{-1}. \tag{8.1.4}$$

8.2 Two applications of the Birkhoff factorization

Singularities of ordinary differential equations

What we have called the Birkhoff factorization was discovered by Birkhoff [11] in 1909 when he was investigating the singularities of differential equations of the form

$$\frac{dv}{dz} = A(z)v(z) \tag{8.2.1}$$

for a $\mathbb{C}^n$-valued function v, where A is a given $(n \times n)$-matrix-valued function which is defined and holomorphic in a neighbourhood U of the origin in the complex plane except for a simple pole at the origin. The problem is to 'change coordinates' by multiplying v by a $GL_n(\mathbb{C})$-valued function T, holomorphic in U, so that the new function $\tilde{v} = Tv$ satisfies a simpler equation. We shall use the Birkhoff factorization to prove

Proposition (8.2.2). *A generic equation* (8.2.1) *can be reduced to the form*

$$\frac{d\tilde{v}}{dz} = z^{-1}K\tilde{v}, \tag{8.2.3}$$

where K is a constant matrix.

The precise meaning of 'generic' is that the residue of A at $z = 0$ is a matrix of which no two distinct eigenvalues differ by an integer. It is not true that every equation of the form (8.2.1) can be reduced to (8.2.3).

Proof. An equation of the form (8.2.1) has a 'solution matrix' X, a multivalued holomorphic function defined in $U - \{0\}$ with values in $GL_n(\mathbb{C})$ which satisfies $dX/dz = AX$. The unique solution of (8.2.1) such that $v(z_0) = v_0$ is expressed in terms of X by

$$v(z) = X(z)X(z_0)^{-1}v_0.$$

8 THE FUNDAMENTAL HOMOGENEOUS SPACE

The solution matrix is unique up to multiplication on the right by a constant invertible matrix. Its many-valuedness can be described by saying that when z travels once anti-clockwise around the origin then $X(z)$ is multiplied on the right by a matrix $M \in GL_n(\mathbb{C})$ called the *monodromy matrix*. A more precise statement is that there is a genuine holomorphic function $\hat{X}$ defined in the set

$$\{\zeta \in \mathbb{C} : e^\zeta \in U\}$$

such that $X(z) = \hat{X}(\log z)$, and

$$\hat{X}(\zeta + 2\pi i) = \hat{X}(\zeta)M.$$

Let us notice that a solution matrix of the equation (8.2.3) is given by $X(z) = z^K = e^{K \log z}$.

Beginning with (8.2.1), let us choose X, and then choose K so that $e^{2\pi i K} = M$. If we define

$$Y(z) = X(z)z^{-K},$$

then Y is a single-valued holomorphic function $U - \{0\} \to GL_n(\mathbb{C})$. The restriction of Y to a small circle $|z| = \varepsilon$ contained in U is a loop in $GL_n(\mathbb{C})$. Birkhoff argued, not quite correctly, that in the generic case, providing K is suitably chosen, this loop has a factorization $Y = Y_+ Y_-$, where Y_+ is holomorphic in U and Y_- is holomorphic in the whole Riemann sphere except for the origin. (One assumes initially that Y_+ and Y_- are holomorphic for $|z| < \varepsilon$ and $|z| > \varepsilon$; but they are then automatically holomorphic wherever Y is.) We can assume also that $Y_-(\infty) = 1$.

Now define $\tilde{v} = Y_+^{-1} v$. The equation satisfied by $\tilde{v}$ is $d\tilde{v}/dz = \tilde{A}\tilde{v}$, where

$$\tilde{A} = Y_+^{-1} A Y_+ - Y_+^{-1} \cdot dY_+/dz.$$

From this we see that $\tilde{A}$ is holomorphic in U except for a simple pole at the origin. But $Y_+^{-1} = Y_- z^K X^{-1}$, and hence

$$\tilde{A} = z^{-1} Y_- K Y_-^{-1} + dY_-/dz \cdot Y_-^{-1}.$$

This shows that $\tilde{A}$ is holomorphic everywhere in the Riemann sphere except for the origin. As $Y_-(z)$ is of the form $1 + y_1 z^{-1} + \ldots$ we know that $Y'_-(z)$ is $O(z^{-2})$ as $z \to \infty$, and so $z\tilde{A}(z) \to K$ as $z \to \infty$. By Liouville's theorem the only possibility is that $\tilde{A}(z) = z^{-1} K$. (In fact it follows also that $Y_-(z) = 1$ for all z.)

It is not sensible to try to repair Birkhoff's argument, as its importance was historical. The simplest way to prove (8.2.2) does not use the factorization theorem: it is better to show directly that when K is suitably chosen the map $Y : U - \{0\} \to GL_n(\mathbb{C})$ extends holomorphically to U. For a full discussion of the subject we refer to Turrettin [148].

The classification of holomorphic vector bundles on the Riemann sphere

The best known application—it would really be better to call it a reformulation—of Birkhoff's theorem is to the classification of holomorphic vector bundles on the Riemann sphere S^2. This was first pointed out by Grothendieck [69].

Let us write $S^2 = U_0 \cup U_\infty$, where $U_0 = S^2 - \{\infty\}$ and $U_\infty = S^2 - \{0\}$. The most obvious bundle on S^2 is the line bundle L which is constructed by attaching $U_0 \times \mathbb{C}$ to $U_\infty \times \mathbb{C}$ by the map

$$(z, \lambda) \mapsto (z, z\lambda).$$

There is also the tensor power L^k for any $k \in \mathbb{Z}$: its attaching function is

$$(z, \lambda) \mapsto (z, z^k \lambda).$$

(If S^2 is regarded as the complex projective line $P(\mathbb{C}^2)$ whose points are the rays in $\mathbb{C}^2$ then L is the 'Hopf bundle' whose fibre at $\xi \in P(\mathbb{C}^2)$ is the line $\xi \subset \mathbb{C}^2$.) Birkhoff's theorem is equivalent to

Proposition (8.2.4). *Any holomorphic vector bundle E on S^2 is isomorphic to a sum $L^{a_1} \oplus \ldots \oplus L^{a_n}$, where the integers $\{a_1, \ldots, a_n\}$ are uniquely determined (apart from their order).*

Proof. The restrictions of E to U_0 and to U_∞ are necessarily trivial, as U_0 and U_∞ are Stein manifolds [66]. So E is obtained by attaching $U_0 \times \mathbb{C}^n$ to $U_\infty \times \mathbb{C}^n$ by means of a holomorphic function

$$\gamma : U_0 \cap U_\infty \to GL_n(\mathbb{C}).$$

By Birkhoff's theorem (8.1.2) we can factorize γ as $\gamma_- \cdot \lambda \cdot \gamma_+$, where γ_+ and γ_- are holomorphic in U_0 and U_∞ respectively, and $\lambda = z^{\mathbf{a}}$ is a homomorphism. If we change coordinates in $U_0 \times \mathbb{C}^n$ by γ_+ and in $U_\infty \times \mathbb{C}^n$ by γ_-^{-1} then we find that E can also be constructed by taking λ as the attaching function. But the bundle defined by λ is $L^{a_1} \oplus \ldots \oplus L^{a_n}$.

8.3 The Grassmannian model of ΩU_n

The group $LGL_n(\mathbb{C})$ acts on the Hilbert space $H^{(n)} = L^2(S^1; \mathbb{C}^n)$, and hence, by Proposition (6.3.1), on the Grassmannian $\mathrm{Gr}(H^{(n)})$. The subspaces W of the form γH_+ for $\gamma \in LGL_n(\mathbb{C})$ have the property that $zW \subset W$, as the action of γ commutes with multiplication by the scalar-valued function z. It turns out that the orbit of H_+ under $LGL_n(\mathbb{C})$ is essentially characterized by this property. (We shall usually write γH_+ for $M_\gamma(H_+)$ and zW for $M_z(W)$ when we do not need to emphasize that γ and z are operators.)

Definition (8.3.1). $\mathrm{Gr}^{(n)}$ *denotes the closed subset of $\mathrm{Gr}(H^{(n)})$ consisting of subspaces W such that $zW \subset W$.*

This is the Grassmannian model of the loop space. Its crucial property is

8 THE FUNDAMENTAL HOMOGENEOUS SPACE

Theorem (8.3.2). *The group $L_{\frac{1}{2}}U_n$ acts transitively on $\mathrm{Gr}^{(n)}$, and the isotropy group of H_+ is the group U_n of constant loops.*

We recall (see Proposition (6.3.3)) that $L_{\frac{1}{2}}U_n$ is the commutant of the multiplication operator M_z in $U_{\mathrm{res}}(H^{(n)})$.

It is obvious that $\gamma H_+ = H_+$ if and only if $\gamma \in L^+GL_n(\mathbb{C})$, so the assertion in Theorem (8.3.2) about the isotropy group is a kind of 'maximum modulus' principle. When $n = 1$ it is the statement that a map $S^1 \to \mathbb{T}$ which extends to a non-vanishing holomorphic function in the disc is constant.

Proof of (8.3.2). The first step is to see that if W belongs to $\mathrm{Gr}^{(n)}$ then zW has codimension n in W. Consider the commutative diagram

$$\begin{array}{ccc} zW & \to & W \\ \downarrow & & \downarrow \\ zH_+ & \to & H_+ \end{array}$$

where the horizontal maps are inclusions and the vertical ones are orthogonal projections. The two vertical maps are Fredholm, and clearly have the same index (equal to the virtual dimension of W). But $zH_+ \to H_+$ is also Fredholm, with index $-n$. It follows that $zW \to W$ is Fredholm, and its index must also be $-n$ in view of the formula $\chi(AB) = \chi(A) + \chi(B)$ for the index of a composite [34]. Thus $\dim(W/zW) = n$.

Now let $\{w_1, \ldots, w_n\}$ be an orthonormal basis for $W \ominus zW$, the orthogonal complement of zW in W. As in the proof of (6.1.1) we put the vector-valued functions w_i side by side to form an $(n \times n)$-matrix-valued function γ on S^1. We then find that $\gamma(\theta)$ is a unitary matrix for almost all $\theta \in S^1$, i.e. γ belongs to $L_{\mathrm{meas}}U_n$. To see this, let us write

$$w_k(\theta) = \sum_m w_{km} e^{im\theta},$$

with $w_{km} \in \mathbb{C}^n$. Then

$$\langle w_k(\theta), w_\ell(\theta) \rangle = \sum_{m,r} \langle w_{km}, w_{\ell r} \rangle e^{i(r-m)\theta}$$

$$= \sum_p \langle w_k, z^p w_\ell \rangle_H e^{ip\theta}$$

$$= \delta_{k\ell},$$

where for the moment we have written $\langle \ , \ \rangle_H$ for the inner product in $H^{(n)}$ to distinguish it from the inner product in $\mathbb{C}^n$.

The multiplication operator M_γ is therefore a unitary operator in $H^{(n)}$, and by its construction it satisfies

$$M_\gamma(H_+ \ominus z^k H_+) = W \ominus z^k W$$

8.3 THE GRASSMANNIAN MODEL OF ΩU_n

for all k. To deduce that $M_\gamma(H_+) = W$ we must prove that $\bigcap z^k W = 0$. Suppose that w belongs to $\bigcap z^k W$, and $\|w\| = 1$. Then $z^{-k}w$ belongs to W for all k. Because the projection $\mathrm{pr}_-: W \to H_-$ is a compact operator, one can find a convergent subsequence of $\{\mathrm{pr}_-(z^{-k}w)\}$, converging, say, to $v \in H$. Clearly $\|v\| = 1$, as $z^k \mathrm{pr}_-(z^{-k}w) \to w$. But $\|v\|^2 = \lim \langle v, z^{-k}w \rangle$, and for any $v, w \in H^{(n)}$ we have $\langle v, z^{-k}w \rangle \to 0$ as $k \to \infty$; a contradiction, proving that $\bigcap z^k W = 0$.

To see that M_γ belongs to U_{res}, we observe that its $(H_+ \to H_-)$ component factorizes

$$H_+ \to W \xrightarrow{\mathrm{pr}_-} H_-,$$

and is therefore Hilbert–Schmidt. The $(H_- \to H_+)$ component is also Hilbert–Schmidt because M_γ is unitary.

We have now proved that $L_{\frac{1}{2}} U_n$ acts transitively on $\mathrm{Gr}^{(n)}$. But any $\gamma \in L_{\frac{1}{2}} U_n$ such that $\gamma H_+ = H_+$ must preserve the n-dimensional subspace $H_+ \ominus z H_+$, and is completely determined by its action there. It therefore belongs to U_n.

Theorem (8.3.2) shows that $\Omega_{\frac{1}{2}} U_n = L_{\frac{1}{2}} U_n / U_n$ can be identified with $\mathrm{Gr}^{(n)}$ as a set, and justifies the name 'Grassmannian model'. We shall return to the topological aspect of the correspondence later. Meanwhile we shall determine which subsets of $\Omega_{\frac{1}{2}} U_n$ correspond to the four subspaces $\mathrm{Gr}_0^{(n)}, \mathrm{Gr}_1^{(n)}, \mathrm{Gr}_\omega^{(n)}, \mathrm{Gr}_\infty^{(n)}$—where $\mathrm{Gr}_\alpha^{(n)}$ denotes $\mathrm{Gr}^{(n)} \cap \mathrm{Gr}_\alpha(H^{(n)})$—and we shall derive the factorization theorems of Section 8.1.

Proposition (8.3.3). *In the correspondence* $\mathrm{Gr}^{(n)} \leftrightarrow \Omega_{\frac{1}{2}} U_n$

(i) $\mathrm{Gr}_0^{(n)}$ *corresponds to* $\Omega_{\mathrm{pol}} U_n$,
(ii) $\mathrm{Gr}_1^{(n)}$ *corresponds to* $\Omega_{\mathrm{rat}} U_n$,
(iii) $\mathrm{Gr}_\omega^{(n)}$ *corresponds to* $\Omega_{an} U_n$,
(iv) $\mathrm{Gr}_\infty^{(n)}$ *corresponds to* ΩU_n.

We recall that the groups of polynomial, rational, real-analytic and smooth loops have been mentioned in Section 3.5, and the corresponding subspaces of the Grassmannian were defined in Section 7.2.

Proof. In one direction, if W belongs to $\mathrm{Gr}_\alpha^{(n)}$, where $\alpha = 0, 1, \omega$, or ∞, then we must show that $W \ominus zW$ consists of functions of the corresponding kind. Let us consider the smooth case. If W belongs to Gr_∞ then by (7.2.1) the images of both projections $W \to H_-$ and $(zW)^\perp \to H_+$ consist of smooth functions. So a function in $W \ominus zW$ has smooth projections on to both H_+ and H_-, and is therefore smooth. The argument for Gr_0 and Gr_ω is identical.

In the case of Gr_1, to say that the image of the projection $W \to H_-$ consists of rational functions is the same as to say that there exists a polynomial $p(z)$ such that $p(z)W \subset H_+$. This is not automatically true when $W \in \mathrm{Gr}_1$, but it is true under the additional hypothesis that $zW \subset W$. For one can take for p the minimal polynomial of the transformation induced by M_z on the finite dimensional space $W/W \cap H_+$. Similarly, for the image of $(zW)^\perp \to H_+$ to consist of rational functions we need a polynomial $q(z^{-1})$ such that $q(z^{-1})W^\perp \subset H_-$. This is equivalent to $\bar{q}(z)H_+ \subset W$, and one can take for $\bar{q}$ the minimum polynomial of M_z on $H_+/W \cap H_+$.

In the converse direction, it is obvious that the action of $L_{\mathrm{pol}}U_n$ preserves Gr_0. And $L_{\mathrm{rat}}U_n$ preserves Gr_1 because the existence of polynomials $p(z)$ and $\bar{q}(z)$ such that

$$p(z)W \subset H_+ \quad \text{and} \quad \bar{q}(z)H_+ \subset W$$

is obviously sufficient as well as necessary for W to belong to $\mathrm{Gr}_1^{(n)}$. The smooth loop group LU_n preserves Gr_∞ by Proposition (7.2.1), and $L_{\mathrm{an}}U_n$ preserves Gr_ω by the corresponding characterization of Gr_ω. This completes the proof of (8.3.3).

Remarks.
(i) The reason there is no simple model for the *continuous* loops $\Omega_{\mathrm{cts}}U_n$ is that the positive and negative frequency parts of a continuous function are not necessarily continuous. In other words, in contrast with the behaviour of the four classes of function just discussed, if C is the space of continuous functions on S^1 it is not true that $C = C_+ \oplus C_-$, where $C_\pm = C \cap H_\pm$. Suppose, for example, that f is the function already mentioned in Section 6.3, defined by

$$f(\theta) = \sum_{k>1} \frac{\sin k\theta}{k \log k}. \tag{8.3.4}$$

This function is continuous. But $f = f_+ + f_-$, where

$$f_+(\theta) = \frac{1}{2i} \sum_{k>1} \frac{e^{ik\theta}}{k \log k},$$

and f_+ is unbounded in the neighbourhood of $\theta = 0$.

(ii) $\mathrm{Gr}_0^{(n)}$ is not dense in $\mathrm{Gr}^{(n)}$, despite the fact that $\mathrm{Gr}_0(H)$ is dense in $\mathrm{Gr}(H)$. For we saw in Section 3.5 that a loop in U_n cannot be polynomial unless its determinant is of the form z^k. We shall see in Section 8.10, however, that $\mathrm{Gr}_0^{(n)}$ has the same homotopy type as $\mathrm{Gr}^{(n)}$.

The Grassmannian model for ΩU_n gives us at once the first of the three basic factorization theorems for loops. For the complex group $L_{\frac{1}{2}}GL_n(\mathbb{C})$ acts on $\mathrm{Gr}^{(n)}$ as well as $L_{\frac{1}{2}}U_n$, and the stabilizer of H_+ in $L_{\frac{1}{2}}GL_n(\mathbb{C})$ is clearly the closed subgroup $L_{\frac{1}{2}}^+ GL_n(\mathbb{C})$ of loops γ which are the boundary values of holomorphic maps

$$\gamma: \{z \in \mathbb{C} : |z| < 1\} \to GL_n(\mathbb{C}).$$

8.4 THE BIRKHOFF AND BRUHAT DECOMPOSITIONS

Because $L_{\frac{1}{2}}U_n$ acts transitively on $\text{Gr}^{(n)}$ we have

Proposition (8.3.5). *The group $L_{\frac{1}{2}}GL_n(\mathbb{C})$ is the product*
$$L_{\frac{1}{2}}U_n \cdot L_{\frac{1}{2}}^+ GL_n(\mathbb{C}).$$
Furthermore, exactly the same factorization property holds for smooth, real-analytic, rational, and polynomial loops.

Remark. The proposition is false for the continuous loop group, as we see from the unique factorization $e^f = e^{f_- - f_+} e^{2f_+}$, where f is the function of (8.3.4).

We have still not quite proved Theorem (8.1.1). It remains to show that the multiplication map $\Omega U_n \times L^+ \to LGL_n(\mathbb{C})$ is a diffeomorphism. For this it is enough to prove the smoothness of the map
$$u : LGL_n(\mathbb{C}) \to \Omega U_n$$
which assigns to a loop its unitary component. The map u factorizes as $\gamma \mapsto \tilde{\gamma} \mapsto u(\gamma)$, where

(i) $\gamma \mapsto \tilde{\gamma}$ is defined by projecting the columns $(\gamma_1, \ldots, \gamma_n)$ of $\gamma \in LGL_n(\mathbb{C})$ on to $(zW)^\perp$, where $W = \gamma H_+$; and

(ii) $\tilde{\gamma} \mapsto u(\gamma)$ is defined by orthonormalizing the basis $(\tilde{\gamma}_1, \ldots, \tilde{\gamma}_n)$ of $W \ominus zW$.

The second of these maps is obviously smooth. The first is smooth because $LGL_n(\mathbb{C})$ acts smoothly on the smooth Grassmannian, whose topology, in turn, is designed to ensure the smoothness of the map
$$C^\infty \times \text{Gr}_\infty(H) \to C^\infty$$
$$(f, W) \mapsto f_W$$
which assigns to a smooth function f on the circle and a subspace $W \in \text{Gr}_\infty(H)$ the projection f_W of f on to W.

Notice that the preceding argument does *not* apply to the group $L_{\frac{1}{2}}GL_n(\mathbb{C})$. Using once again the function f of (8.3.4) we observe that for any $t \in \mathbb{R}$ the loop $e^{tf} \in L_{\frac{1}{2}}$ factorizes as $e^{t(f_- - f_+)} e^{2tf_+}$. But the map $t \mapsto e^{tf}$ is smooth, while the set $\{e^{t(f_- - f_+)}\}_{t \in \mathbb{R}}$ has the discrete topology on $L_{\frac{1}{2}}U_n$ because $f_- - f_+$ is unbounded.

8.4 The stratification of $\text{Gr}^{(n)}$: the Birkhoff and Bruhat decompositions.

For the rest of this chapter, except for the Appendix, we shall be concerned only with the smooth Grassmannian $\text{Gr}_\infty(H^{(n)})$ and its subspace $\text{Gr}_\infty^{(n)}$, and shall have no use for the Hilbert manifold $\text{Gr}(H^{(n)})$. We shall therefore change notation by dropping the subscript, and shall write $\text{Gr}(H^{(n)})$ and $\text{Gr}^{(n)}$ for the smooth spaces.

8 THE FUNDAMENTAL HOMOGENEOUS SPACE

Because $\mathrm{Gr}^{(n)}$ can be identified with $LGL_n(\mathbb{C})/L^+GL_n(\mathbb{C})$, Birkhoff's theorem (8.1.2) amounts to the description of the orbits of the action of $L^-GL_n(\mathbb{C})$ on $\mathrm{Gr}^{(n)}$: it asserts that each L^--orbit contains a point of the form $z^{\mathbf{a}}H_+^{(n)}$, unique up to the order of $\{a_1, \ldots, a_n\}$. We shall prove a slightly more precise result. If $N^- = N^-GL_n(\mathbb{C})$ is the subgroup of L^- consisting of loops γ such that $\gamma(\infty)$ is upper triangular with 1s on the diagonal then we shall show that each orbit of N^- on $\mathrm{Gr}^{(n)}$ contains a unique point of the form $z^{\mathbf{a}}H_+^{(n)}$. In fact we shall show that the orbits of N^- are precisely the intersections of $\mathrm{Gr}^{(n)}$ with the strata of $\mathrm{Gr}(H)$ defined in Chapter 7.

The proof we are about to give is very elementary and explicit, but it is nevertheless rather tedious. For that reason we shall mention in advance the following geometrical description of what will ultimately be proved. The fixed points of the rotation action of $\mathbb{T}$ on ΩG (see (8.1.4)) are easily seen to be the homomorphisms $\lambda: S^1 \to U_n$, corresponding to the subspaces $\lambda \cdot H_+ \in \mathrm{Gr}^{(n)}$. But the action of $\mathbb{T}$ extends to an action of the semigroup $\mathbb{C}_{\leq 1}^\times$ (see Section 7.6), and for any $W \in \mathrm{Gr}^{(n)}$ the point $R_u W$ tends to a fixed-point of the $\mathbb{T}$-action as $u \to 0$ in $\mathbb{C}_{\leq 1}^\times$. It turns out that if $R_u W \to \lambda \cdot H_+$ then W belongs to the L^--orbit of $\lambda \cdot H_+$. (Notice that if $\gamma \in L^-$ then $R_u \gamma$ tends to the constant loop $\gamma(\infty)$ as $u \to 0$.)

The stratification of $\mathrm{Gr}(H)$ was defined by regarding H as $L^2(S^1; \mathbb{C})$, whereas in this chapter we are concerned with $H^{(n)} = L^2(S^1; \mathbb{C}^n)$. In fact all we need is a Hilbert space with an orthonormal basis indexed by the integers. It is surprisingly convenient, however, to identify $H^{(n)}$ with H in the way explained in Section 6.5, and to think of its elements sometimes as vector-valued functions of z and sometimes as scalar-valued functions of ζ, the two being related by the formulae (6.5.1). Thus $H = H^{(n)}$ has the orthonormal basis $\{\zeta^k\}_{k \in \mathbb{Z}}$, and the definition of $\mathrm{Gr}^{(n)}$ can be rewritten

$$\mathrm{Gr}^{(n)} = \{W \in \mathrm{Gr}(H) : \zeta^n W \subset W\}.$$

Recall that $\mathrm{Gr}(H)$ is the union of disjoint strata Σ_S, where W belongs to Σ_S if S is the set of integers s such that W contains an element of order s. If $W \in \mathrm{Gr}^{(n)}$ belongs to Σ_S then obviously

$$S + n \subset S.$$

Sets $S \in \mathscr{S}$ satisfying this condition are completely determined by giving the complement S^* of $S + n$ in S, which must consist of n elements, one in each congruence class modulo n. They correspond precisely to the homomorphisms from $\mathbb{T}$ into the maximal torus of U_n: to the homomorphism $z^{\mathbf{a}}$ there corresponds the set $S_{\mathbf{a}}$ such that $S_{\mathbf{a}}^*$ is

$$\{na_1, na_2 + 1, na_3 + 2, \ldots, na_n + n - 1\}.$$

8.4 THE BIRKHOFF AND BRUHAT DECOMPOSITIONS

The subspace H_{S_a} spanned by $\{\zeta^k\}_{k \in S_a}$ is $z^a H_+^{(n)}$. Thus the strata of $\mathrm{Gr}(H)$ which meet $\mathrm{Gr}^{(n)}$ can be indexed by the homomorphisms z^a. We shall write Σ_a for $\Sigma_{S_a} \cap \mathrm{Gr}^{(n)}$, and H_a for H_{S_a}.

Notice that as a group of operators the subgroup N^- of $LGL_n(\mathbb{C})$ is the intersection of $LGL_n(\mathbb{C})$ with the lower triangular subgroup $\mathcal{N}^-$ of (7.3.3).†

Proposition (8.4.1). *The orbit of H_a under N^- is Σ_a. It can be identified with the subgroup L_a^- of N^-, where $L_a^- = N^- \cap z^a L_1^- z^{-a}$.*

Proof. The strata of $\mathrm{Gr}(H)$ are the orbits of $\mathcal{N}^-$ by Proposition (7.3.3). As the group N^- is contained in $\mathcal{N}^-$ each orbit of N^- is certainly contained in some Σ_a.

If $W \in \Sigma_a$ then the projection $W \to H_a$ is an isomorphism. Let w_i be the inverse image of $z^{a_i}\varepsilon_i$, where $\{\varepsilon_1, \ldots, \varepsilon_n\}$ is the standard basis of $\mathbb{C}^n$. The functions w_i are smooth; furthermore $\{z^k w_i : 1 \le i \le n \text{ and } k \ge 0\}$ is a basis for W^{fin} because its projection is a basis for H_a^{fin}; and W^{fin} is a dense subspace of the space W^{sm} of smooth functions in W, because H_a^{fin} is dense in H_a^{sm}. We know from (8.3.3) that the evaluation map $W^{\mathrm{sm}} \to \mathbb{C}^n$ at any point z of S^1 is surjective, for W^{sm} contains n functions which form the columns of a smooth map $S^1 \to U_n$. It follows that $w_1(z), \ldots, w_n(z)$ are independent in $\mathbb{C}^n$ for any $z \in S^1$, and so $w = (w_1, \ldots, w_n)$ is an element of $LGL_n(\mathbb{C})$, and $w(H_+) = W$.

Now

$$w(\varepsilon_i) = z^{a_i}\varepsilon_i + \text{(lower terms)}, \qquad (8.4.2)$$

where 'lower terms' refers to the lexicographic ordering of the basis elements $\{z^k \varepsilon_j\} = \{\zeta^m\}$ of $H^{(n)} = H$. So

$$wz^{-a}(\varepsilon_i) = \varepsilon_i + \text{(lower terms)}.$$

In other words, $\gamma = wz^{-a}$ is the boundary value of a holomorphic map from the hemisphere $|z| > 1$ to the $n \times n$ matrices, and $\gamma(\infty)$ is upper triangular. Furthermore $\gamma(H_a) = W$. But the determinant of γ cannot vanish when $|z| > 1$, for if it did then $\det(\gamma)$ would have non-zero winding number on S^1, contradicting the fact that H_a and W have the same virtual dimension.

Finally, the loop γ belongs to $z^a L_1^- z^{-a}$ as well as to N^-. For the basis elements occurring in the 'lower terms' in (8.4.2) are not only lower than $z^{a_i}\varepsilon_i$ but also belong to $H_a^\perp = z^a H_-$. This means that when the operator

---

† Our terminology concerning upper and lower triangular matrices is a little muddled, for when discussing infinite matrices indexed by $\mathbb{Z} \times \mathbb{Z}$ we regard the ith row as above the jth if $i > j$, while it is customary to do the opposite with finite matrices. We have decided to tolerate the anomaly that the positive Borel subgroup of $GL_n(\mathbb{C})$ is taken to be the lower triangular matrices in this chapter, but the upper triangular matrices everywhere else in the book.

z^{-a} is applied to equation (8.4.2) we have

$$z^{-a}\gamma z^a(\varepsilon_i) = \varepsilon_i + \text{(element of } H_-\text{)}.$$

We have now proved Proposition (8.4.1), but to complete the proof of the Birkhoff factorization theorem (8.1.2) we must show that the multiplication map $L_1^- \times L^+ \to LGL_n(\mathbb{C})$ is a diffeomorphism on to a dense open subset of the identity component. This is, however, very easy. The map $\gamma \mapsto \gamma H_+$ from $LGL_n(\mathbb{C})$ to the smooth Grassmannian is smooth. On the open subset of $LGL_n(\mathbb{C})$ where γH_+ belongs to the coordinate chart $U_\mathbb{N}$, i.e. where γH_+ is the graph of an operator $T_\gamma : H_+ \to H_-$, we have $\gamma = \gamma_- \gamma_+$, where the columns of γ_- are $\{\varepsilon_i + T_\gamma \varepsilon_i\}$. Thus γ_- (and hence also γ_+) depends smoothly on γ.

The preceding argument shows also that $\text{Gr}^{(n)}$, with its subspace topology in the smooth Grassmannian $\text{Gr}(H)$, is locally homeomorphic to L_1^- and hence is a smooth manifold. It is also a smooth submanifold of $\text{Gr}(H)$: it is easy to see that in the coordinate patch $U_\mathbb{N}$ of $\text{Gr}(H)$, identified with a space $\mathcal{D}(H_+; H_-)$ of linear maps $H_+ \to H_-$, there is an open subset which is the product of $U_\mathbb{N} \cap \text{Gr}^{(n)} \cong L_1^-$ with the linear subspace $\mathcal{D}(zH_+; H_-)$.

The proof of (8.4.1) shows that the group L_a^- is diffeomorphic to the homogeneous space $N^-/{}_a N^-$, where ${}_a N^-$ is $N^- \cap z^a L^+ z^{-a}$, a finite dimensional group which is the stabilizer of H_a in N^-. This means that the multiplication map

$$L_a^- \times {}_a N^- \to N^- \tag{8.4.3}$$

is a diffeomorphism.

The splitting (8.4.3) evidently arises from a splitting of the basis elements of the Lie algebra of N^- into two subsets which span the two subgroups on the left. The fact that such a splitting of the Lie algebra induces a splitting of the group is well-known and elementary for a finite dimensional simply connected nilpotent group ([20] Chapter 3 Section 9.5); and (8.4.3) is really a finite dimensional result, because L_a^- contains, for large q, the normal subgroup N_q^- of N^- consisting of loops γ such that $\gamma(z) - 1$ vanishes to order q at $z = \infty$, and N^-/N_q^- is finite dimensional. (N^-/N_q^- is nilpotent because N_r^-/N_{r+1}^- is abelian for $r > 0$.) At present the splitting which is of most interest to us is a slight variant of (8.4.3), namely

$$L_a^- \times L_a^+ \xrightarrow{\cong} z^a L_1^- z^{-a}, \tag{8.4.4}$$

where $L_a^+ = N^+ \cap z^a L_1^- z^{-a}$, and N^+ is the subgroup of L^+ consisting of loops γ such that $\gamma(0)$ is lower triangular with ones on the diagonal.

8.4 THE BIRKHOFF AND BRUHAT DECOMPOSITIONS

Theorem (8.4.5).
(i) *The map $\gamma \mapsto \gamma H_\mathbf{a}$ defines a diffeomorphism between $z^\mathbf{a} L_1^- z^{-\mathbf{a}}$ and a contractible open neighbourhood $U_\mathbf{a}$ of $H_\mathbf{a}$ in $\mathrm{Gr}^{(n)}$.*
(ii) *The stratum $\Sigma_\mathbf{a}$ is a contractible closed submanifold of $U_\mathbf{a}$, of complex codimension*

$$d(\mathbf{a}) = \sum_{i<j} |a_i - a_j| - v(\mathbf{a}),$$

where $v(\mathbf{a})$ is the number of pairs i, j with $i < j$ but $a_i > a_j$.
(iii) *The orbit of $H_\mathbf{a}$ under N^+ is a complex cell $C_\mathbf{a}$, of complex dimension $d(\mathbf{a})$, which meets $\Sigma_\mathbf{a}$ transversally in the single point $H_\mathbf{a}$. The splitting (8.4.4) defines a diffeomorphism*

$$\Sigma_\mathbf{a} \times C_\mathbf{a} \to U_\mathbf{a}.$$

(iv) *The union of the cells $C_\mathbf{a}$ is $\mathrm{Gr}_0^{(n)}$; in fact $C_\mathbf{a}$ is the intersection of $\mathrm{Gr}^{(n)}$ with the cell $C_{S_\mathbf{a}}$ of Gr_0.*

Remarks.
(i) We shall call the cells $C_\mathbf{a}$ the *Bruhat cells* of $\mathrm{Gr}^{(n)}$. Notice that part (iv) of (8.4.5) implies the Bruhat factorization theorem (8.1.3).
(ii) The stratum $\Sigma_\mathbf{a}$ is contractible, and diffeomorphic to $L_\mathbf{a}^-$. But we cannot assert, as in the analogous finite dimensional theorem, that $L_\mathbf{a}^-$ is diffeomorphic to its Lie algebra by the exponential map, for the exponential map of $L_\mathbf{a}^-$ is not surjective. Goodman and Wallach have given the following example of an element γ of $N^- SL_2(\mathbb{C})$ which does not belong to the image of exp:

$$\gamma = \begin{pmatrix} 1 + 2z^{-2} & 4z^{-1} \\ z^{-1} + z^{-3} & 1 + 2z^{-2} \end{pmatrix}.$$

This cannot be of the form $\exp(\xi)$, for

$$\gamma(i) = \begin{pmatrix} -1 & -4i \\ 0 & -1 \end{pmatrix}$$

does not belong to the image of exp in $SL_2(\mathbb{C})$.

Proof of (8.4.5). Little more needs to be said. The contractibility of $U_\mathbf{a}$ and $\Sigma_\mathbf{a}$ follows from the contractibility of the groups L_1^- and $N_\mathbf{a}^-$; these consist of holomorphic functions in the disc $|z| > 1$, and can be contracted by the homomorphisms $\gamma \mapsto \gamma_t$ (for $0 \leq t \leq 1$), where $\gamma_t(z) = \gamma(t^{-1}z)$.

The orbit $C_\mathbf{a}$ is contractible by the same argument (applied to the disc $|z| < 1$). It is a cell because the exponential map of the nilpotent group $L_\mathbf{a}^+$ is a diffeomorphism.

The proof that $C_\mathbf{a}$ is the intersection of $\mathrm{Gr}^{(n)}$ with $C_{S_\mathbf{a}}$ is exactly like that of (8.4.1)—cf. (7.4.1).

It remains to calculate the dimension $d(\mathbf{a})$ of the group $L_\mathbf{a}^+$. Because conjugation by $z^\mathbf{a}$ multiplies the $(i, j)^{\text{th}}$ entry of a matrix by $z^{a_i - a_j}$, we find that $L_\mathbf{a}^+$ is an open subset of the matrix-valued functions (f_{ij}) such that

$$f_{ii} = 1,$$
$$f_{ij} \text{ belongs to } zH_+ \cap z^{a_i - a_j} H_- \text{ if } i < j, \text{ and}$$
$$f_{ij} \text{ belongs to } H_+ \cap z^{a_i - a_j} H_- \text{ if } i > j.$$

This leads at once to the above formula for $d(\mathbf{a})$.

One thing lacking from the preceding Proposition, in comparison with (7.3.3) and (7.4.1), is a description of the closures of the strata $\Sigma_\mathbf{a}$ and cells $C_\mathbf{a}$. We shall content ourselves with the slightly weaker statement in our next proposition.

At the beginning of our discussion we asked about the orbits of L^- and L^+ on $\text{Gr}^{(n)}$, rather than of N^- and N^+. It is clear that the orbit of $H_\mathbf{a}$ under L^- contains $H_{\sigma\mathbf{a}}$ for every permutation σ of $\{1, \ldots, n\}$, where $\sigma\mathbf{a} = (a_{\sigma(1)}, \ldots, a_{\sigma(n)})$. The orbit does not contain $H_\mathbf{b}$ for any other $\mathbf{b}$, as one sees from the fact that the action of L^- on W does not change the dimension of $W \cap z^k H_-$. The orbit is therefore the union of the strata $\Sigma_{\sigma\mathbf{a}}$; we shall denote it by $\Sigma_{|\mathbf{a}|}$. The orbit of $H_\mathbf{a}$ under L^+, similarly, is the union of the cells $C_{\sigma\mathbf{a}}$, and will be denoted by $C_{|\mathbf{a}|}$. Notice that if $a_1 \geq a_2 \geq \ldots \geq a_n$ then $\Sigma_\mathbf{a}$ is a dense open subset of $\Sigma_{|\mathbf{a}|}$, and if $a_1 \leq a_2 \leq \ldots \leq a_n$ then $C_\mathbf{a}$ is a dense open subset of $C_{|\mathbf{a}|}$.

The set of multi-indices $\mathbf{a}$ such that $a_1 \geq a_2 \geq \ldots \geq a_n$ can be identified with the set of conjugacy classes of homomorphisms from S^1 into U_n. We shall order such multi-indices by prescribing $\mathbf{a} \leq \mathbf{b}$ if

$$a_1 + a_2 + \ldots + a_k \leq b_1 + b_2 + \ldots + b_k \quad \text{for} \quad 1 \leq k < n,$$

and

$$a_1 + a_2 + \ldots + a_n = b_1 + b_2 + \ldots + b_n.$$

The disposition of the sets $\Sigma_{|\mathbf{a}|}$ and $C_{|\mathbf{a}|}$ is summarized in the following Proposition, where all indices $\mathbf{a}$ are assumed to be written in decreasing order $a_1 \geq a_2 \geq \ldots \geq a_n$.

Proposition (8.4.6).

(i) *The orbits $\{\Sigma_{|\mathbf{a}|}\}$ of L^- on $\text{Gr}^{(n)}$ are indexed by the conjugacy classes of homomorphisms $S^1 \to U_n$. The set $\Sigma_{|\mathbf{a}|}$ is a locally closed submanifold of $\text{Gr}^{(n)}$ of codimension $d(\mathbf{a})$. Furthermore $\Sigma_{|\mathbf{a}|}$ lies in the closure of $\Sigma_{|\mathbf{b}|}$ if and only if $\mathbf{a} \geq \mathbf{b}$.*

(ii) *The orbits $\{C_{|\mathbf{a}|}\}$ of L^+ on $\text{Gr}_0^{(n)}$ are indexed in the same way, and $C_{|\mathbf{a}|}$ is a locally closed submanifold of $\text{Gr}^{(n)}$ of dimension $d(\tilde{\mathbf{a}})$, where $\tilde{\mathbf{a}} = (a_n, a_{n-1}, \ldots, a_1)$. The closure of $C_{|\mathbf{b}|}$ contains $C_{|\mathbf{a}|}$ if and only if $\mathbf{a} \leq \mathbf{b}$.*

8.4 THE BIRKHOFF AND BRUHAT DECOMPOSITIONS   135

(iii) $C_{|\mathbf{a}|}$ meets $\Sigma_{|\mathbf{b}|}$ if and only if $\mathbf{a} \geq \mathbf{b}$, and $C_{|\mathbf{a}|}$ meets $\Sigma_{|\mathbf{a}|}$ transversally in the set $\Lambda_{\mathbf{a}}$ of homomorphisms $S^1 \to U_n$ which are conjugate to $z^{\mathbf{a}}$.

Remark. $\Lambda_{\mathbf{a}}$ is a generalized flag manifold of the form $U_n/U_{n_1} \times \ldots \times U_{n_r}$; its dimension is $\nu(\mathbf{a})$ in the notation of (8.4.5).

Proof. What needs to be proved is the assertions about the ordering, and about the intersection of $C_{|\mathbf{a}|}$ with $\Sigma_{|\mathbf{a}|}$.

For any decreasing multi-index $\mathbf{a}$ and any integer p let us define

$$\delta_p(\mathbf{a}) = \sum_{k=1}^{n} (p - a_k)_+,$$

where a_+ means a if $a \geq 0$ and 0 otherwise. It is easy to check that, providing $\Sigma a_i = \Sigma b_i$,

$$\mathbf{a} \leq \mathbf{b} \Leftrightarrow \delta_p(\mathbf{a}) \leq \delta_p(\mathbf{b}) \text{ for all } p,$$

and that

$$W \in \Sigma_{|\mathbf{a}|} \Leftrightarrow \dim(W \cap z^p H_-) = \delta_p(\mathbf{a}) \text{ for all } p,$$

and

$$W \in C_{|\mathbf{a}|} \Leftrightarrow \dim(W/W \cap z^p H_+) = \delta_p(\mathbf{a}) \text{ for all } p.$$

From this it follows that $\mathbf{a} \leq \mathbf{b}$ if $\Sigma_{|\mathbf{b}|}$ is in the closure of $\Sigma_{|\mathbf{a}|}$, or $C_{|\mathbf{a}|}$ is in the closure of $C_{|\mathbf{b}|}$, or $C_{|\mathbf{b}|}$ meets $\Sigma_{|\mathbf{a}|}$.

It also follows that if $W \in C_{|\mathbf{a}|} \cap \Sigma_{|\mathbf{a}|}$ then

$$W = (W \cap z^p H_-) \oplus (W \cap z^p H_+)$$

for all p, and hence that

$$W \ominus zW = \bigoplus_{i=1}^{r} A_i z^{k_i},$$

where $\mathbb{C}^n = A_1 \oplus \ldots \oplus A_r$ is some orthogonal decomposition of $\mathbb{C}^n$. This implies that $W = \lambda H_+$, where $\lambda : S^1 \to U_n$ is the homomorphism defined by

$$\lambda(z) = z^{k_1} \oplus \ldots \oplus z^{k_r}$$

with respect to the decomposition $A_1 \oplus \ldots \oplus A_r$. Thus $C_{|\mathbf{a}|} \cap \Sigma_{|\mathbf{a}|} = \Lambda_{\mathbf{a}}$. The intersection is transversal because the tangent space to $\mathrm{Gr}^{(n)}$ at $H_{\mathbf{a}}$ can be identified with the Lie algebra $\Omega_{\mathbf{a}}$ of $z^{\mathbf{a}} L_1^- z^{-\mathbf{a}}$, and then the tangent spaces to $\Sigma_{|\mathbf{a}|}$, $C_{|\mathbf{a}|}$ and $\Lambda_{\mathbf{a}}$ correspond to the intersection of $\Omega_{\mathbf{a}}$ with the factors of the decomposition $L\mathfrak{g}_{\mathbb{C}} = L_1^- \mathfrak{g}_{\mathbb{C}} \oplus L_1^+ \mathfrak{g}_{\mathbb{C}} \oplus \mathfrak{g}_{\mathbb{C}}$. (Here $\mathfrak{g}_{\mathbb{C}} = \mathfrak{gl}_n(\mathbb{C})$.)

We have proved the 'only if' part of the three assertions about the ordering. For the converse let us first consider a pair $\mathbf{a} < \mathbf{b}$ where

8 THE FUNDAMENTAL HOMOGENEOUS SPACE

$\mathbf{b} = \mathbf{a} + \mathbf{e}_{pq}$ for some $p < q$, where

$$\mathbf{e}_{pq} = (0, \ldots, 0, 1, 0, \ldots, 0, -1, 0, \ldots, 0)$$

with 1 and -1 in the pth and q^{th} places. We shall show that there is a complex projective line $\mathbb{C} \cup \infty$ in $\text{Gr}^{(n)}$ which joins $H_{\mathbf{a}}$ to $H_{\mathbf{b}'}$, where $\mathbf{b}'$ is $\mathbf{b}$ with its pth and q^{th} elements interchanged, and which lies, except for its end-points $\{0, \infty\}$ in $\Sigma_{|\mathbf{a}|} \cap C_{|\mathbf{b}|}$.

There is an embedding i_{pq} of $SL_2(\mathbb{C})$ in $LGL_n(\mathbb{C})$ which takes

$$\begin{pmatrix} a & b \\ c & d \end{pmatrix}$$

to (f_{ij}), where (f_{ij}) differs from the identity matrix only in the pth and q^{th} rows and columns, and

$$\begin{pmatrix} f_{pp} & f_{pq} \\ f_{qp} & f_{qq} \end{pmatrix} = \begin{pmatrix} a & bz \\ cz^{-1} & d \end{pmatrix}.$$

Consider the orbit of $H_{\mathbf{a}}$ under the action of $SL_2(\mathbb{C})$ induced by i_{pq}. The stabilizer consists of the lower triangular matrices, so the orbit is a standard projective line $S^2 \cong \mathbb{C} \cup \infty$. All of its points except the point at infinity form the orbit of the strictly upper triangular matrices in $SL_2(\mathbb{C})$. These matrices map into L_1^-, so $S^2 - \{\infty\}$ is contained in $\Sigma_{|\mathbf{a}|}$. But the point at infinity is $i_{pq}(A).H_{\mathbf{a}}$, where

$$A = \begin{pmatrix} 0 & 1 \\ -1 & 0 \end{pmatrix};$$

and this is $H_{\mathbf{b}'}$, which belongs to $\Sigma_{|\mathbf{b}|}$. Thus the closure of $\Sigma_{|\mathbf{a}|}$ contains $\Sigma_{|\mathbf{b}|}$. On the other hand $S^2 - \{0\}$ is the orbit of $H_{\mathbf{b}'}$ under the strictly lower triangular matrices, and belongs to $C_{|\mathbf{b}|}$. So the closure of $C_{|\mathbf{b}|}$ contains $C_{|\mathbf{a}|}$, and $C_{|\mathbf{b}|}$ meets $\Sigma_{|\mathbf{a}|}$.

That completes the proof in the case $\mathbf{b} = \mathbf{a} + \mathbf{e}_{pq}$. In the general case the first two assertions about the ordering follow because whenever $\mathbf{a} < \mathbf{b}$ one can get from $\mathbf{a}$ to $\mathbf{b}$ by successively adding multi-indices of the form $\mathbf{e}_{pq}$. (See [108] (1.15).) We shall leave the third assertion to the reader.

8.5 The Grassmannian model for the other classical groups

The Grassmannian description we have given of the loop space of U_n can be modified very easily to treat the orthogonal and symplectic groups.

The orthogonal group

The group O_n consists of the real matrices in U_n, so ΩO_n is a submanifold of ΩU_n.

Proposition (8.5.1). *A subspace $W \in \text{Gr}^{(n)}$ corresponds to a loop in O_n if*

8.5 THE MODEL FOR THE OTHER CLASSICAL GROUPS

and only if it belongs to
$$\mathrm{Gr}_{\mathbb{R}}^{(n)} = \{W \in \mathrm{Gr}^{(n)} : \bar{W}^\perp = zW\}.$$

All the spaces $W \in \mathrm{Gr}_{\mathbb{R}}^{(n)}$ have virtual dimension 0, but it follows from (8.5.1) that $\mathrm{Gr}_{\mathbb{R}}^{(n)}$ has two connected components, which in fact are distinguished by the parity of the dimension of the kernel of the projection $W \to H_+$. (That will appear in Chapter 12: the space $\mathcal{J}(H)$ considered in Section 12.4 is closely related to $\mathrm{Gr}_{\mathbb{R}}^{(n)}$.)

Before proving (8.5.1) let us notice that $\mathrm{Gr}_{\mathbb{R}}^{(n)}$ is a *complex* submanifold of $\mathrm{Gr}^{(n)}$. For the map $W \mapsto z^{-1}\bar{W}^\perp$ is a holomorphic involution on $\mathrm{Gr}(H^{(n)})$: in the coordinate patch consisting of graphs of operators $T: H_+ \to H_-$ it is represented by the complex linear map $T \to -z^{-1}\bar{T}^*$. The condition of (8.5.1) asserts that W is very nearly an *isotropic* subspace of $H^{(n)}$ for the complex bilinear form B on $H^{(n)}$ defined by $B(\xi, \eta) = \langle \bar{\xi}, \eta \rangle$: more precisely, the radical of W with respect to B is zW.

Notice that we must now avoid identifying $H^{(n)}$ with $H^{(1)}$, for that does not respect the real subspaces.

Proof of (8.5.1). First suppose that γ is a loop in the complex orthogonal group $O_n(\mathbb{C})$. Then the multiplication operator M_γ preserves the complex bilinear form B on $H^{(n)}$, so it commutes with the operation $W \mapsto \bar{W}^\perp$ of forming the orthogonal complement with respect to B. As H_+ satisfies $\bar{H}_+^\perp = zH_+$, so does γH_+.

Conversely, if $\bar{W}^\perp = zW$, then $W \ominus zW = W \cap \bar{W}$, and so $W \ominus zW$ is the complexification of a real n-dimensional subspace of $L^2(S^1; \mathbb{R}^n)$, and we can find an orthonormal basis for it consisting of real functions. Thus $W = \gamma H_+$ for some loop γ in O_n.

Proposition (8.5.1) gives us two factorization theorems immediately. Because $\mathrm{Gr}_{\mathbb{R}}^{(n)}$ is a homogeneous space of $LO_n(\mathbb{C})$ on which ΩO_n acts transitively, we have

Proposition (8.5.2). *The multiplication map*
$$\Omega O_n \times L^+ O_n(\mathbb{C}) \to LO_n(\mathbb{C})$$
is a diffeomorphism, where $L^+ O_n(\mathbb{C})$ denotes the loops which are boundary values of holomorphic maps
$$\{z \in \mathbb{C} : |z| < 1\} \to O_n(\mathbb{C}).$$

Secondly, any element W of $\mathrm{Gr}_{\mathbb{R}}^{(n)}$ in a suitable neighbourhood of H_+ is transversal to H_-, i.e. $W \cap H_- = 0$ and $W + H_- = H^{(n)}$. For such spaces W the intersection $W \cap zH_-$ is n-dimensional, and $W = (W \cap zH_-) \oplus zW$. We know then (see the proof of (8.4.1)) that any basis $\{w_1, \ldots, w_n\}$ for $W \cap zH_-$ forms the columns of a loop $\gamma_- \in L^- GL_n(\mathbb{C})$ such that

8 THE FUNDAMENTAL HOMOGENEOUS SPACE

$\gamma = \gamma_- \gamma_+$, with $\gamma_+ \in L^+GL_n(\mathbb{C})$. If the basis is chosen orthogonal with respect to B—which is possible because the restriction of B to $W \cap zH_-$ is necessarily nondegenerate—then γ_- belongs to $L^-O_n(\mathbb{C})$, because (see the proof of (8.3.2))

$$\langle \overline{w_k(e^{i\theta})}, w_\ell(e^{i\theta}) \rangle = \sum_{p,q \leq 0} \langle \bar{w}_{kp}, w_{\ell q} \rangle e^{+ip\theta + iq\theta}$$

$$= \sum_{r \geq 0} \langle \bar{w}_k, z^r w_\ell \rangle_H e^{-ir\theta}$$

$$= B(w_k, w_\ell).$$

This gives us

Proposition (8.5.3). *The multiplication map*

$$L_1^- O_n(\mathbb{C}) \times L^+ O_n(\mathbb{C}) \to LO_n(\mathbb{C})$$

is a diffeomorphism on to an open subset of $LO_n(\mathbb{C})$.†

We could go on to derive Birkhoff and Bruhat decomposition theorems, but we shall postpone that until the next section.

The symplectic group

The group Sp_n is the subgroup of all elements u in U_{2n} which preserve a nondegenerate skew form on $\mathbb{C}^{2n}$. Equivalently, it consists of the unitary transformations u which are quaternionic-linear when $\mathbb{C}^{2n}$ is identified with $\mathbb{H}^n$. If $J: \mathbb{C}^{2n} \to \mathbb{C}^{2n}$ is the antilinear map representing multiplication by the quaternion j, then u belongs to Sp_n if and only if $uJ = Ju$. The complexification of Sp_n is the subgroup $Sp_n(\mathbb{C})$ of all elements of $GL_{2n}(\mathbb{C})$ which preserve the skew complex-bilinear form S defined by $S(\xi, \eta) = \langle J\xi, \eta \rangle$.

Corresponding to (8.5.1) we have

Proposition (8.5.4). *A subspace $W \in \text{Gr}^{(2n)}$ corresponds to a loop in Sp_n if and only if it belongs to*

$$\text{Gr}^{(2n)}_{\emptyset} = \{W \in \text{Gr}^{(2n)} : (JW)^\perp = zW\}.$$

The proof is identical to that of (8.5.1); and the result implies two factorization theorems precisely analogous to (8.5.2) and (8.5.3).

8.6 The Grassmannian model for a general compact Lie group

In studying ΩG for a compact semisimple group G one may as well assume that the centre of G is trivial. For if $\tilde{G}$ is a covering group of G then the manifold $\Omega \tilde{G}$ is just the union of some of the connected

† The open subset is dense in the group of null-homotopic loops in $O_n(\mathbb{C})$. These form two of the four connected components of $LO_n(\mathbb{C})$.

8.6 THE MODEL FOR A GENERAL COMPACT GROUP

components of ΩG. If the centre is trivial then G is the identity component of the group of automorphisms of its Lie algebra $\mathfrak{g}$.

The most obvious Hilbert space on which LG acts is $H^{\mathfrak{g}} = L^2(S^1; \mathfrak{g}_\mathbb{C})$. This is, essentially, its adjoint representation. We shall identify $H^{\mathfrak{g}}$ with $H^{(n)}$, where n is the dimension of G. Thus we are regarding LG as a subgroup of LU_n by the adjoint representation of G on $\mathbb{C}^n$. The loop space ΩG is a submanifold of ΩU_n, which can be identified with a submanifold of $\mathrm{Gr}(H^{\mathfrak{g}})$.

Definition (8.6.1). $\mathrm{Gr}^{\mathfrak{g}}$ *is the subset of* $\mathrm{Gr}(H^{\mathfrak{g}})$ *consisting of subspaces* W *such that*
 (i) $zW \subset W$,
 (ii) $\bar{W}^\perp = zW$, *and*
 (iii) W^{sm} *is a Lie algebra.*

Here W^{sm} denotes the subspace of smooth functions in W, which we know is dense. To say that it is a Lie algebra means simply that it is closed under the bracket operation defined pointwise for $\mathfrak{g}_\mathbb{C}$-valued functions.

Theorem (8.6.2). *The action of* $LG_\mathbb{C}$ *on* $\mathrm{Gr}(H^{\mathfrak{g}})$ *preserves* $\mathrm{Gr}^{\mathfrak{g}}$, *and if the centre of G is trivial then* $\gamma \mapsto \gamma H_+$ *defines a diffeomorphism* $\Omega G \to \mathrm{Gr}^{\mathfrak{g}}$.

Proof. The first statement is obvious, as any group acts on its own Lie algebra by Lie algebra automorphisms. (The condition (ii) arises as in (8.5.1) because the adjoint action preserves the Killing form, so that $G_\mathbb{C}$ is contained in the orthogonal group $O(\mathfrak{g}_\mathbb{C})$.)

Conversely, suppose that W satisfies the conditions of (8.6.1). From condition (ii) we have $W \ominus zW = W \cap \bar{W}$. We know that $W \ominus zW$ consists of smooth functions. For any point z of the circle the evaluation map $e_z : W \cap \bar{W} \to \mathfrak{g}_\mathbb{C}$ at z is an isomorphism, and must be an isomorphism of Lie algebras. It also commutes with complex conjugation. If γ is defined by $\gamma(z) = e_z e_1^{-1}$ then γ is a loop in the group of automorphisms of $\mathfrak{g}$. For a group with trivial centre this means that γ belongs to ΩG. By our usual argument we have $\gamma H_+ = W$.

As in the preceding section we can deduce from (8.6.2) that the multiplication $\Omega G \times L^+ G_\mathbb{C} \to LG_\mathbb{C}$ is a diffeomorphism, and that the multiplication $L_1^- G_\mathbb{C} \times L^+ G_\mathbb{C} \to LG_\mathbb{C}$ is a diffeomorphism on to a dense open subset of the identity component. We shall now go further and determine the stratification and cell decomposition of ΩG corresponding to the theorems proved earlier for U_n.

Let us choose a maximal torus T of G, and a system of positive roots. Then we can define the nilpotent subgroups $N_0^\pm$ of $G_\mathbb{C}$ whose Lie algebras are spanned by the root vectors of $\mathfrak{g}_\mathbb{C}$ corresponding to the positive (resp.

negative) roots. We can also define subgroups $N^\pm$ of $L^\pm = L^\pm G_\mathbb{C}$: $N^\pm$ consists of the loops $\gamma \in L^\pm$ such that $\gamma(0) \in N_0^\pm$ (resp. $\gamma(\infty) \in N_0^\pm$). Thus $L_1^\pm \subset N^\pm \subset L^\pm$.

The result we wish to prove is

Theorem (8.6.3).
(i) $\text{Gr}^\mathfrak{a} \cong \Omega G$ is the union of strata Σ_λ indexed by the lattice $\check{T}$ of homomorphisms $\lambda : \mathbb{T} \to T$.

(ii) Σ_λ is the orbit of $\lambda.H_+$ under N^-. It is a locally closed contractible complex submanifold of finite codimension d_λ in $\text{Gr}^\mathfrak{a}$, and it is diffeomorphic to $L_\lambda^- = N^- \cap \lambda.L_1^-.\lambda^{-1}$.

(iii) The orbit of $\lambda.H_+$ under N^+ is a complex cell C_λ of dimension d_λ. It is diffeomorphic to $L_\lambda^+ = N^+ \cap \lambda.L_1^-.\lambda^{-1}$, and meets Σ_λ transversally in the single point $\lambda.H_+$.

(iv) The orbit of $\lambda.H_+$ under $\lambda.L_1^-.\lambda^{-1}$ is an open subset U_λ of $\text{Gr}^\mathfrak{a}$. The multiplication $L_\lambda^+ \times L_\lambda^- \to \lambda.L_1^-.\lambda^{-1}$ defines a diffeomorphism $C_\lambda \times \Sigma_\lambda \to U_\lambda$.

(v) The union of the cells C_λ is $\text{Gr}_0^\mathfrak{a} \cong \Omega_{\text{pol}} G$.

Once again the stratification of $\text{Gr}^\mathfrak{a}$ will be induced by that of $\text{Gr}(H^\mathfrak{a})$. To define the latter we must choose an orthonormal basis $\{\zeta_k\}$ of $H^\mathfrak{a}$ indexed by $\mathbb{Z}$. We shall do this so that H_+ is spanned by $\{\zeta_k\}$ for $k \geq 0$, and $z\zeta_k = \zeta_{k+n}$. Thus $\{\zeta_0, \ldots, \zeta_{n-1}\}$ is a basis for $\mathfrak{g}_\mathbb{C}$. We shall choose it to consist of eigenvectors of the action of T. Each vector ζ_i (for $0 \leq i < n$) then has a weight with respect to the action of T: either zero or a root of G. We choose the order of the ζ_i so that ζ_i precedes ζ_j whenever the difference between the weight of ζ_j and the weight of ζ_i is a sum of positive roots. In particular, $\{\zeta_0, \ldots, \zeta_{m-1}\}$, where $m = \frac{1}{2}(n - \ell)$ and ℓ is the rank of G, are the negative root vectors, and span the Lie algebra of N_0^-, while $\{\zeta_{m+\ell}, \ldots, \zeta_{n-1}\}$ span the Lie algebra of N_0^+.

With respect to the basis $\{\zeta_k\}$ the group N^- acts on $H^\mathfrak{a}$ by elements of the lower triangular subgroup $\mathcal{N}^-$ of (7.3.3).

The strata and cells of $\text{Gr}(H^\mathfrak{a})$ are indexed by subsets S of $\mathbb{Z}$. Among them are the sets S_λ corresponding to the lattice $\check{T}$ of homomorphisms $\mathbb{T} \to T$: these are defined by

$$H_{S_\lambda} = \lambda.H_+.$$

We shall write H_λ for H_{S_λ}. Our proof of (8.6.3) depends on the following lemma.

Lemma (8.6.4). *The strata Σ_S and cells C_S of $\text{Gr}(H^\mathfrak{a})$ which meet $\text{Gr}^\mathfrak{a}$ are precisely the Σ_{S_λ} and C_{S_λ} for $\lambda \in \check{T}$.*

We shall postpone the proof of the lemma to the end of this section, and shall proceed with the proof of (8.6.3).

8.6 THE MODEL FOR A GENERAL COMPACT GROUP

Proof of (8.6.3). Let us define
$$\Sigma_\lambda = \Sigma_{S_\lambda} \cap \mathrm{Gr}^{\mathfrak{g}}.$$
The lemma tells us that $\mathrm{Gr}^{\mathfrak{g}}$ is the union of the Σ_λ. Because N^- is contained in $\mathcal{N}^-$ it is clear that the orbit of H_λ is contained in Σ_λ. Thus the main point is to show that L_λ^- acts transitively on Σ_λ. Now from (8.4.1) we know that any $W \in \Sigma_\lambda$ can be expressed *uniquely* as γH_λ, where γ belongs to $L_\lambda^- GL_n(\mathbb{C})$. As L_λ^- is the intersection of $L_\lambda^- GL_n(\mathbb{C})$ with $LG_\mathbb{C}$, it is enough to show that γ belongs to $LG_\mathbb{C}$. The construction of γ in (8.4.1) can be formulated as follows. If W belongs to Σ_λ then $W \in zH_\lambda^\perp$ has dimension n, and the evaluation map $e_z : W \cap zH_\lambda^\perp \to \mathfrak{g}_\mathbb{C}$ at each point $z \in S^1$ is an isomorphism. The orthogonal projection $\mathrm{pr} : W \cap zH_\lambda^\perp \to H_\lambda \ominus zH_\lambda$ is also an isomorphism. Now $\gamma(z)$ is the composite

$$\mathfrak{g}_\mathbb{C} \xrightarrow{M_\lambda} H_\lambda \ominus zH_\lambda \xrightarrow{\mathrm{pr}^{-1}} W \cap zH_\lambda^\perp \xrightarrow{e_z} \mathfrak{g}_\mathbb{C}.$$

Each of the three spaces here is a Lie algebra—in the case of $W \cap zH_\lambda^\perp$, notice that $(zH_\lambda^\perp)^{\mathrm{sm}} = \lambda \bar{H}_+^{\mathrm{sm}}$. Furthermore each map is a homomorphism of Lie algebras—for pr is induced by the projection of $(zH_\lambda^\perp)^{\mathrm{sm}}$ on to $zH_\lambda^\perp \ominus H_\lambda^\perp$, and $(H_\lambda^\perp)^{\mathrm{sm}}$ is an ideal in $(zH_\lambda^\perp)^{\mathrm{sm}}$. So $\gamma(z)$ is a homomorphism of Lie algebras, and hence belongs to $G_\mathbb{C}$, as we want.

The rest of the proof is exactly the same as for $GL_n(\mathbb{C})$, and we shall say no more about it.

The orbits of L^- and L^+ on $\mathrm{Gr}^{\mathfrak{g}}$ and $\mathrm{Gr}_0^{\mathfrak{g}}$ are obtained, again just as before, by grouping the Σ_λ and C_λ together into pieces $\Sigma_{|\lambda|}$ and $C_{|\lambda|}$ indexed by the *conjugacy classes* of homomorphisms $\mathbb{T} \to G$, or equivalently by the set of orbits $\check{T}/W$ of the Weyl group W of G acting on the lattice $\check{T}$. This set $\check{T}/W$ can be ordered by prescribing $|\lambda| \leq |\mu|$ if the convex hull of the orbit $W \cdot \lambda$ is contained in the convex hull of $W \cdot \mu$. (Here $\check{T}$ is regarded as a lattice in the vector space $\mathfrak{t}$.) Without any further discussion we record

Proposition (8.6.5).

(i) $\Sigma_{|\lambda|}$ *intersects* $C_{|\lambda|}$ *transversally in the set* Λ_λ *of homomorphisms* $\mathbb{T} \to G$ *which are conjugate to* λ.

(ii) $\Sigma_{|\lambda|}$ *is contained in the closure of* $\Sigma_{|\mu|}$
$C_{|\mu|}$ *is contained in the closure of* $C_{|\lambda|}$ $\bigg\}$ *if and only if* $|\lambda| \geq |\mu|$.

$C_{|\lambda|}$ *meets* $\Sigma_{|\mu|}$

Because we have a cell decomposition of $\mathrm{Gr}_0^{\mathfrak{g}}$ whose cells are all of even dimension the fundamental group $\pi_1(\mathrm{Gr}_0^{\mathfrak{g}})$ must be trivial. Now $\mathrm{Gr}_0^{\mathfrak{g}}$ is the polynomial loop space $\Omega_{\mathrm{pol}} G$. The fundamental group of ΩG is the second homotopy group $\pi_2(G)$. If we show that $\Omega_{\text{pol}}G$ is homotopy equivalent to ΩG then we shall have a proof of the well-known but important fact that $\pi_2(G)$ is zero for any compact Lie group G. This proof is in essence the same as Bott's Morse theory proof [14].

Proposition (8.6.6). *The inclusion* $\text{Gr}_0^{\mathfrak{g}} \to \text{Gr}^{\mathfrak{g}}$, *or equivalently* $\Omega_{\text{pol}}G \to \Omega G$, *is a homotopy equivalence.*

Corollary (8.6.7). *The homotopy group* $\pi_2(G)$ *is zero.*

Proof of (8.6.6). The idea is that $\text{Gr}_0^{\mathfrak{g}}$ and $\text{Gr}^{\mathfrak{g}}$ have corresponding stratifications by homotopy equivalent subsets.

Let us arrange the elements of the lattice $\check{T}$ in a sequence beginning with 0 so that if Σ_μ is contained in the closure of Σ_λ then λ precedes μ (which we shall denote by $\lambda \leq \mu$). Let $\text{Gr}^{\leq \lambda}$ denote the union of all the open sets U_μ of $\text{Gr}^{\mathfrak{g}}$ such that $\mu \leq \lambda$, and $\text{Gr}^{<\lambda}$ the union of the U_μ such that $\mu < \lambda$. We shall also write $\text{Gr}_0^{\leq \lambda}$ and $\text{Gr}_0^{<\lambda}$ for the corresponding parts of $\text{Gr}_0^{\mathfrak{g}}$. It is enough for us to prove that $\text{Gr}_0^{\leq \lambda} \to \text{Gr}^{\leq \lambda}$ is a homotopy equivalence for all λ. (Cf. [114] Appendix.)

Now $\text{Gr}^{\leq \lambda}$ is the union of U_λ and $\text{Gr}^{<\lambda}$, and the intersection of the two sets is $U_\lambda - \Sigma_\lambda$. (The fact that all points of U_λ belong to strata $\leq \lambda$ follows from the corresponding fact for $\text{Gr}(H)$ proved in (7.3.3).) The set U_λ is contractible, while $U_\lambda - \Sigma_\lambda$ is diffeomorphic to $\Sigma_\lambda \times (C_\lambda - \{H_\lambda\})$ and so homotopy equivalent to $C_\lambda - \{H_\lambda\}$.

The space $\text{Gr}_0^{\leq \lambda}$ is likewise the union of $U_{\lambda,0}$ and $\text{Gr}_0^{<\lambda}$, whose intersection is $U_{\lambda,0} - \Sigma_{\lambda,0}$. Whereas U_λ was diffeomorphic to $\lambda \cdot L_1^- \cdot \lambda^{-1}$, the set $U_{\lambda,0}$ is homeomorphic to $\lambda \cdot L_{1,\text{pol}}^- \cdot \lambda^{-1}$— we do not know that $\text{Gr}_0^{\mathfrak{g}}$ is a manifold—and hence to $\Sigma_{\lambda,0} \times C_\lambda$. Furthermore $U_{\lambda,0}$ and $\Sigma_{\lambda,0}$ are contractible for the same reason as U_λ, and $U_{\lambda,0} - \Sigma_{\lambda,0}$ is homotopy equivalent to $U_\lambda - \Sigma_\lambda$. We can conclude by induction that $\text{Gr}_0^{\leq \lambda}$ is homotopy equivalent to $\text{Gr}^{\leq \lambda}$. (We are using the fact that if $X = U \cup V$ and $X' = U' \cup V'$, where U, V, U', V' are open subsets, then a map $f: X \to X'$ is a homotopy equivalence if it induces homotopy equivalences $U \to U'$, $V \to V'$, and $U \cap V \to U' \cap V'$. Cf. [67] (16.24).)

Proof of (8.6.4). We end this section with the postponed proof of Lemma (8.6.4).

Let us define an action of the circle $\mathbb{T}$ on $L\mathfrak{g}$ as follows. Choose a homomorphism $\rho = \exp(\dot\rho): \mathbb{T} \to T$ such that $\dot\rho$ belongs to the positive Weyl chamber in $\mathfrak{t}$, i.e. $\langle \alpha, \dot\rho \rangle > 0$ for every root α. For any sufficiently large integer q the centralizer of $\rho(e^{2\pi i/q})$ in G will be T. The action of $\mathbb{T}$ which we want is got by simultaneously rotating with speed q and conjugating by ρ:

$$\mathbb{T} \times L\mathfrak{g} \to L\mathfrak{g}$$

takes (u, ξ) to $S_u\xi$, where

$$S_u\xi(z) = \rho(u)\xi(u^{-q}z)\rho(u)^{-1}.$$

This action extends to the Hilbert space $H^\mathfrak{a}$, and is diagonal with respect to the basis $\{\zeta_k\}$. In fact $S_u\zeta_k = u^{m_k}\zeta_k$, where m_k increases monotonically with k providing q is large enough. The action induces an action on $\text{Gr}(H^\mathfrak{a})$ which extends to an action of $\mathbb{C}^\times_{\leq 1}$ (see Section 7.6). Furthermore the action preserves $\text{Gr}^\mathfrak{a}$, and also the strata Σ_S. For any $W \in \text{Gr}(H^\mathfrak{a})$ the point $S_u W$ tends to a limit as $u \to 0$, the limit being necessarily fixed under the $\mathbb{T}$-action and contained in the same stratum as W. (For the strata are characterized by (7.5.4), and S_u simply multiplies the Plücker coordinate $\pi_S(W)$ by $u^{m(S)}$, where $m(S)$ increases monotonically with S and tends to ∞ as $\ell(S) \to \infty$.) But the only fixed points of the action on $\text{Gr}^\mathfrak{a}$ are the spaces λH_+ with $\lambda \in \check{T}$. Indeed if γH_+ is fixed for some $\gamma \in \Omega G$ then

$$\rho(u)\gamma(u^{-q}z)\gamma(u^{-q})^{-1}\rho(u)^{-1} = \gamma(z) \tag{8.6.8}$$

for all $u, z \in \mathbb{T}$. Putting $u = e^{2\pi i/q}$ we find that $\gamma(z)$ commutes with $\rho(e^{2\pi i/q})$, and so $\gamma(z)$ belongs to T. Equation (8.6.8) then reduces to the condition that $\gamma: \mathbb{T} \to T$ is a homomorphism. Thus every stratum contains a point λH_+, and is therefore of the form Σ_{S_λ}.

The argument for the cells C_S is essentially the same: one considers $S_u W$ as $u \to \infty$.

8.7 The homogeneous space LG/T and the periodic flag manifold

We saw in Section 2.8 that the most important homogeneous space of a compact group G is G/T, where T is a maximal torus of G. We have already mentioned that the analogue of G/T for a loop group is LG/T rather than the more natural space $\Omega G = LG/G$ which we have been studying so far in this chapter. We shall now give a rapid account of LG/T.

We know already that LG/T is a complex manifold, for it can be identified with $LG_\mathbb{C}/B^+$, where B^+ consists of the elements $\gamma_0 + \gamma_1 z + \ldots$ of $L^+ G_\mathbb{C}$ such that γ_0 belongs to the positive Borel subgroup B_0^+ of $G_\mathbb{C}$. The main property of LG/T is that it is stratified by the orbits of N^-, and the strata are indexed by the affine Weyl group W_{aff}. This group was defined in Section 5.1. It is the semidirect product $W \tilde{\times} \check{T}$, where W is the Weyl group of G and $\check{T}$ is the lattice of homomorphisms $\mathbb{T} \to T$. We can regard W_{aff} as a subset of LG/T, for $W_{\text{aff}} = (N_T \cdot \check{T})/T$, where N_T is the normalizer of T in G. The following Proposition is essentially a restatement of the proof of (5.1.2).

Proposition (8.7.1). *The set of fixed points of the rotation action of $\mathbb{T}$ on LG/T is W_{aff}.*

144   8 THE FUNDAMENTAL HOMOGENEOUS SPACE

The properties of the stratification of LG/T are listed in

Theorem (8.7.2).

(i) *The complex manifold $Y = LG/T = LG_\mathbb{C}/B^+$ is the union of strata Σ_w indexed by $w \in W_{\mathrm{aff}}$.*

(ii) *The stratum Σ_w is the orbit of w under N^-, and is a locally closed contractible complex submanifold of Y whose codimension is the length $\ell(w)$ of w. It is diffeomorphic to $N_w^- = N^- \cap wN^-w^{-1}$.*

(iii) *Σ_w is a closed subset of the open subset U_w of Y, where $U_w = w \cdot \Sigma_1$. The action of $A_w = N^+ \cap wN^-w^{-1}$ defines a diffeomorphism*

$$A_w \times \Sigma_w \to U_w.$$

(iv) *The orbit of w under A_w is a complex cell C_w of dimension $\ell(w)$ which intersects Σ_w transversally at w. The union of the cells C_w is $Y_{\mathrm{pol}} = L_{\mathrm{pol}}G/T$.*

(v) *If $\ell(w') = \ell(w) + 1$ then $\Sigma_{w'}$ is contained in the closure of Σ_w if and only if $w' = ws$, where $s \in W_{\mathrm{aff}}$ is the reflection corresponding to a simple affine root.*

Here the length $\ell(w)$ is defined as the dimension of A_w, i.e. as the number of positive affine roots α such that $w \cdot \alpha$ is negative.

The most important part of the content of Proposition (8.7.2) is the pair of factorization theorems

$$LG_\mathbb{C} = \bigcup_{w \in W_{\mathrm{aff}}} N^- w B^+ \tag{8.7.3a}$$

and

$$L_{\mathrm{pol}}G_\mathbb{C} = \bigcup_{w \in W_{\mathrm{aff}}} N_{\mathrm{pol}}^+ w B_{\mathrm{pol}}^+. \tag{8.7.3b}$$

These follow from (8.6.3) together with the Bruhat decomposition of the finite dimensional group

$$G_\mathbb{C} = \bigcup_{w \in W} N_0^- w B_0^+ = \bigcup_{w \in W} N_0^+ w B_0^+. \tag{8.7.4}$$

Granting (8.7.4) the proof of (8.7.2) presents nothing new. It does not seem worth giving the details. For a proof of (8.7.4) we refer to Bourbaki [20] Chapter 6 Section 2. We shall, however, mention the crucial point in the proof of part (v) of (8.7.2), although it is identical with the finite dimensional case.

If α is a simple affine root of LG then there is (see (5.2.4)) an associated homomorphism $i_\alpha: SL_2(\mathbb{C}) \to LG_\mathbb{C}$ which maps the torus $\mathbb{T}$ of $SL_2(\mathbb{C})$ into T. This gives us a map $i_\alpha: S^2 \to Y$, where $S^2 = \mathbb{C} \cup \infty$ is $SL_2(\mathbb{C})/\mathbb{T}$. If $w' = ws_\alpha$ in W_{aff}, where s_α is the reflection corresponding to α, and $\ell(w') = \ell(w) + 1$, then the map

$$z \mapsto i_\alpha(z) \cdot w$$

8.7 THE HOMOGENEOUS SPACE LG/T

from S^2 to Y defines a holomorphic curve in Y linking w to w'. This curve lies in Σ_w except for the point $i_\alpha(\infty) \cdot w = w'$, and so the closure of Σ_w contains $\Sigma_{w'}$.

In the case of U_n there is a geometrical model for LU_n/T which we shall now describe: it is analogous to the Grassmannian model of ΩU_n.

Definition (8.7.5). $\mathrm{Fl}^{(n)}$ consists of all sequences $\{W_k\}_{k \in \mathbb{Z}}$ of subspaces of $H^{(n)}$ such that
  (i) each W_k belongs to $\mathrm{Gr}(H^{(n)})$,
  (ii) $W_{k+1} \subset W_k$ for each k, and $\dim(W_k/W_{k+1}) = 1$,
  (iii) $zW_k = W_{k+n}$.

It is natural to refer to the points of $\mathrm{Fl}^{(n)}$ as *periodic flags*. We shall also consider the subspace $\mathrm{Fl}_0^{(n)}$ of $\mathrm{Fl}^{(n)}$ consisting of flags such that each W_k belongs to Gr_0.

In this section it is now once again convenient to identify $H^{(n)}$ with $H = L^2(S^1; \mathbb{C})$. Then the flag $\{\zeta^k H_+\}$ is a canonical base-point in $\mathrm{Fl}^{(n)}$; its stabilizer in $LGL_n(\mathbb{C})$ is $B^+ GL_n(\mathbb{C})$. The following proposition is proved in the same way as (8.3.2), and even follows from it.

Proposition (8.7.6). *The group LU_n acts transitively on $\mathrm{Fl}^{(n)}$, and the stabilizer of $\{\zeta^k H_+\}$ is the maximal torus T of U_n.*

This means that $\mathrm{Fl}^{(n)} \cong LU_n/T \cong LGL_n(\mathbb{C})/B^+$. Evidently $\mathrm{Fl}^{(n)}$ is fibred over $\mathrm{Gr}^{(n)}$ by the map $\{W_k\} \mapsto W_0$, the fibre being the finite dimensional flag manifold $U_n/T = \mathrm{Fl}(\mathbb{C}^n)$.

The orbits of N^- and N^+ provide a stratification of $\mathrm{Fl}^{(n)}$ and a cellular subdivision of $\mathrm{Fl}_0^{(n)}$. We shall not pursue this any further, however, beyond explaining how the strata and cells are indexed by the affine Weyl group W_{aff} of LU_n. In the present case W_{aff} is the semidirect product of the symmetric group S_n with the lattice $\check{T} \cong \mathbb{Z}^n$, on which S_n acts by permuting the factors. When W_{aff} is identified with a subgroup of LU_n it acts on $H^{(n)}$ by permuting the basis elements $\{\zeta^k\}$. In fact it can be identified with the group of all permutations π of $\mathbb{Z}$ with the property

$$\pi(k+n) = \pi(k) + n \tag{8.7.7}$$

for all k.

Proposition (8.7.8). *The strata and the cells of $\mathrm{Fl}^{(n)}$ are indexed by the affine Weyl group W_{aff} of LU_n.*

Proof. We must show that any flag $\{W_k\}$ belongs to the orbit of $\pi\{H_k\}$, where $H_k = \zeta^k H_+$, for some permutation π satisfying (8.7.7). Now each W_k belongs to some stratum Σ_{S_k} of the Grassmannian, and $S_k - S_{k+1}$ has exactly one element, say s_k. The desired permutation is given by $\pi(k) = s_k$. For $k = 0, 1, 2, \ldots, n-1$ we choose a vector $w_k \in H^{(n)}$

spanning W_k/W_{k+1} which is of order s_k. Just as in the proof of (8.3.2) we find that $\{w_0, w_1, \ldots, w_{n-1}\}$ are the columns of a loop γ such that $\gamma\{H_k\} = \{W_k\}$; and $\gamma\pi^{-1}$ belongs to N^-.

The argument for the cells is precisely analogous.

8.8 Bott periodicity

We have already mentioned in Section 6.4 the Bott periodicity theorem, which asserts that the infinite unitary group $U = \bigcup_n U_n$ is homotopy equivalent to its second loop space $\Omega^2 U$. Another formulation is that ΩU is homotopy equivalent to

$$\mathrm{Gr}_0(H) = \bigcup_{p \leq q} \mathrm{Gr}(\zeta^p H_+/\zeta^q H_+).$$

(This formulation is well-known [17]: $\mathrm{Gr}_0(H)$ is the standard model for the space which algebraic topologists call $\mathbb{Z} \times BU$.) The theory which we have built up incorporates a proof of the theorem, for the identification of ΩU_n with a subspace of $\mathrm{Gr}(H)$ is precisely the Bott map.

Before explaining the proof, let us recall that we showed in (8.6.6) that the polynomial loop group $\Omega_{\mathrm{pol}} U_n$ is homotopy equivalent to the smooth loop group ΩU_n (and hence, by very standard arguments, to the continuous loop group $\Omega_{\mathrm{cts}} U_n$). It is a much more elementary fact, but can be proved by the same argument (8.6.6), that $\mathrm{Gr}_0(H)$ is homotopy equivalent to $\mathrm{Gr}(H)$. So Bott's theorem is a consequence of the following.

Proposition (8.8.1). *The inclusion*

$$\Omega_{\mathrm{pol}} U_n = \mathrm{Gr}_0^{(n)} \to \mathrm{Gr}_0(H)$$

induces an isomorphism of homotopy groups up to dimension $2n - 2$.

Proof. It is enough to consider the identity components of the two spaces.

$\mathrm{Gr}_0(H)$ is the union of cells C_S indexed by subsets S of $\mathbb{Z}$. If S satisfies

$$n + k + \mathbb{N} \subset S \subset k + \mathbb{N} \tag{8.8.2}$$

for some k then any $W \in C_S$ is sandwiched between $\zeta^{k+n} H_+$ and $\zeta^k H_+$. This implies that $\zeta^n W \subset W$, and hence that W belongs to $\mathrm{Gr}_0^{(n)}$. The cell C_S is therefore completely contained in $\mathrm{Gr}_0^{(n)}$. (Although we shall not need the fact, it may be worth remarking that in the notation of (8.4.5) the cells $C_\mathbf{a}$ of $\mathrm{Gr}_0^{(n)}$ obtained in this way are those such that

$$a_i = \pm 1 \text{ or } 0 \text{ for each } i,$$

and

if $a_i = 1$ and $a_j = -1$ then $i < j$.)

8.8 BOTT PERIODICITY

To prove (8.8.1) we must show that every cell C_S of Gr_0 such that S does not satisfy (8.8.2) has complex dimension $\geq n$, and that the same applies to every cell $C_S^{(n)} = C_S \cap Gr_0^{(n)}$ of $Gr_0^{(n)}$ when $n + S \subset S$ but S does not satisfy (8.8.2).

The dimension of C_S is given (if S has virtual cardinal zero) by

$$\ell(S) = \sum_{k \geq 0} (k - s_k) \tag{8.8.3}$$

by (7.4.1). The condition (8.8.2) is easily seen to be equivalent to

$$s_m = m \quad \text{when} \quad m = s_0 + n.$$

So if (8.8.2) does not hold then $k - s_k \geq 1$ when $k \leq s_0 + n$, and

$$\ell(S) \geq -s_0 + (s_0 + n) = n,$$

as we want.

The argument is essentially the same for the cells $C_S^{(n)}$. We leave it to the reader to check that the formula of (8.4.5) for the dimension can be rewritten in the same form as (8.8.3), except that for $C_S^{(n)}$ the sum must be taken only over the n values of k for which s_k does not belong to $n + S$. There are two cases. If $s_0 \leq -n$ the result is clear. If not, then s_k does not belong to $n + S$ when $k \leq s_0 + n$, because $s_k < k$; the preceding argument then applies.

Remark. The possibility of proving the Bott periodicity theorem by the above method was first pointed out in the announcement [57] of Garland and Raghunathan.

8.9 ΩG as a Kähler manifold: the energy flow

In this section we shall look at the homogeneous space ΩG afresh, thinking of it somewhat more geometrically.

Let us choose a positive definite invariant inner product $\langle \ , \ \rangle$ on the Lie algebra $\mathfrak{g}$. In Chapter 4 we introduced a skew form ω on $L\mathfrak{g}$, defined by

$$\omega(\xi, \eta) = \frac{1}{2\pi} \int_0^{2\pi} \langle \xi(\theta), \eta'(\theta) \rangle \, d\theta.$$

This defines a left-invariant closed 2-form ω on LG; because it is invariant under conjugation by constant loops, and because it vanishes when ξ or η is constant, it defines an invariant closed 2-form ω on the homogeneous space $\Omega G = LG/G$. If ξ is a non-zero element of the tangent space $\Omega \mathfrak{g} = L\mathfrak{g}/\mathfrak{g}$ to ΩG at its base-point then there is always some $\eta \in \Omega \mathfrak{g}$ such that $\omega(\xi, \eta) \neq 0$. (One can take $\eta = \xi'$.) We shall therefore think of ω as defining a *symplectic* structure on ΩG.

8 THE FUNDAMENTAL HOMOGENEOUS SPACE

If X is a finite dimensional symplectic manifold with symplectic form ω then to each smooth function $F: X \to \mathbb{R}$ there corresponds a so-called *Hamiltonian vector field* ξ_F on X characterized by

$$\omega_x(\xi_F(x), \eta) = dF(x; \eta), \tag{8.9.1}$$

where $x \in X$ and η is a tangent vector to X at x. In the infinite-dimensional case the existence of ξ_F is not automatic, but *at most* one vector field ξ_F can satisfy (8.9.1). On the other hand it is easy to see that if X is simply connected and ξ is a vector field on X which leaves ω invariant then $\xi = \xi_F$ for some smooth function F. (For de Rham's theorem applies, and $(x; \eta) \mapsto \omega_x(\xi(x), \eta)$ is a closed 1-form on X.) We shall make use of two examples of this construction.

We consider first the *energy* function $\mathscr{E}: \Omega G \to \mathbb{R}$ defined by

$$\mathscr{E}(\gamma) = \frac{1}{4\pi} \int_0^{2\pi} \|\gamma(\theta)^{-1} \gamma'(\theta)\|^2 \, d\theta. \tag{8.9.2}$$

(Here and elsewhere we always use notation as if a loop γ were a matrix-valued function. Thus $\gamma(\theta)^{-1} \gamma'(\theta)$ denotes the element of $\mathfrak{g}$ got by left-translating to the origin the tangent vector $\gamma'(\theta)$ to G at $\gamma(\theta)$. The formula (8.9.2) is written for $\gamma \in LG$, but is invariant under both left and right multiplication by elements of G.)

Proposition (8.9.3). *The Hamiltonian vector field on ΩG which corresponds to $\mathscr{E}$ is the generator of the flow defined by rotating the loops.*

Note. In this proposition ΩG is regarded as a homogeneous space LG/G rather than as a subset of LG. As a subset it would not be preserved by rotation. If one wants to regard ΩG as a subset of LG then the action on $\gamma \in \Omega G$ of the rotation R_α through the angle α must be defined by

$$(R_\alpha \gamma)(\theta) = \gamma(\theta - \alpha) \gamma(-\alpha)^{-1}. \tag{8.9.4}$$

Proof of (8.9.3). For an infinitesimal change $\delta\gamma$ in γ the change in $\mathscr{E}$ is

$$d\mathscr{E}(\gamma; \delta\gamma) = \frac{1}{2\pi} \int \langle \gamma^{-1} \gamma', \delta(\gamma^{-1} \gamma') \rangle.$$

On the other hand the value at γ of the vector field corresponding to rotation is γ' (modulo the action of a constant element of $\mathfrak{g}$). So what we have to show is that

$$\frac{1}{2\pi} \int \langle \gamma^{-1} \gamma', \delta(\gamma^{-1} \gamma') \rangle = \omega_\gamma(\gamma', \delta\gamma)$$

$$= \frac{1}{2\pi} \int \langle \gamma^{-1} \gamma', (\gamma^{-1} \delta\gamma)' \rangle.$$

8.9 THE ENERGY FLOW ON ΩG

This is true because
$$(\gamma^{-1}\delta\gamma)' = \gamma^{-1}\delta\gamma' - \gamma^{-1}\gamma'\gamma^{-1}\delta\gamma$$
$$= \delta(\gamma^{-1}\gamma') + [\gamma^{-1}\gamma', \gamma^{-1}\delta\gamma],$$
while
$$\langle\gamma^{-1}\gamma', [\gamma^{-1}\gamma', \gamma^{-1}\delta\gamma]\rangle = \langle[\gamma^{-1}\gamma', \gamma^{-1}\gamma'], \gamma^{-1}\delta\gamma\rangle = 0.$$

Corollary (8.9.5). *The critical points of the energy function $\mathscr{E}$ on ΩG are the loops γ which are homomorphisms $\gamma: S^1 \to G$.*

Proof. The critical points are precisely the stationary points of the corresponding Hamiltonian flow. From the formula (8.9.4) we find that $R_\alpha \gamma = \gamma$ for all α if and only if γ is a homomorphism.

Our other example of a Hamiltonian flow is even simpler.

Proposition (8.9.6). *The flow on ΩG generated by $\xi \in \mathfrak{g}$ corresponds to the Hamiltonian function $F_\xi: \Omega G \to \mathbb{R}$ given by*
$$F_\xi(\gamma) = \frac{1}{2\pi} \int_0^{2\pi} \langle \xi, \gamma'(\theta)\gamma(\theta)^{-1}\rangle \, d\theta.$$

The proof of this is completely straightforward. Cf. [8].

Remark (8.9.7). Combining (8.9.3) and (8.9.6) we see that the Hamiltonian function corresponding to the twisted rotational flow on ΩG which was considered in the proof of (8.6.4) is the *tilted energy* given by
$$\tilde{\mathscr{E}}(\gamma) = \frac{q}{4\pi} \int_0^{2\pi} \left\|\gamma(\theta)^{-1}\gamma'(\theta) - \frac{1}{q}\rho\right\|^2 d\theta.$$

This function has isolated critical points: they are the homomorphisms $\gamma: S^1 \to T$.

It is our object to investigate the Morse theory of the energy function on ΩG, i.e. the trajectories of the gradient flow of $\mathscr{E}$. For this to make sense, the manifold ΩG must be given a Riemannian structure. There are very many invariant Riemannian metrics on ΩG, but the choice is fixed for us by the fact that ΩG is a *complex* manifold: in Section 8.6 we proved that ΩG can be regarded as $LG_\mathbb{C}/L^+G_\mathbb{C}$, and is locally diffeomorphic to $L_1^- G_\mathbb{C}$.

Proposition (8.9.8). *The complex structure and the symplectic structure of ΩG are compatible, and combine to make ΩG into a Kähler manifold.*

This means that if T_γ is the real tangent space to ΩG at γ, and J_γ is the automorphism of T_γ which corresponds to multiplication by i in terms of the complex structure of ΩG, then
  (i) $\omega_\gamma(J_\gamma\xi, J_\gamma\eta) = \omega_\gamma(\xi, \eta)$ for all $\xi, \eta \in T_\gamma$,

8 THE FUNDAMENTAL HOMOGENEOUS SPACE

(ii) $(\xi, \eta) \mapsto g_\gamma(\xi, \eta) = \omega_\gamma(\xi, J_\gamma \eta)$ is a *positive definite inner product* on T_γ.

The Kähler form on T_γ is then given by
$$(\xi, \eta) \mapsto g_\gamma(\xi, \eta) + i\omega_\gamma(\xi, \eta).$$

Proof. Because both the complex structure and the form ω are invariant, it suffices to prove (i) and (ii) when $\gamma = 1$. If ξ is expanded as $\xi = \sum \xi_k z^k$, with $\xi_k \in \mathfrak{g}_\mathbb{C}$, then the action of J_1 on ξ multiplies ξ_k by i when $k < 0$ and by $-i$ when $k > 0$. (The constant term ξ_0 is to be disregarded, as we are really working in $L\mathfrak{g}/\mathfrak{g}$.) Now
$$\omega(\xi, \eta) = i \sum_{k \in \mathbb{Z}} k \langle \xi_{-k}, \eta_k \rangle,$$
where the inner product has been extended to a complex bilinear form on $\mathfrak{g}_\mathbb{C}$. This gives us both (i) and (ii), for
$$g(\xi, \xi) = \omega(\xi, J_1 \xi) = 2 \sum_{k>0} k \langle \xi_{-k}, \xi_k \rangle \geq 0.$$

For a function F on a Kähler manifold the gradient flow of F is related to its Hamiltonian flow simply by applying the operators J_γ in the tangent bundle: the two flows are in fact the real and imaginary parts of a flow parametrized by $\mathbb{C}$. Now we have already seen in Section 7.6 and Section 8.6 that the rotation-action of $\mathbb{T}$ on the Grassmannian $\text{Gr}(H)$ and on ΩG extends to a holomorphic action of the semigroup $\mathbb{C}^\times_{\leq 1}$. (Notice that we have by now considered three different actions of $\mathbb{T}$ on H. The action studied in Section 7.6 came from the identification of H with $L^2(S^1; \mathbb{C})$. The one we are concerned with now comes from $H \cong H^\mathfrak{g} = L^2(S^1; \mathfrak{g}_\mathbb{C})$. There is also the twisted version of this which was used in Section 8.6, where the rotation was combined with conjugation by the elements of a one-parameter subgroup of G. All three are diagonal with respect to the standard orthonormal basis of H, and in each case $R_u \in \mathbb{T}$ multiplies the kth basis vector by a power of u which increases with k and tends to ∞ as $k \to \infty$.) We can now reinterpret the discussion in Sections 7.6 and 8.6, and in particular the proof of (8.6.4).

First let us recall [115] that a vector field on an infinite dimensional manifold does not always possess trajectories, and even when it does they need not be unique. The theory which we have developed can be summarized as follows.

Theorem (8.9.9).

(i) *There is a downwards trajectory of the energy function emanating from every point γ of ΩG. It is given by*
$$t \mapsto \gamma_t = R_{e^{-t}} \gamma$$
for $t \geq 0$.

8.9 CLASSICAL AND QUANTUM-MECHANICAL ENERGY

(ii) *The loop γ_t is real-analytic when $t > 0$, and it converges to a homomorphism $\gamma_\infty : S^1 \to G$ as $t \to \infty$.*

(iii) *There is an upwards trajectory $t \mapsto \gamma_t$ defined for non-zero time $\varepsilon < t \leq 0$ if and only if γ is real-analytic. It is defined for all $t \leq 0$ if and only if γ is a polynomial loop, and in that case γ_t converges to a homomorphism $\gamma_{-\infty}$ as $t \to -\infty$.*

(iv) *The sets $\Sigma_{|\lambda|}$ and $C_{|\lambda|}$ of Proposition (8.6.5) are the ascending and descending manifolds of the critical level Λ_λ of the energy; i.e.*

$$\gamma_t \in \Sigma_{|\lambda|} \Leftrightarrow \gamma_t \to \gamma_\infty \in \Lambda_\lambda \quad \text{as} \quad t \to \infty.$$

$$\gamma_t \in C_{|\lambda|} \Leftrightarrow \gamma_t \to \gamma_{-\infty} \in \Lambda_\lambda \quad \text{as} \quad t \to -\infty.$$

It is interesting to specialize this result to the case $G = \mathbb{T}$. The identity component of $\Omega \mathbb{T}$ can be identified with the vector space of smooth functions $f : S^1 \to \mathbb{R}$. We have

$$\mathscr{E}(f) = \frac{1}{4\pi} \int_0^{2\pi} |f'(\theta)|^2 \, d\theta.$$

The downwards gradient flow of $\mathscr{E}$ is $\{f_t\}$, where $\{f_t\}$ is obtained by solving the parabolic pseudo-differential equation

$$\frac{\partial f_t}{\partial t} = (-\Delta)^{\frac{1}{2}} f_t, \tag{8.9.10}$$

where $\Delta = (\partial/\partial\theta)^2$. If f is expanded as $\Sigma \, a_k e^{ik\theta}$, then

$$f_t(\theta) = \Sigma \, a_k e^{-|k|t + ik\theta}.$$

The assertions (i), (ii), (iii) of (8.9.9) are quite clear in this case. But for a general group the equation corresponding to (8.9.10) is non-linear; and although our results are very plausible it would probably be rather difficult to prove them by direct methods.

Classical and quantum-mechanical energy

Combining the isomorphism $\Omega G \cong \text{Gr}^\Omega$ with the Plücker embedding of the Grassmannian (see Sections 7.5 and 7.7) we obtain a holomorphic embedding

$$\pi : \Omega G \to P(\mathscr{H}).$$

The rotation action of the circle on H^Ω induces an action on $\mathscr{H}$ generated by an unbounded hermitian operator $i(d/d\theta)$. Let us think of ΩG as a classical state-space, and $P(\mathscr{H})$ as the corresponding quantum state-space. For any loop γ let us choose a unit vector Ω_γ belonging to the ray $\pi(\gamma)$ in $\mathscr{H}$.

152   8 THE FUNDAMENTAL HOMOGENEOUS SPACE

Proposition (8.9.11). *We have*

$$\mathscr{E}(\gamma) = \left\langle \Omega_\gamma, \mathrm{i}\frac{\mathrm{d}}{\mathrm{d}\theta}\Omega_\gamma \right\rangle,$$

where the classical energy $\mathscr{E}(\gamma)$ is defined using the Killing form on $\mathfrak{g}$.

Notice that the right-hand-side is the expected value of the quantum energy operator $\mathrm{i}\,(\mathrm{d}/\mathrm{d}\theta)$ in the state Ω_γ.

Proof. The result follows from the fact that the canonical Kähler structure of $P(\mathscr{H})$ induces the Kähler structure on ΩG corresponding to the Killing form on $\mathfrak{g}$. This, in turn, is true because the Kähler form of $P(\mathscr{H})$ restricts to the standard $U_{\mathrm{res}}(H^\mathfrak{a})$-invariant form on $\mathrm{Gr}(H^\mathfrak{a})$, which corresponds to the basic Lie algebra 2-cocycle of $U_{\mathrm{res}}(H^\mathfrak{a})$. The latter restricts to the basic 2-cocycle of $\Omega\mathfrak{u}_n$, and so—via the embedding of G in U_n by the adjoint action—to the 2-cocycle of ΩG associated to the Killing form.

Remark. It is instructive to derive (8.9.11) more directly from the formula (7.8.4), i.e. to prove that

$$\mathscr{E}(\gamma) = \mathrm{trace}\left\{\mathrm{i}\frac{\mathrm{d}}{\mathrm{d}\theta}\cdot(\gamma J\gamma^{-1} - J)\right\},$$

where J is the Hilbert transform of (6.3.2). It is easy to check that the operator on the right has the kernel

$$-\frac{1}{2\pi}\frac{\partial}{\partial\theta}\{\cot\tfrac{1}{2}(\theta - \phi)\cdot\gamma(\phi)^{-1}(\gamma(\theta) - \gamma(\phi))\}$$

at $(\theta, \phi) \in S^1 \times S^1$. When $\theta = \phi$ this reduces to

$$\frac{1}{4\pi}\gamma(\theta)\gamma''(\theta),$$

from which (8.9.11) follows by taking the trace and integrating by parts.

8.10 ΩG and holomorphic bundles

The factorization theorems for loops give us a description of points of ΩG as holomorphic bundles on the Riemann sphere $S^2 = \mathbb{C} \cup \infty$. (We shall write $S^2 = D_0 \cup D_\infty$, where $D_0 = \{z : |z| \leq 1\}$ and $D_\infty = \{z : |z| \geq 1\}$.)

Proposition (8.10.1). *A point of ΩG is the same thing as an isomorphism class of pairs (P, τ), where P is a holomorphic principal $G_\mathbb{C}$-bundle on S^2, and τ is a trivialization of P over D_∞. The stratum to which (P, τ) belongs—in the sense of (8.6.5)—is simply the isomorphism class of P.*

8.10 ΩG AND HOLOMORPHIC BUNDLES

Here a 'trivialization over D_∞' means a smooth cross-section of $P \mid D_\infty$ which is holomorphic over the interior of D_∞.

Proof of (8.10.1). Given (P, τ) we choose a trivialization σ of $P \mid D_0$. The transition function between σ and τ over the intersection $S^1 = D_0 \cap D_\infty$ is an element of $LG_\mathbb{C}$. But σ is indeterminate up to multiplication by an element of $L^+G_\mathbb{C}$, so we have an element of $LG_\mathbb{C}/L^+G_\mathbb{C} = \Omega G$.

Conversely, given $\gamma \in LG$, we can choose a factorization $\gamma = \gamma_- \cdot \lambda \cdot \gamma_+$, where $\lambda: S^1 \to G$ is a homomorphism. Because λ extends to a holomorphic map $\lambda: \mathbb{C}^\times \to G_\mathbb{C}$ it defines a holomorphic bundle P_λ on S^2 (see Section 8.2(ii)) which is canonically trivial over $S^2 - \{\infty\}$ and $S^2 - \{0\}$. We assign to γ the pair (P_λ, τ), where $\tau = \gamma_+ \cdot \tau_\infty$, and τ_∞ is the canonical trivialization of $P_\lambda \mid D_\infty$.

A complex manifold X is completely described by giving the set of holomorphic maps $M \to X$ for every complex manifold M. So ΩG is completely described as a complex manifold by the following simple generalization of (8.10.1), which was pointed out to us by Atiyah (see [5]).

Proposition (8.10.2). *A holomorphic map $M \to \Omega G$, for any complex manifold M, is the same thing as an isomorphism class of pairs (P, τ), where P is a holomorphic principal $G_\mathbb{C}$-bundle on $M \times S^2$, and τ is a trivialization of $P \mid M \times D_\infty$.*

Proof. First suppose that we are given (P, τ). By Proposition (8.10.1) we know that $m \mapsto (P, \tau) \mid (m \times S^2)$ defines a map $f: M \to \Omega G$. We must show that f is holomorphic. But for any $m \in M$ we can suppose that P is trivial over a set of the form $V \times \tilde{D}_0$, where V is a neighbourhood of m in M and $\tilde{D}_0$ is $\{z \in S^2 : |z| \leq r\}$ for some $r > 1$. Then $f \mid V$ can be represented by the transition function between τ and a trivialization over $V \times \tilde{D}_0$. This is a smooth map

$$V \times \{z: 1 \leq |z| \leq r\} \to G_\mathbb{C}$$

which is holomorphic for $1 < |z| < r$. Its restriction to $V \times S^1$ is therefore a holomorphic map $V \to LG_\mathbb{C}$, as we want.

Conversely, to obtain a pair (P, τ) from each map $f: M \to \Omega G$ it is enough to define a pair (P, τ) over $\Omega G \times S^2$ itself. The definition of P as a smooth bundle is clear: it is constructed by attaching trivial bundles on $\Omega G \times D_0$ and $\Omega G \times D_\infty$ by the clutching function given by the evaluation map $\varepsilon: \Omega G \times S^1 \to G$. The resulting bundle has a canonical trivialization over $\Omega G \times D_\infty$. If we had taken ΩG to be the space of *real-analytic* loops instead of smooth ones it would be clear that P was a holomorphic bundle, and the proof would be complete: for in that case ε would extend to a holomorphic map $W_0 \cap W_\infty \to G_\mathbb{C}$, where W_0 and W_∞ are suitable open neighbourhoods of $\Omega G \times D_0$ and $\Omega G \times D_\infty$ in $\Omega G \times S^2$. But to treat the *smooth* loops ΩG requires more care.

We recall from (8.6.3) that ΩG is covered by open sets U_λ such that $U_\lambda \cong \Sigma_\lambda \times C_\lambda$ as a complex manifold, and $C_\lambda \subset \Omega_{\text{pol}} G$. Thus C_λ is a space of holomorphic maps $\mathbb{C}^\times \to G_\mathbb{C}$. Composing the projection $U_\lambda \to C_\lambda$ with the evaluation map gives us a holomorphic map $U_\lambda \times \mathbb{C}^\times \to G_\mathbb{C}$ which defines a holomorphic bundle P_λ on $U_\lambda \times S^2$. Proposition (8.6.3) implies that P_λ is *canonically* isomorphic to $P \mid U_\lambda \times S^2$ as a smooth bundle. But the bundles P_λ fit together to define a holomorphic bundle on $\Omega G \times S^2$. That is the case because the canonical smooth isomorphism between the restrictions of P_λ and P_μ to $(U_\lambda \cap U_\mu) \times S^2$ is easily seen to be holomorphic when $|z| < 1$ and when $|z| > 1$, and is therefore holomorphic everywhere.

If M is a compact manifold then the trivialization τ of $P \mid M \times D_\infty$ is unique up to the action of a single element of $L^+ G_\mathbb{C}$, for any holomorphic map $M \to L^+ G_\mathbb{C}$ is constant. That gives us

Proposition (8.10.3). *If M is a compact complex manifold with a base point m_0 then the set of base-point-preserving holomorphic maps $M \to \Omega G$ can be identified with the set of isomorphism classes of $G_\mathbb{C}$-bundles P on $M \times S^2$ which are trivial over $m_0 \times S^2$ and $M \times D_\infty$.*

This result is the starting point of a circle of interesting ideas for which we refer the reader to Atiyah [5]. We shall content ourselves with pointing out one immediate corollary.

Proposition (8.10.4). *If M is a compact complex manifold with a base-point then each connected component of the space of base-point-preserving holomorphic maps $M \to \Omega G$ is finite dimensional.*

This follows from (8.10.3), because the space of all holomorphic bundles of a given topological type on a compact manifold is finite dimensional. (See Mumford and Fogarty [119].)

Proposition (8.10.4) reveals a striking difference between ΩG or $\text{Gr}^{(n)}(H)$ and more familiar infinite dimensional complex manifolds such as the projective space $P(H)$ or the Grassmannian $\text{Gr}(H)$ of Chapter 7. For example, if $v \in H$ represents the base-point in $P(H)$ then

$$(z_0, z_1) \mapsto [z_0 v + z_1 w]$$

is a family of base-point preserving holomorphic maps $S^2 \to P(H)$ which is parametrized by the space of non-zero vectors w orthogonal to v. A similar family can be defined for $\text{Gr}(H)$.

8.11 The homogeneous space associated to a Riemann surface: the moduli spaces of vector bundles

Throughout the last three chapters we have always polarized the Hilbert space $H = L^2(S^1; \mathbb{C})$ as $H_+ \oplus H_-$, where H_+ is the space of boundary

8.11 MODULI SPACES OF VECTOR BUNDLES

values of holomorphic functions in the disc $|z|<1$. A rather natural generalization of this procedure is to replace the disc by some other Riemann surface whose boundary is a circle.

Suppose then that X is a compact Riemann surface with a distinguished point x_∞ and a given local parameter around x_∞. We shall write the local parameter as z^{-1}: thus z is a holomorphic map from a neighbourhood of x_∞ to a neighbourhood of ∞ in the Riemann sphere. We shall assume that $z(x_\infty) = \infty$, and that z is an isomorphism between a neighbourhood of x_∞ and the region $|z| > \frac{1}{2}$ on the Riemann sphere. The standard circle S^1 can then be identified with the circle $|z| = 1$ around x_∞ on X. We shall denote the part of X where $|z|>1$ by X_∞, and the complement of the region where $|z| \geq 1$ by X_0. Thus

$$\bar{X}_0 \cap \bar{X}_\infty = S^1.$$

Associated to these data there is a subspace $H_X^{(n)}$ of $H^{(n)} = L^2(S^1; \mathbb{C}^n)$ analogous to $H_+^{(n)}$. It is the closed subspace of $H^{(n)}$ consisting of the boundary values of holomorphic maps $X_0 \to \mathbb{C}^n$.

Proposition (8.11.1). *The space $H_X^{(n)}$ belongs to* Gr, *and has virtual dimension* $-ng$, *where g is the genus of X.*

We shall postpone the proof of this, and shall go on to characterize the orbit of $H_X^{(n)}$ under the complex loop group $LGL_n(\mathbb{C})$ in analogy with the description of $\text{Gr}^{(n)}$ in (8.3.1) as $\{W \in \text{Gr}(H^{(n)}): zW \subset W\}$.

Let A_X denote the ring of rational functions on X which are holomorphic everywhere except for a pole of arbitrary order at x_∞. Using the local parameter z we think of A_X as a ring of functions of z. As such it acts by multiplication operators on the Hilbert space $H^{(n)}$, and hence on $\text{Gr}(H^{(n)})$.

Definition (8.11.2).

$$\text{Gr}^{(n),X} = \{W \in \text{Gr}(H^{(n)}): A_X.W \subset W\}.$$

If X is the Riemann sphere then A_X is the polynomial ring $\mathbb{C}[z]$, and then $\text{Gr}^{(n),X}$ coincides with $\text{Gr}^{(n)}$. We shall prove that $\text{Gr}^{(n),X}$ is always a homogeneous space of $LGL_n(\mathbb{C})$.

Let us recall, to begin with, that A_X is a finitely generated algebra which is filtered by the order of the pole at x_∞. It contains, up to a scalar multiple, exactly one element of each sufficiently large degree. More precisely, if $A_X^{(d)}$ is the vector space of functions in A_X with poles of orders $\leq d$, then the Weierstrass 'gap' theorem ([68] p. 273) asserts that $A_X^{(d)}$ has dimension $d + 1 - g$, where g is the genus of X, providing $d \geq 2g$.

Proposition (8.11.3). $\text{Gr}^{(n),X}$ *is the orbit of $H_X^{(n)}$ under $LGL_n(\mathbb{C})$.*

Proof. Suppose that W belongs to $\text{Gr}^{(n),X}$. The subspace W^{fin} of

8 THE FUNDAMENTAL HOMOGENEOUS SPACE

functions of finite order in W, which is dense in W by Proposition (7.3.2), is a module over A_X. Like A_X the module W^{fin} has an increasing filtration $\{W^{(k)}\}$ by the order of the pole at x_∞. We know that $\dim(W^{(k)}/W^{(k-1)}) = n$ for all large k, and it follows that W^{fin} is a finitely generated A_X-module. It is obviously torsion-free, and hence projective (because A_X is a Dedekind ring). This means that it is the module of algebraic sections of an algebraic vector bundle E on $X - \{x_\infty\}$ whose fibre E_x at x is $W^{\text{fin}}/\mathfrak{a}_x W^{\text{fin}}$, where $\mathfrak{a}_x = \{f \in A_X : f(x) = 0\}$. As a holomorphic bundle E is necessarily trivial, for there are no non-trivial holomorphic vector bundles on an affine curve. Thus there are holomorphic sections $w_1, \ldots, w_n$ in W whose values at any point $x \in X - \{x_\infty\}$ span the fibre E_x. In particular the matrix $(w_1(z), \ldots, w_n(z))$ is invertible at each point z of the circle, and defines a loop $\gamma : S^1 \to GL_n(\mathbb{C})$ such that $\gamma H_X^{(n)} = W$.

The stabilizer of $H_X^{(n)}$ in $LGL_n(\mathbb{C})$ is obviously the group $L_X^+ GL_n(\mathbb{C})$ of loops which are the boundary values of holomorphic maps $X_0 \to GL_n(\mathbb{C})$. So the preceding proposition gives us

$$\text{Gr}^{(n), X} \cong LGL_n(\mathbb{C})/L_X^+ GL_n(\mathbb{C}). \tag{8.11.4}$$

The proof of the proposition shows also that a point of $\text{Gr}^{(n), X}$ can be identified with an isomorphism class of pairs (E, α), where E is a holomorphic vector bundle on X and α is a trivialization of $E \mid X_\infty$ which extends smoothly to $\bar{X}_\infty$. The natural action of $L^- GL_n(\mathbb{C})$ on $\text{Gr}^{(n), X}$ permutes the trivializations α transitively, so we have the following generalization of the Birkhoff factorization theorem.

Proposition (8.11.5). *The set of double cosets*

$$L^- GL_n(\mathbb{C}) \backslash LGL_n(\mathbb{C}) / L_X^+ GL_n(\mathbb{C})$$

is the set of isomorphism classes of n-dimensional holomorphic vector bundles on X.

It is in fact more convenient and usual to consider the double coset space $L_1^- \backslash L / L_X^+$. This is the set of isomorphism classes of bundles E with a chosen identification of the fibre at x_∞ with $\mathbb{C}^n$. It is a better space than $L^- \backslash L / L_X^+$ because L_1^- acts freely on a dense open set $\mathcal{U}$ of $L/L_X^+ = \text{Gr}_X^{(n)}$, and is a contractible group. The quotient space, $L_1^- \backslash \mathcal{U}$, which is homotopy equivalent to $\mathcal{U}$, is the *moduli space* of n-dimensional bundles in the sense of Mumford [119]. (Cf. also Atiyah and Bott [7].)

It is interesting to consider the homotopy type of the group L_X^+ and the homogeneous space $\text{Gr}_X^{(n)}$. Any loop γ belonging to L_X^+ has winding number zero, i.e. is contractible to a point, for if $\det(\gamma)$ is the boundary value of a holomorphic function in X_0 then its winding number is the number of zeros of $\det(\gamma)$ in X_0. On the other hand L_X^+ is not connected, for $\det(\gamma)$ has a well-defined integral winding number around each

8.11 MODULI SPACES OF VECTOR BUNDLES

non-trivial loop in X_0. In fact the group of components of L_X^+ is given by

$$\pi_0(L_X^+) \cong \mathbb{Z}^{2g} \cong \mathrm{Hom}(\pi_1(X_0); \mathbb{Z}) \cong H^1(X; \mathbb{Z}),$$

where g is the genus of X. This is the group of homotopy classes of maps $X_0 \to \mathbb{C}^\times$, or equivalently of maps $X_0 \to GL_n(\mathbb{C})$. An even stronger result is true, namely

Proposition (8.11.6).

(i) *The group $L_X^+ GL_n(\mathbb{C})$ is homotopy equivalent to the group of continuous maps $X_0 \to GL_n(\mathbb{C})$.*

(ii) *The space $\mathrm{Gr}_X^{(n)}$ is homotopy equivalent to the space of base-point preserving maps $X \to BGL_n(\mathbb{C})$, i.e. to the space of topological $\mathbb{C}^n$-bundles on X (with the fibre at x_∞ identified with $\mathbb{C}^n$).*

In part (ii) of this proposition $BGL_n(\mathbb{C})$ denotes the classifying space of the group $GL_n(\mathbb{C})$. Thinking of the space of maps $X \to BGL_n(\mathbb{C})$ as the 'space' of vector bundles on X can be justified from various points of view. The essential point is that for a given bundle E on X the space of pairs (f, α), where $f: X \to BGL_n(\mathbb{C})$ and $\alpha: E \to f^* E_{\mathrm{univ}}$ is an isomorphism between E and the pull-back of the universal bundle E_{univ} on $BGL_n(\mathbb{C})$, is contractible. This implies that the space of maps $\mathrm{Map}(X; BGL_n)$ is homotopy equivalent to the 'space' or 'realization' of the category of vector bundles on X in the sense of [128]. More explicitly, the space of maps has one connected component for each isomorphism class of bundles on X, and the component corresponding to a bundle E has the homotopy type of $B\,\mathrm{Aut}(E)$, the classifying space of the 'gauge group' $\mathrm{Aut}(E)$ of all automorphisms of E. For more information about this space we refer to Atiyah and Bott [7].

Assertion (ii) above follows immediately from (i). For L/L_X^+ is homotopy equivalent to the fibre of $BL_X^+ \to BL$, i.e. to the fibre of

$$\mathrm{Map}(X_0; BGL_n) \to \mathrm{Map}(S^1; BGL_n).$$

The cofibration sequence $S^1 \to X_0 \to X$ shows that this is $\mathrm{Map}(X; BGL_n)$.

Proof of (i). In order not to go too far afield we shall content ourselves with indicating how the statement follows simply from the results of [130].

The proof is by induction on n. Let us first consider the case $n = 1$. We have to prove that $\mathrm{Hol}(X_0; \mathbb{C}^\times)$ is homotopy equivalent to $\mathrm{Map}(X_0; \mathbb{C}^\times)$, where Hol denotes the holomorphic maps which extend smoothly to $\bar{X}_0$, and Map denotes continuous maps. (It is permissible, and more convenient, to replace $\mathrm{Map}(X_0; \mathbb{C}^\times)$ by $\mathrm{Map}(\bar{X}_0; \mathbb{C}^\times)$.) The exact sequence of groups

$$0 \to \mathbb{Z} \to \mathrm{Hol}(X_0; \mathbb{C}) \xrightarrow{\exp} \mathrm{Hol}(X_0; \mathbb{C}^\times) \to H^1(X_0; \mathbb{Z}) \quad (8.11.7)$$

shows that each connected component of $\text{Hol}(X_0; \mathbb{C}^\times)$ has the homotopy type of a circle, as is also the case for the continuous maps. It is therefore enough to prove that the right hand map in (8.11.7) is surjective. That is true because its cokernel is $H^1(X_0; \mathcal{O})$, where $\mathcal{O}$ is the sheaf of germs of holomorphic functions on X_0. This group vanishes because X_0 is a Stein manifold.

For the inductive step we consider the holomorphic fibration

$$GL_{n-1,1} \to GL_n \to P^{n-1},$$

where $GL_{n-1,1}$ is the group of echelon matrices which is the stabilizer of a one-dimensional subspace of $\mathbb{C}^n$, and P^{n-1} is $(n-1)$-dimensional complex projective space. To prove that $\text{Hol}(X_0; GL_n)$ is homotopy equivalent to $\text{Map}(X_0; GL_n)$ it is enough to show that the sequence

$$\text{Hol}(X_0; GL_{n-1,1}) \to \text{Hol}(X_0; GL_n) \to \text{Hol}(X_0; P^{n-1}) \quad (8.11.8)$$

is a fibration (e.g. to show that it has local cross-sections) and also that the inclusions

$$\text{Hol}(X_0; GL_{n-1,1}) \to \text{Map}(\bar{X}_0; GL_{n-1,1})$$

and

$$\text{Hol}(X_0; P^{n-1}) \to \text{Map}(\bar{X}_0; P^{n-1}) \quad (8.11.9)$$

are equivalences; for the sequence of spaces of continuous maps analogous to (8.11.8) is trivially a fibration. Now $GL_{n-1,1}$ is isomorphic to $GL_{n-1} \times \mathbb{C}^\times \times \mathbb{C}^{n-1}$ as a complex manifold, so $\text{Hol}(X_0; GL_{n-1,1})$ is equivalent to the product

$$\text{Hol}(X_0; GL_{n-1}) \times \text{Hol}(X_0; \mathbb{C}^\times),$$

and hence to $\text{Map}(X_0; GL_{n-1,1})$ by the inductive hypothesis. On the other hand it is proved in [130] that the map (8.11.9) is an equivalence. (Strictly speaking, the proof in [130] applies to the holomorphic maps $f: X_0 \to P^{n-1}$ which extend *real-analytically* to X_0, for it supposes that the homogeneous coordinates $f_1, \ldots, f_n$ of f have only finitely many zeros. But it is easy to see that the space of sets of n subsets $S_1, \ldots, S_n$ of $\bar{X}_0$ which have empty intersection and no points of accumulation in X_0 is homotopy equivalent to the space $Q^{(n)}(\bar{X}_0)$ of [130] by the map $S_i \mapsto S_i \cap \bar{X}_0^\varepsilon$, where $\bar{X}_0^\varepsilon$ is obtained from $\bar{X}_0$ by deleting a collar of small width ε around the boundary.)

It remains to explain why (8.11.8) is a fibration. The second map is surjective because a holomorphic map $X_0 \to P^{n-1}$ is the same thing as a holomorphic line sub-bundle L of $X \times \mathbb{C}^n$. To lift the map to GL_n is to find an isomorphism between the exact sequence

$$L \to X_0 \times \mathbb{C}^n \to (X_0 \times \mathbb{C}^n)/L$$

8.11 MODULI SPACES OF VECTOR BUNDLES

and the trivial sequence

$$X_0 \times \mathbb{C} \to X_0 \times \mathbb{C}^n \to X_0 \times \mathbb{C}^{n-1}.$$

This can be done because X_0 is a Stein manifold. In view of the surjectivity and the fact that the total space of the bundle is a group, it is enough to prove that (8.11.8) has a local cross-section near a constant map $X_0 \to P^{n-1}$. But that is obvious because $GL_n \to P^{n-1}$ has holomorphic cross-sections.

Proof of (8.11.1). We conclude this section by giving the omitted proof of Proposition (8.11.1). We shall actually prove a slightly more general result.

Proposition (8.11.10). *Let E be an n-dimensional holomorphic vector bundle on X with a given trivialization in a neighbourhood of X_∞. Let W be the closed subspace of $H^{(n)} = L^2(S^1; \mathbb{C}^n)$ consisting of functions which are the boundary values of holomorphic sections of E over X_0. Then W belongs to $\mathrm{Gr}_\omega(H^{(n)})$, and its virtual dimension is*

$$\dim H^0(X; \mathscr{E}) - \dim H^1(X; \mathscr{E}) - n,$$

where $\mathscr{E}$ is the sheaf of holomorphic sections of E. In fact $H^0(X; \mathscr{E})$ and $H^1(X; \mathscr{E})$ are respectively the kernel and cokernel of the orthogonal projection $W \to zH_+$.

Proof. We observe first that the projection $\mathrm{pr}: W \to H_-$ factorizes

$$W \xrightarrow{R_{\rho^{-1}}} H^{(n)} \xrightarrow{\mathrm{pr}} H_- \xrightarrow{R_\rho} H_-,$$

for some ρ such that $0 < \rho < 1$. Here R_ρ is the operator of Section 7.6 such that $R_\rho z^k = \rho^{-k} z^k$; the operator $R_{\rho^{-1}}: W \to H^{(n)}$ is bounded because it assigns to the boundary value of a holomorphic section ϕ of $\mathscr{E}$ over X the function $z \mapsto \phi(\rho z)$, i.e. the value of ϕ on a circle slightly inside the boundary of X_0. The operator $R_\rho: H_- \to H_-$ is compact, so the projection $\mathrm{pr}: W \to H_-$ is compact. It follows that the projection $W \to H_+$ has closed image.

Now let U_0 and U_∞ be open sets of X slightly larger than $\bar{X}_0$ and $\bar{X}_\infty$. Because U_0 and U_∞ are Stein manifolds the kernel and cokernel of the map

$$\mathscr{E}(U_0) \oplus \mathscr{E}(U_\infty) \to \mathscr{E}(U_0 \cap U_\infty)$$

taking (ϕ_0, ϕ_∞) to $\phi_0 - \phi_\infty$ can be identified with $H^0(X; \mathscr{E})$ and $H^1(X; \mathscr{E})$. Passing to the direct limit as U_0 and U_∞ shrink towards $\bar{X}_0$ and $\bar{X}_\infty$ we find that the same groups are the kernel and cokernel of

$$W^{\mathrm{an}} \oplus zH_-^{\mathrm{an}} \to (H^{(n)})^{\mathrm{an}},$$

and hence of

$$\mathrm{pr}: W^{\mathrm{an}} \to zH_+^{\mathrm{an}}. \qquad (8.11.11)$$

(Here W^{an} denotes the set of real-analytic functions in W, and so on.) The kernel of (8.11.11) is the same as that of $\mathrm{pr}: W \to zH_+$, for an element of the kernel of the latter is the common boundary value of holomorphic functions in X_0 and X_∞. But we know that $W \to zH_+$ has closed image, so its cokernel must also coincide with the cokernel of (8.11.11). This essentially completes the proof: it remains only to observe that W belongs to Gr_ω because it is of the form $R_\rho \tilde{W}$ for some $\rho < 1$, where $\tilde{W}$ is the analogue of W constructed from the circle $|z| = \rho$ on X.

8.12 Appendix: Scattering theory

The Grassmannian interpretation of loop groups arises in the approach to 'scattering theory' developed by Lax and Phillips [99]. We shall give a very brief account of its role there.

Suppose that we are studying the solutions of a wave equation

$$\frac{\partial^2 \psi}{\partial t^2} - \frac{\partial^2 \psi}{\partial x^2} + \rho(x)\psi = 0, \qquad (8.12.1a)$$

where ψ is a complex-valued function of x and t, and ρ is a positive real-valued function independent of t which vanishes outside a finite interval. We think of the equation as describing waves which are scattered by an obstacle described by ρ. Intuitively it seems plausible that if a solution ψ is fairly well localized in space at time $t = 0$ then after a long period the solution will (in the sense of its energy, to be defined presently) be concentrated mainly in the region where ρ is zero. That is to say, we expect that an arbitrary solution of (8.12.1a) will, for large positive t, be close to a definite solution of the 'unperturbed equation'

$$\frac{\partial^2 \psi}{\partial t^2} - \frac{\partial^2 \psi}{\partial x^2} = 0. \qquad (8.12.1b)$$

We expect that the same thing will be true also for large negative t.

Now let V be the vector space of solutions of (8.12.1a) and V_0 that of (8.12.1b): to begin with we shall consider solutions which have compact support in x for each t. We expect that there will be two isomorphisms $T_\pm: V \to V_0$ which assign to a solution ψ the solutions of the unperturbed equation to which ψ is asymptotic as $t \to \pm\infty$. The composite

$$S = T_+ \circ T_-^{-1}: V_0 \to V_0$$

is called the *scattering matrix* of the original equation; from one point of view it obviously gives a good description of the behaviour of the solutions. (It would not be reasonable to expect T_+ and T_- to be isomorphisms if the equation (8.12.1a) had 'bound states', i.e. if the operator $-\partial^2/\partial x^2 + \rho$ had negative eigenvalues; but that is excluded by the positivity of ρ.)

8.12 APPENDIX: SCATTERING THEORY

Because ρ is independent of t there is a one-parameter group of transformations $\{U_t\}$ of V defined by time-translation. This action preserves the energy-norm

$$\|\psi\|_\rho^2 = \frac{1}{2}\int_{-\infty}^{\infty}\left\{\left(\frac{\partial\psi}{\partial t}\right)^2 + \left(\frac{\partial\psi}{\partial x}\right)^2 + \rho\psi^2\right\}dx;$$

here the integral is taken along any line $t = $ constant. We can complete V using this norm to get a Hilbert space H with a unitary group of transformations; and V_0 can be completed similarly to get H_0. The transformations $T_\pm$ are isometries, and they clearly commute with time translations; so S is a unitary transformation which commutes with time translation.

A solution of (8.12.1b) can be analysed by making a Fourier transformation in t:

$$\psi(t, x) = \int_{-\infty}^{\infty} \psi_\omega(x)e^{i\omega t}\,d\omega. \tag{8.12.2}$$

Here ψ_ω belongs to the two-dimensional space of solutions of

$$\psi_\omega'' + \omega^2\psi_\omega = 0,$$

which can be identified with $\mathbb{C}^2$ by mapping ψ_ω to $(\psi_\omega(0), \psi_\omega'(0))$. Thus H_0 can be identified with the Hilbert space $L^2(\mathbb{R}; \mathbb{C}^2)$ in such a way that the time translation U_t in H_0 is given by multiplication by the function $e^{i\omega t}$, where ω is the coordinate in $\mathbb{R}$. By a simple variant of Proposition (6.1.1) we know that the unitary transformations of H_0 which commute with all U_t are the group of measurable maps $\text{Map}_{\text{meas}}(\mathbb{R}; U_2)$. The scattering matrix S is therefore an element of this group, which is a kind of loop group. (If S corresponds to a map $\sigma: \mathbb{R} \to U_2$ then $\sigma(\omega)$ describes the scattering of waves of frequency ω. As very high-frequency waves are comparatively unaffected by the potential ρ we shall have $\sigma(\omega) \to 1$ as $\omega \to \pm\infty$, which justifies our regarding σ as a loop.)

The relevance of this discussion to the Grassmannian model of loops comes from the theorem that to give an isomorphism between a Hilbert space H with a unitary group $\{U_t\}$ and a standard space $L^2(\mathbb{R}; K)$ with its multiplication group $\{e^{i\omega t}\}$ is precisely the same thing as to prescribe what is called an *outgoing subspace* in H. (Here K is an unspecified auxiliary Hilbert space.) The standard outgoing subspace H_0^+ in H_0 is the closure of the solutions ψ such that $\psi(t, 0) = 0$ for $t < 0$. When H_0 is identified by the Fourier transform (8.12.2) with the $\mathbb{C}^2$-valued functions of ω the space H_0^+ consists of the boundary values of functions holomorphic in the half-plane $\text{Im}(\omega) < 0$.

Definition (8.12.3). *An outgoing subspace in a Hilbert space H with a*

8 THE FUNDAMENTAL HOMOGENEOUS SPACE

one-parameter unitary group $\{U_t\}$ *is a closed subspace W of H such that*

(i) $U_t(W) \subset W$ *when* $t \geq 0$,

(ii) $\bigcap_{t \geq 0} U_t(W) = 0$,

(iii) $\bigcup_{t \leq 0} U_t(W)$ *is dense in H*.

The basic theorem of Lax–Phillips scattering theory [99] is that when the data $\{H, \{U_t\}, W\}$ are given one can construct a Hilbert space K and a canonical isomorphism of the data with the standard data $\{H_0 = L^2(\mathbb{R}; K), \{e^{i\omega t}\}, H_0^+\}$. In other words, instead of giving the two maps $T_\pm : H \to H_0$ it is equally informative simply to prescribe two subspaces $W_\pm$ in H. Intuitively, W_+, which is mapped by T_+ to H_0^+, consists of 'outgoing waves', while W_- consists of 'incoming waves', and is mapped to $(H_0^+)^\perp$ by T_-.

We shall not prove the theorem here. The variant of it which is directly related to loop groups is that where the continuous group $\{U_t\}_{t \in \mathbb{R}}$ is replaced by a discrete group $\{u^k\}_{k \in \mathbb{Z}}$. This is very easy to prove. The standard model is then the space $H_0 = L^2(S^1; K)$, the group is generated by multiplication by z, and H_0^+ has the meaning which is usual in this book. When $\{H, u, W\}$ is given one can determine K as $W \ominus u(W)$. The theorem reduces essentially to the following.

Proposition (8.12.4). *Let K be a Hilbert space, and $U(K)$ its unitary group. Then the measurable loop group $\Omega_{\text{meas}} U(K) = L_{\text{meas}} U(K)/U(K)$ can be identified with the set of closed subspaces W of $L^2(S^1; K)$ such that*

(i) $zW \subset W$,

(ii) $\bigcap_{k \geq 0} z^k W = 0$,

(iii) $\bigcup_{k \leq 0} z^k W$ *is dense in H*.

The proof of (8.3.2) included a proof of this result. In fact the hardest step in proving (8.3.2) was to show that a space $W \in \text{Gr}(H^{(n)})$ satisfies the conditions of (8.12.4).

Remark. We have pointed out that there is no simple model for $\Omega_{\text{cts}} U_n$. The present result, however, shows that there is a—not very explicit—model for the space of measurable loops.

PART II

9

REPRESENTATION THEORY

In this chapter we begin our study of the representation theory of loop groups, to which the rest of the book is devoted. It has two aspects. On the one hand we want to have a uniform description of the set of irreducible representations of a general loop group—at present that can be done only for representations of positive energy—and on the other we are interested in explicit constructions of interesting representations of particular groups. This dichotomy is well known in the case of compact groups. The irreducible representations of a compact group are parametrized by the characters of its maximal torus, modulo the action of the Weyl group, and the Borel–Weil theorem (see Section 2.9) tells us that all the representations can be constructed in a uniform way as spaces of holomorphic sections of line bundles. On the other hand for the particular group U_n we have n fundamental representations given by its action on the exterior powers $\Lambda^k(\mathbb{C}^n)$, and we know that all representations can be constructed from these. (Better still, in that case, we can say that all the irreducible representations arise on spaces of tensors over $\mathbb{C}^n$ satisfying appropriate symmetry conditions, or, equivalently, that they are obtained by decomposing the tensor products $\mathbb{C}^n \otimes \mathbb{C}^n \otimes \ldots \otimes \mathbb{C}^n$ according to the action of the symmetric group. Cf. Weyl [154].) For the orthogonal group O_n we have to include the spin representation as well as the tensor representations to obtain all (projective) irreducible representations. All of these statements, even the relationship between the unitary and symmetric groups (see Proposition (10.6.4) below), have analogues for loop groups.

The present chapter, which is summarized below, is an introductory survey of the representation theory. To put the subsequent theory in perspective it begins with a brief account of some representations of Map$(X; G)$ when X is of dimension greater than one, and of some representations of LG which are not of the class we shall study. The remaining chapters are as follows.

In Chapter 10 we describe very explicitly a projective representation of the group $GL_{\text{res}}(H)$ introduced in Chapter 6. This restricts to give an irreducible representation of $\tilde{L}U_n$ which is called its *basic* representation. One can think of it as an analogue of the representations of U_n on the spaces $\Lambda^k(\mathbb{C}^n)$. The representation space is described in three different

ways:
  (i) as a space of holomorphic sections;
  (ii) as an exterior algebra; and
  (iii) as a sum of symmetric algebras.

The equivalence between the last two descriptions is the correspondence between bosons and fermions which has attracted attention in two-dimensional quantum field theory—cf. [29], [111], [45], [155], and Section 10.7 below.

Chapter 11 returns to the uniform theory, and describes the Borel–Weil theory for loop groups, which is closely analogous to that for compact groups. Most of the assertions in Chapter 9 are proved here.

Chapter 12 is a continuation of Chapter 10. The representation constructed there extends from $GL_{\text{res}}(H)$ to the restricted orthogonal group of the real Hilbert space underlying H: it is the *spin* representation of the latter. To explain this we begin with a fairly detailed description of the spin representation in the finite dimensional case, comparing the known classical constructions. We hope that this is of some interest in its own right: in particular we give an explicit description of the spin group (12.2.10) which does not seem to have been pointed out before. But as far as loop groups are concerned the essential contribution of this chapter is to construct the basic representation of $\tilde{L}O_{2n}$.

Chapter 13 is devoted to what have become known as 'vertex operators', although we prefer the term 'blips'. These arose in quantum field theory [80], from which their name derives, and they have no analogue in finite dimensional representation theory. They provide a very interesting construction of the basic representation of $\tilde{L}G$ when G is a simply laced group (cf. Section 2.5). They enable us to prove in general that positive energy representations of loop groups admit an intertwining action of the group of diffeomorphisms of the circle.

The final Chapter 14 discusses the Kac character formula and the Bernstein–Gelfand–Gelfand resolution. Our treatment of these topics is not what we would have desired. In the literature they have been approached only in a formal algebraic way which in our view fails to bring out their simple geometrical content and their beauty. To carry through a geometrical treatment, however, one would need a theory of complex analysis on the fundamental homogeneous space of Chapter 8 which has still not been developed, though we have little doubt of its feasibility. We have therefore contented ourselves with the compromise of describing what is known and what is probably true, while giving at the same time the standard algebraic proof of the character formula (due to Kac). We have said very little about the combinatorial implications of the

9.1 GENERAL REMARKS ABOUT REPRESENTATIONS

character formula (Macdonald's identity, etc.), although they were the motivation for much of the recent interest in loop groups, and are extensively treated in the literature.

We conclude this introduction by summarizing the present chapter. After the background discussion in Section 9.1 we introduce the fundamental idea of positive energy representations in Section 9.2. Then Section 9.3 lists the main properties of these representations and describes their classification in terms of characters of the maximal torus. On the whole we have preferred wherever possible to use global methods, but the use of Lie algebras can certainly not be avoided, and in that direction much the most important tool is the Casimir operator described in Section 9.4. At the same time we can see with very little extra work how the Lie algebra of the group of diffeomorphisms of the circle (called by physicists the 'Virasoro algebra') automatically acts on positive energy representations of loop groups. In fact this action can be integrated to give an intertwining action of the diffeomorphism group itself, but we shall prove that in Section 13.4 by completely different methods. The chapter concludes with an appendix describing the classical and well-known facts about the representations of finite and infinite dimensional Heisenberg groups. These are an essential background for the loop group theory.

9.1 General remarks about representations

Terminology

By a representation of a topological group G we shall always mean a complete locally convex complex topological vector space E on which G acts $\mathbb{C}$-linearly and continuously in the sense that $(g, \xi) \mapsto g \cdot \xi$ is a continuous map $G \times E \to E$. We are primarily interested in unitary representations on Hilbert spaces, but it is convenient to work in a more general framework. Thus the natural action of the circle group $\mathbb{T}$ on S^1 by rotation gives us representations of $\mathbb{T}$ on any of the following vector spaces:

the smooth functions $S^1 \to \mathbb{C}$,
the continuous functions $S^1 \to \mathbb{C}$,
any of the Banach spaces $L^p(S^1; \mathbb{C})$ for $1 \leq p < \infty$,
the space of distributions on S^1,

but *not* on $L^\infty(S^1; \mathbb{C})$.

We shall not want to regard these representations of $\mathbb{T}$ as interestingly different, so we shall make the definition that representations E and E' of a group G are *essentially equivalent* if there is a continuous linear map

$E \to E'$ which is injective, has dense image, and is equivariant with respect to G.†

A representation is called *irreducible* if it has no closed invariant subspace.

If E is a representation of a Lie group G then a vector $\xi \in E$ is called *smooth* if the map $G \to E$ given by $g \mapsto g \cdot \xi$ is smooth. The smooth vectors form a G-invariant subspace E_{sm} of E on which the Lie algebra of G acts. We cannot hope in general that all vectors in E will be smooth, but we shall say that the representation E is smooth if E_{sm} is *dense* in E.

We shall sometimes consider the *dual* E^* of a representation E, with the contragredient action—if g acts on E by T then it acts on E^* by $(T')^{-1}$, the inverse of the transpose of T. We shall always give E^* the topology of uniform convergence on compact subsets. Then it is locally convex and complete, and $(E^*)^*$ is canonically isomorphic to E (see [19] Chapter 4, Section 2.3). The *antidual* $\bar{E}^*$ means the complex conjugate of E^*, i.e. the space of continuous antilinear maps $E \to \mathbb{C}$.

Representations of $Map(X; G)$.

Although this book is concerned with loop groups we shall make some general introductory remarks about the representation theory of the group of maps $Map(X; G)$, where X is a compact manifold. Essentially only one interesting irreducible representation of this group is known when $\dim(X) > 1$.

If $S = \{x_1, \ldots, x_p\}$ is a finite subset of X then evaluation at the points of S gives us a homomorphism

$$\varepsilon_S : Map(X; G) \to G^p.$$

So if $E_1, \ldots, E_p$ are irreducible representations of G we can make $Map(X; G)$ act irreducibly on $E_1 \otimes \ldots \otimes E_p$ by means of ε_S. More generally, because the group $Map(X; G)$ is a kind of product of a family of copies of G indexed by the points of X one can expect the irreducible representations of $Map(X; G)$ to be 'continuous tensor products'

$$\bigotimes_{x \in X} E_x$$

of families of irreducible representations of G. The representation theory of $Map(X; G)$ has been developed from this point of view by Gelfand and his coworkers [59], [151]. (A short survey of their work can be found in [58].) We shall say only a little about it here.

† The notion of essential equivalence must be treated with caution. As we have defined it it is not an equivalence relation. Worse, a reducible representation can be essentially equivalent to an irreducible one. (E.g. the infinite symmetric group $S_\infty = \cup S_n$ acts on the sequence spaces ℓ^1 and ℓ^2 by permuting the coordinates. The inclusion $\ell^1 \to \ell^2$ is an essential equivalence. But ℓ^2 is irreducible, while in ℓ^1 the sequences with sum 0 form an invariant subspace.)

9.1 REPRESENTATIONS OF MAP(X; G)

If V is a representation of G then $V^X = \mathrm{Map}(X; V)$ is a representation of $\mathrm{Map}(X; G)$. It is, of course, highly reducible, for the maps $X \to V$ which vanish on an arbitrary closed subset Y of X form an invariant subspace. Now the symmetric algebra and exterior algebra functors have the 'exponential' property

$$S(E_1 \oplus E_2) \cong S(E_1) \otimes S(E_2)$$
$$\Lambda(E_1 \oplus E_2) \cong \Lambda(E_1) \otimes \Lambda(E_2),$$

and so if one thinks of V^X as a 'sum' of copies of V indexed by the points of X then it is reasonable to regard $S(V^X)$ and $\Lambda(V^X)$ as continuous tensor products of the desired type. Unfortunately they are even more reducible than V^X. In some cases, however, one can modify the action of $\mathrm{Map}(X; G)$ on them so as to obtain an irreducible representation.

A symmetric algebra $S(E)$ can be identified with the ring of polynomial functions on the dual E^* of E. In that light it is clear that not only linear but also *affine* transformations of E^* induce linear transformations of $S(E)$. It turns out that by combining the initial linear action of $\mathrm{Map}(X; G)$ on $E = V^X$ with a suitable translation component one can sometimes find an affine action on E^* which induces an irreducible representation of $\mathrm{Map}(X; G)$ on $S(E)$. To be more precise, we should assume V is unitary and must choose a measure on X so that E acquires an L^2 inner product. Then the Hilbert space completion $\hat{S}(E)$ of $S(E)$—see Section 9.5 below—is the space of square-summable holomorphic functions on E^* for the Gaussian measure defined by the inner product of E. We make the affine transformation f of E^* given by $f(\xi) = A\xi + b$ act unitarily on $\hat{S}(E)$ by defining

$$(f^{-1} \cdot \phi)(\xi) = e^{\langle b, A\xi \rangle - \frac{1}{2}\|b\|^2} \phi(f \cdot \xi).$$

A very important variant of this construction is when G is a compact Lie group, X is a compact manifold, and $E = \Omega^1(X; \mathfrak{g}_{\mathrm{C}})$ is the space of smooth 1-forms on X with values in the Lie algebra of G. Equivalently, E is the space of *connections* in the trivial principal G-bundle on X. As such there is an affine action of $\mathrm{Map}(X; G)$ on E:

$$(g, \alpha) \mapsto g\alpha g^{-1} + dg \cdot g^{-1}. \tag{9.1.1}$$

This can be transferred to E^*, and the induced unitary action of $\mathrm{Map}(X; G)$ on $\hat{S}(E)$ is known [59] to be irreducible when $\dim(X) \geq 4$. We shall refer to this as the *Gelfand representation*. It seems to be the *only* known irreducible representation of $\mathrm{Map}(X; G)$ when G is compact and $\dim(X) > 1$, except for representations which are degenerate in the sense that they factorize through the restriction map $\mathrm{Map}(X; G) \to \mathrm{Map}(Y; G)$ for some proper subspace Y of X.

It is natural to look for representations of $\mathrm{Map}(X; G)$ which involve all the points of X symmetrically. This property can be formulated as follows: If $\phi: X \to X$ is a diffeomorphism then for each representation E of Map$(X; G)$ we can form a new representation $\phi^* E$ by composing the old action with the automorphism of Map$(X; G)$ induced by ϕ. We can ask whether $\phi^* E \cong E$ for some transitive group of diffeomorphisms ϕ of X. The Gelfand representation has the property that $\phi^* E \cong E$ for all measure-preserving diffeomorphisms ϕ. The representations of loop groups which we shall presently be studying are invariant under *all* diffeomorphisms of the circle. No such representations are known when $\dim(X) > 1$.

In the case of a loop group LG the Gelfand representation is highly reducible. Frenkel [47] has pointed out the following very interesting equivalent description of it, which relates it to the 'regular representation' of LG.

Let $P_{\text{cts}}G$ denote the space of continuous paths $[0, 2\pi] \to G$ which begin at the identity element of G. For each $\lambda > 0$ there is a measure μ_λ on $P_{\text{cts}}G$ called the Wiener measure. It is characterized by the fact that the evaluation map $\varepsilon_t : P_{\text{cts}}G \to G$ at time t carries μ_λ to the measure on G whose density is the fundamental solution of the heat equation on G at time λt. The continuous loop group $L_{\text{cts}}G$ acts on $P_{\text{cts}}G$ by $(\gamma, \pi) \mapsto \gamma * \pi$, where (cf. (4.3.5))

$$(\gamma * \pi)(t) = \gamma(t)\pi(t)\gamma(0)^{-1}. \tag{9.1.2}$$

The measure μ_λ is quasi-invariant under LG, which therefore acts unitarily on the Hilbert space $L^2(P_{\text{cts}}G; \mu_\lambda)$. The orbits of $L_{\text{cts}}G$ on $P_{\text{cts}}G$ correspond to the conjugacy classes in G: to a conjugacy class ω corresponds the set $P_{\text{cts}}^\omega G$ of paths which end in ω. The representation $L^2(P_{\text{cts}}G; \mu_\lambda)$ is therefore the direct integral of the family of representations $L^2(P_{\text{cts}}^\omega G; \mu_\lambda)$, of which the member with $\omega = 1$ can be called the 'regular representation'. It is easy to see that the components $L^2(P_{\text{cts}}^\omega G; \mu_\lambda)$ are themselves still highly reducible.

Now in Section 4.3 we observed that the diffeomorphism between $\Omega^1(S^1; \mathfrak{g})$ and PG given by the indefinite integral, i.e. by $\xi \mapsto \pi$, where $\pi'\pi^{-1} = \xi$, was equivariant with respect to the affine action (9.1.1) of LG on $\Omega^1(S^1, \mathfrak{g})$ and the action (9.1.2) on PG. Formally the Wiener measure μ_λ on $P_{\text{cts}}G$ is

$$e^{-\lambda^{-1}\mathscr{E}(\pi)} \, d(\pi'\pi^{-1}),$$

where

$$\mathscr{E}(\pi) = \frac{1}{4\pi} \int_0^{2\pi} \|\pi'\pi^{-1}\|^2 \, d\theta$$

is the energy of π. So formally μ_λ corresponds to the natural Gaussian measure on $\Omega^1(S^1; \mathfrak{g})$ with variance λ. Frenkel's observation is that this

9.2 THE POSITIVE ENERGY CONDITION

statement can be made precise, so that the Gelfand representation of LG on $\hat{S}(\Omega^1(S^1; \mathfrak{g}_{\mathbb{C}}))$ is actually equivalent to its action on $L^2(P_{\text{cts}}G; \mu_\lambda)$.

With that we conclude our general survey, and turn to the remarkable class of representations of loop groups which we shall study from now on, which we call the representations of *positive energy*. In terms of the preceding description the crucial point is that the homogeneous spaces $P^\omega G \cong LG/Z_\omega$ mentioned above (where Z_ω is the centralizer in G of an element of ω) are naturally *complex* manifolds. Instead of considering the spaces of continuous functions on them we can consider the spaces of holomorphic sections of various line bundles, and that gives us a large class of irreducible projective representations of LG.

9.2 The positive energy condition

We shall now define the class of representations E of loop groups LG which we shall study from now on. They will first of all be symmetric with respect to rigid rotations of the circle, i.e. in the notation of the preceding section we have $\phi^* E \cong E$ when ϕ is a rotation. But we shall make a stronger assumption: we shall assume that the group $\mathbb{T}$ of rotations acts on E by operators R_θ which intertwine with the action of LG, i.e. are such that

$$R_\theta U_\gamma R_\theta^{-1} = U_{R_\theta \gamma}, \qquad (9.2.1)$$

where U_γ is the action of $\gamma \in LG$ on E, and $R_\theta \gamma$ denotes γ rotated through the angle θ, i.e. $R_\theta \gamma(\theta') = \gamma(\theta' - \theta)$.

Actually we must of necessity study *projective* representations of LG, i.e. those such that

$$U_\gamma U_{\gamma'} = c(\gamma, \gamma') U_{\gamma \gamma'}$$

for $\gamma, \gamma' \in LG$, with $c(\gamma, \gamma') \in \mathbb{C}^\times$, or, more precisely, representations of central extensions $\tilde{L}G$ of LG by $\mathbb{C}^\times$. But we saw in Chapter 4 that the action of $\mathbb{T}$ on LG lifts (essentially uniquely) to an action on $\tilde{L}G$. The intertwining property (9.2.1) is best expressed by saying that we have a representation of the semidirect product $\mathbb{T} \tilde{\times} \tilde{L}G$. We shall often refer to it, however, simply as a *symmetric* representation of $\tilde{L}G$.

An action of the circle $\mathbb{T}$ on a topological vector space E is roughly the same thing as a $\mathbb{Z}$-grading of E. For if $E(k)$ is the closed subspace of E where R_θ acts as $e^{-ik\theta}$ then the algebraic direct sum

$$\check{E} = \bigoplus_{k \in \mathbb{Z}} E(k) \qquad (9.2.2)$$

is a dense subspace of E, which we shall call the space of vectors of *finite energy*. We shall say that the action of $\mathbb{T}$ on E has *positive energy* if $E(k) = 0$ when $k < 0$, or equivalently, if R_θ is represented by $\exp(-iA\theta)$

where A is an operator with positive spectrum. Correspondingly we shall say that a symmetric representation of a loop group has positive energy if the associated action of $\mathbb{T}$ has positive energy. (It might be better to use 'positive energy' to mean that $E(k) = 0$ for $k < k_0$ for some k_0; but the distinction is not important, for we can always multiply the action of $\mathbb{T}$ on E by a character of $\mathbb{T}$.)

The most obvious symmetric representations of loop groups, for example the natural action of LU_n on the Hilbert space $L^2(S^1; \mathbb{C}^n)$, are not of positive energy. On the other hand there seems to be no explicitly known *irreducible* symmetric representation of a loop group which is not of positive energy or of negative energy. (The complex conjugate of a representation of positive energy is of negative energy.)

The positive energy condition involves the canonical parametrization of the circle. If ϕ is a diffeomorphism of the circle and E is a representation of positive energy it is natural to ask whether $\phi^* E$ is of positive energy. That turns out to be true, though it is not at all obvious. In fact we shall see (Section 13.4 below) that the action of $\mathbb{T}$ on a representation of positive energy always extends canonically to a projective action of $\mathrm{Diff}^+(S^1)$, the group of orientation-preserving diffeomorphisms of the circle. It even extends to an action of the identity component of the group of automorphisms of LG: that is a strong version of the fact that the representations form a *discrete* set.

Another remark of the same kind which is quite easy to prove and will be useful technically is the following. Let us choose a finite covering homomorphism $G \to G_1 \times G_2 \times \ldots \times G_p$, where G_1 is a torus and the remaining G_i are simple groups. Then LG is a finite covering of a subgroup of finite index in $\prod LG_i$, and there is an action of $\mathbb{T}^p$ on LG which is compatible with the product action on $\prod LG_i$ which rotates each LG_i separately. We shall see (see Remark (11.1.5)(i)) that $\mathbb{T}^p$ acts on every positive energy representation E of LG, and hence that the graded vector space E actually admits a finer multigrading indexed by $\mathbb{Z}^p$.

A representation of a loop group $\tilde{L}G$ which is of positive energy is by definition a representation of $\mathbb{T} \tilde{\times} \tilde{L}G$. But we have

Proposition (9.2.3). *An irreducible unitary representation of $\mathbb{T} \tilde{\times} \tilde{L}G$ which is of positive energy is also irreducible as a representation of $\tilde{L}G$.*

Remark. The hypothesis of unitarity is not actually needed here, as we shall see that *all* representations of positive energy are essentially unitary.

Proof. Suppose that E is an irreducible unitary representation of $\mathbb{T} \tilde{\times} \tilde{L}G$. If it were not irreducible under $\tilde{L}G$ then we could find a

9.2 THE POSITIVE ENERGY CONDITION

bounded self-adjoint operator $T: E \to E$ commuting with the action of $\tilde{L}G$ but not with the action of $\mathbb{T}$.

Define $T_q: E \to E$ by

$$T_q(\xi) = \frac{1}{2\pi} \int_0^{2\pi} e^{iq\theta} R_\theta T R_\theta^{-1}(\xi) \, d\theta.$$

Then T_q is bounded and commutes with $\tilde{L}G$, and $T_{-q} = T_q^*$. Furthermore T_q maps $E(k)$ to $E(k+q)$ for each k. As T does not commute with $\mathbb{T}$ we must have $T_q \neq 0$ for at least one $q < 0$. Let m be the lowest energy which occurs in E, i.e. the smallest m such that $E(m) \neq 0$. Then T_q must annihilate $E(m)$. But $E(m)$ generates E under $\tilde{L}G$, for E is irreducible under $\mathbb{T} \tilde{\times} \tilde{L}G$, and $E(m)$ is $\mathbb{T}$-invariant. So T_q annihilates all of E: a contradiction.

Although the $\mathbb{T}$-action on a representation of positive energy is supposed to be given we know from Schur's lemma that for an irreducible representation of $\tilde{L}G$ the $\mathbb{T}$-action is determined by the $\tilde{L}G$-action up to multiplication by a character of $\mathbb{T}$. A characterization of positive energy representations in terms of the action of $\tilde{L}G$ is given by

Proposition (9.2.4). *A representation of $\tilde{L}G$ is irreducible and of positive energy if and only if it is generated by a smooth cyclic vector ξ which is an eigenvector of the Lie algebra $B^- \mathfrak{g}_\mathbb{C}$.*

Here $B^- \mathfrak{g}_\mathbb{C}$ denotes the subalgebra of $L\mathfrak{g}_\mathbb{C}$ consisting of all elements of the form $\sum_{k \geq 0} a_k z^{-k}$ with $a_k \in \mathfrak{g}_\mathbb{C}$ and $a_0 \in \mathfrak{b}_0^-$, where $\mathfrak{b}_0^-$ is the Borel subalgebra of $\mathfrak{g}_\mathbb{C}$ corresponding to the negative roots.

A vector ξ as in (9.2.4) is usually called a *lowest weight vector*. Proposition (9.2.4) will be proved in Chapter 11. (Cf. (11.2.4).) We shall see there that an equivalent formulation of (9.2.4) is

Proposition (9.2.4'). *A representation E of $\tilde{L}G$ is irreducible and of positive energy if and only if it is generated by a $\tilde{T}$-invariant ray whose orbit under $\tilde{L}G$ in the projective space $P(E)$ defines a holomorphic map $LG/T \to P(E)$.*

There is an interesting relationship between 'energy' in the quantum-mechanical sense used in this section—i.e. as the eigenvalue of the infinitesimal rotation $i(d/d\theta)$—and the classical energy $\mathscr{E}(\gamma)$ of a loop γ, defined (cf. (8.9.2)) by

$$\mathscr{E}(\gamma) = \frac{1}{4\pi} \int_0^{2\pi} \|\gamma(\theta)^{-1} \gamma'(\theta)\|^2 \, d\theta.$$

Proposition (9.2.5). *Suppose that $\mathcal{H}$ is an irreducible unitary positive-energy representation of $\tilde{L}G$ of level h. Let $\Omega \in \mathcal{H}$ be a unit lowest weight vector, and for $\gamma \in \tilde{L}G$ let $\Omega_\gamma = \gamma . \Omega$. Then*

$$\left\langle \Omega_\gamma, \mathrm{i}\frac{\mathrm{d}}{\mathrm{d}\theta}\Omega_\gamma \right\rangle = h\mathscr{E}(\gamma).$$

In other words, the expectation value of $\mathrm{i}(\mathrm{d}/\mathrm{d}\theta)$ in the state Ω_γ is $h\mathscr{E}(\gamma)$.

For the meaning of 'level' we refer to p. 177. The inner product on $\mathfrak{g}$ used in defining $\mathscr{E}(\gamma)$ is the same as the one which defines the extension $\tilde{L}G$. Notice that $\langle \Omega_\gamma, \mathrm{i}(\mathrm{d}/\mathrm{d}\theta)\Omega_\gamma \rangle$ depends only on the image of γ in LG.

Proposition (9.2.5) is simply a reformulation of (4.9.4). A particular case has already been given as Proposition (8.9.11).

Variants of the positive energy condition

(i) *Replacing the circle by the line (cf. Section 6.9)*

The group ΛG of smooth maps $\mathbb{R} \to G$ with compact support can be regarded as a subgroup of LG by identifying S^1 with $\mathbb{R} \cup \infty$ by stereographic projection (i.e. $e^{i\theta} \in S^1 \leftrightarrow 2\tan\frac{1}{2}\theta \in \mathbb{R}$). The natural definition of positive energy for a representation E of ΛG is that the action of ΛG intertwines with an action $t \mapsto T_t$ of the group of translations of $\mathbb{R}$, and that $T_t = \exp(-\mathrm{i}tA)$, where A is an operator with positive spectrum. With this definition we have

Proposition (9.2.6). *Any representation of positive energy of LG restricts to a representation of positive energy of ΛG.*

Remark. We do not know whether every positive energy representation of ΛG comes from a representation of LG, but we know of no counter example. (Cf. the remark at the end of Section 10.7.)

Proof. The group $PSL_2(\mathbb{R})$ of automorphisms of the real projective line, regarded as a group of diffeomorphisms of S^1, contains both the rotation group $\mathbb{T}$ and the group of translations of $\mathbb{R}$. Any representation of LG of positive energy intertwines with an action of $\mathrm{Diff}^+(S^1)$, and in particular with one of $PSL_2(\mathbb{R})$. (The result for $PSL_2(\mathbb{R})$ is much more elementary—see Remark (11.1.5)(ii).) But the only irreducible representations of $PSL_2(\mathbb{R})$ which are of positive energy are its actions on the spaces of holomorphic differentials of various degrees on the upper half-plane. The translation subgroup of $PSL_2(\mathbb{R})$ clearly acts positively on these. A little more precisely, we shall see that any positive energy representation of $\tilde{L}G$ is a sum of irreducible ones. Each irreducible

9.2 VARIANTS OF THE POSITIVE ENERGY CONDITION

representation, in turn, can be identified (see (11.1.2)) with a subspace of a completion of the symmetric algebra $S(N^-\mathfrak{g}_\mathbb{C}^*)$, compatibly with the action of $PSL_2(\mathbb{R})$. But the translations of $\mathbb{R}$ act positively on $S(N^-\mathfrak{g}_\mathbb{C}^*)$.

(ii) *Twisted actions and twisted loop groups*

In Section 8.6 we saw how it is sometimes useful to replace the natural rotation action of $\mathbb{T}$ on LG by the 'twisted' action of $\tilde{\mathbb{T}} = \mathbb{R}/2\pi q\mathbb{Z}$ got by simultaneously rotating and conjugating: for $\alpha \in \tilde{\mathbb{T}}$ and $\gamma \in LG$ we define

$$\tilde{R}_\alpha \gamma(\theta) = \exp(\alpha A)\gamma(\theta - \alpha)\exp(-\alpha A),$$

where $A \in \mathfrak{g}$ is such that $\exp(2\pi qA) = 1$. If A belongs to the fundamental alcove in $\mathfrak{t}$, which we shall assume, then the positive and negative eigenspaces of the twisted action of $\tilde{\mathbb{T}}$ on $L\mathfrak{g}_\mathbb{C}$ are $N^+\mathfrak{g}_\mathbb{C}$ and $N^-\mathfrak{g}_\mathbb{C}$. One can define a notion of 'positive energy' with respect to the twisted action just as for the usual action. But we have

Proposition (9.2.7). *A representation is of positive energy for the twisted action if and only if it is of positive energy in the usual sense.*

This will be proved below.

One use of (9.2.7) is in connection with twisted loop groups. We recall from Section 3.7 that if β is an automorphism of G of finite order q the twisted loop group $L_{(\beta)}G$ consists of the smooth maps $\gamma: \mathbb{R} \to G$ such that $\gamma(\theta + 2\pi) = \beta(\gamma(\theta))$. The circle group $\tilde{\mathbb{T}} = \mathbb{R}/2\pi q\mathbb{Z}$ evidently acts on $L_{(\beta)}G$, and permits us to define representations of positive energy. We shall not pursue the study of these, but it should be noticed that if the automorphism β is inner, so that $L_{(\beta)}G \cong LG$, then the natural rotation action of $\tilde{\mathbb{T}}$ on $L_{(\beta)}G$ corresponds to a twisted action on LG. If β is not inner but becomes inner in a larger (connected) group G' containing G then (9.2.7) tells us that a representation of positive energy of LG' restricts to one of $L_{(\beta)}G$.

One particular case is of interest to us. We often make use of the embedding of $L\mathbb{T}$ in LU_n as a maximal abelian subgroup (see Sections 3.6 and 6.5). This is not compatible with the obvious rotation action of $\mathbb{T}$. But the rotation action on $L\mathbb{T}$ corresponds to that on the isomorphic group $L_{(\beta)}\mathbb{T}^n$, where β is the automorphism of $\mathbb{T}^n$ given by cyclic permutation of the factors. As β is inner in U_n we conclude that representations of positive energy of LU_n restrict to representations of positive energy of $L\mathbb{T}$.

Proof of (9.2.7). The theorem of complete reducibility of positive energy representations applies, with the same proof (see Section 11.2), to ones of positive energy in the twisted sense. So it is enough to consider irreducible representations. But then the criterion (9.2.4) is valid for both types.

9.3 The classification and main properties of representations of positive energy

For the rest of this chapter we shall make the convention that 'representation' means 'smooth representation of positive energy'. The restriction to representations of positive energy is essential for our approach. The requirement of smoothness (i.e. that smooth vectors are dense) is almost certainly unnecessary because automatically satisfied, but we know how to prove that only for LU_n, by an indirect argument sketched in Section 11.4.

We shall now list the most important properties of the representations. The proofs will mostly be given in Chapter 11.

Theorem (9.3.1). *Up to essential equivalence, every representation of LG*
   (i) *is projective,*
   (ii) *is completely reducible, i.e. a discrete direct sum of irreducible representations,*
   (iii) *is unitary,*
   (iv) *extends to a holomorphic projective representation of $LG_\mathbb{C}$,*
   (v) *admits a (projective) intertwining action of $\mathrm{Diff}^+(S^1)$.*

Properties (ii), (iii) and (iv) are familiar as properties of representations of compact groups which are not possessed by the representations of any non-compact locally compact groups such as $SL_2(\mathbb{R})$. It is quite striking that they hold in the present situation.

To be more explicit about the role of essential equivalence, assertion (ii) means that any representation E has irreducible closed subspaces E_α such that the natural map

$$\bigoplus E_\alpha \to E$$

is injective with dense image; (iii) means that E has a dense invariant subspace on which there is a positive definite invariant Hermitian form; and a more precise statement of (iv) is that E is canonically sandwiched between holomorphic representations of $LG_\mathbb{C}$ on antidual spaces:

$$E_0 \hookrightarrow E \hookrightarrow \bar{E}_0^*.$$

To describe all positive energy representations it is enough, in view of (9.3.1) (ii) to classify the irreducible ones. They are parametrized by their 'lowest weights', just like representations of compact groups. Let us consider a particular central extension $\tilde{L}G$, and let $\mathbb{T}_1 \times T \times \mathbb{T}_0$ be a maximal torus of $\mathbb{T} \tilde{\times} \tilde{L}G$. (Here $\mathbb{T}_1$ is the rotation group, T is a maximal torus of G, and $\mathbb{T}_0$ is the kernel of the central extension.) An arbitrary representation E of $\mathbb{T} \tilde{\times} \tilde{L}G$ can be decomposed (up to essential

9.3 REPRESENTATIONS OF POSITIVE ENERGY

equivalence) as

$$\bigoplus E_{(n,\lambda,h)}, \tag{9.3.2}$$

where $E_{(n,\lambda,h)}$ is the part of E where $\mathbb{T}_1 \times T \times \mathbb{T}_0$ acts by the character $(n, \lambda, h) \in \mathbb{Z} \times \hat{T} \times \mathbb{Z}$. The characters which occur in this decomposition are called the *weights* of E. If E is irreducible then $\mathbb{T}_0$ must act by scalars (by Schur's lemma), so only one value of h can occur. This is called the *level* of the representation. The weights of any representation are obviously permuted by the normalizer of $\mathbb{T}_1 \times T \times \mathbb{T}_0$, and hence by the affine Weyl group $W_{\text{aff}} = N(\mathbb{T}_1 \times T \times \mathbb{T}_0)/(\mathbb{T}_1 \times T \times \mathbb{T}_0)$ (see Section 5.1). We recall that W_{aff} is the semidirect product of the Weyl group W of G and the lattice $\check{T} = \text{Hom}(\mathbb{T}; T)$. From the formula (4.9.5) we find that an element $\xi \in \check{T}$ acts on a weight (n, λ, h) by

$$\xi \cdot (n, \lambda, h) = (n + \lambda(\xi) + \tfrac{1}{2}h\|\xi\|^2, \lambda + h\xi^*, h), \tag{9.3.3}$$

where $\|\xi\|^2$ is defined by the inner product corresponding to the extension $\tilde{L}G$, and ξ^* is the image of ξ under the map $\check{T} \to \hat{T}$ defined by the inner product.

The decomposition (9.3.2) is a refinement of the decomposition according to the action of $\mathbb{T}_1$, i.e. the *energy* decomposition

$$E = \bigoplus E(n)$$

discussed in Section 9.2. We shall prove

Proposition (9.3.4). *Every irreducible representation is of finite type, i.e. each $E(n)$ is of finite dimension.*

A fortiori each weight-space $E_{(n,\lambda,h)}$ is of finite dimension.

Now let us recall from Section 5.1 that the characters α of $\mathbb{T}_1 \times T$ arising in the decomposition of the adjoint representation of LG are called the *roots* of LG. We can regard them as characters of $\mathbb{T}_1 \times T \times \mathbb{T}_0$ which are trivial on $\mathbb{T}_0$. As usual we shall often think of roots and weights as linear forms on the Lie algebra $\mathbb{R} \oplus \mathfrak{t} \oplus \mathbb{R}$ of $\mathbb{T}_1 \times T \times \mathbb{T}_0$. A root is classified as positive or negative according to its value on a preferred 'alcove' in the Lie algebra of $\mathbb{T}_1 \times T$. We recall that the roots α are of the form (n, α), where $n \in \mathbb{Z}$ and α is either 0 or a root of G. For each root $\alpha = (n, \alpha)$ with $\alpha \neq 0$ there is a coroot h_α in $\mathfrak{t} \oplus \mathbb{R} \subset \mathbb{R} \oplus \mathfrak{t} \oplus \mathbb{R}$. In fact

$$h_\alpha = \left(h_\alpha, -\frac{n}{2}\|h_\alpha\|^2\right),$$

where the inner product is the one which defines the central extension $\tilde{L}G$.

If G is simply connected—cf. Remark (vi) below—then the classification of the irreducible representations is given by

Theorem (9.3.5).

(i) *Every irreducible representation E has a unique 'lowest weight'* $\lambda = (n, \lambda, h)$ *characterized by the fact that* $\lambda - \alpha$ *is not a weight of E for any positive root* α.

(ii) *The lowest weight λ is* antidominant *in the sense that* $\lambda(h_\alpha) \leq 0$ *for each positive coroot* h_α.

(iii) *The isomorphism classes of irreducible representations of* $\mathbb{T} \tilde{\times} \tilde{L}G$ *are parametrized precisely by the set of antidominant weights.*

Remarks.

(i) As already mentioned in Section 2.7 we work with antidominant weights instead of the more usual dominant ones because we have arranged for our representations to have *lowest* weight vectors rather than highest weight vectors.

(ii) The condition that (n, λ, h) is antidominant can be written explicitly as

$$-\tfrac{1}{2}h\|h_\alpha\|^2 \leq \lambda(h_\alpha) \leq 0 \qquad (9.3.6)$$

for each positive root α, where $\| \ \|^2$ is the inner product which defines the extension $\tilde{L}G$. Because the h_α span $\mathfrak{t}$ this shows that *there are only finitely many irreducible representations of each level.* It also follows that the only representation with $h = 0$ is the trivial representation, i.e. that *all representations of LG are projective* and of positive level.

(iii) The condition that (n, λ, h) is antidominant does not involve n. But the irreducible representations with antidominant weights (n, λ, h) and (m, λ, h) are equivalent as representations of $\tilde{L}G$, and differ only by multiplication by the character $R_\alpha \mapsto e^{i(m-n)\alpha}$ of the circle of rotations. So one may as well restrict attention to antidominant weights of the form $(0, \lambda, h)$.

(iv) If $h > 0$ and the inner product $\langle \ , \ \rangle$ is positive-definite, then the characters $\lambda \in \hat{T}$ such that $(0, \lambda, h)$ is antidominant constitute a fundamental domain for the affine action of W_{aff} on $\hat{T}$ for which $\xi \in \check{T} \subset W_{\text{aff}}$ acts by translation by $h\xi^*$. (This follows from the same argument (5.1.4) (ii) which proves that an alcove is a fundamental domain of the action of W_{aff} on $\mathfrak{t}$.)

(v) If G is *simple* then the antidominant weights $(0, \lambda, h)$ can be described more explicitly. The condition (9.3.6) then reduces to the fact that $-\lambda$ belongs to the simplex cut off from the simplicial cone of dominant weights (in the usual finite dimensional sense of Section 2.7) by the hyperplane $\{\mu : \langle \mu, \alpha_0 \rangle = h\}$, where α_0 is the highest root. We may as well assume that the inner product is the basic one of Section 4.4. Then if we write

$$\omega_0 = (0, 0, 1),$$

$$\omega_i = (0, -\omega_i, \langle \omega_i, \alpha_0 \rangle) \quad (1 \leq i \leq \ell),$$

where the ω_i are the fundamental weights of G (i.e. $\omega_i(h_{\alpha_j}) = \delta_{ij}$), it is easy to see that the antidominant weights are precisely the positive integral combinations of $\omega_0, \ldots, \omega_\ell$. We call $\omega_0, \ldots, \omega_\ell$ the *fundamental* weights of LG.

(vi) The restriction to simply connected groups G in (9.3.5) is not very important. The Proposition and its proof remain true without change for representations of the *identity component* of LG in all cases, and then it is usually easy to extend to representations of the whole group. In the case of LU_n the representations of each level still correspond to the orbits of W_{aff} on $\hat{T}$: cf. Example (ii) below. The same applies when G is a torus, as we shall prove in (9.5.11).

Examples.
(i) If $G = SU_n$ we write characters of the maximal torus in the form
$$\text{diag}\{u_1, \ldots, u_n\} \mapsto u_1^{\lambda_1} u_2^{\lambda_2} \ldots u_n^{\lambda_n}$$
where the λ_i are integers, and the multi-index $(\lambda_1, \ldots, \lambda_n)$ is determined only up to the addition of an integral multiple of $(1, 1, \ldots, 1)$.

We find that $(0; \lambda_1, \ldots, \lambda_n; h)$ is antidominant if and only if $\lambda_1 \leq \lambda_2 \leq \ldots \leq \lambda_n$ and $\lambda_n - \lambda_1 \leq h$. The fundamental weights are the n possible antidominant weights of level 1:
$$\omega_i = (0, -\varpi_i, 1)$$
for $0 \leq i < n$, where
$$-\varpi_i = (0, \ldots, 0, 1, \ldots, 1)$$
with $n - i$ zeros.

(ii) If $G = U_n$ we write characters of the maximal torus as for SU_n, but there is no indeterminacy in the multi-index $\lambda = (\lambda_1, \ldots, \lambda_n)$. Representations of level h of the identity component $(\tilde{L}U_n)^0$ still correspond to the λs such that $\lambda_1 \leq \ldots \leq \lambda_n$ and $\lambda_n - \lambda_1 \leq h$. But now the standard inner product induces an isomorphism $\check{T} \to \hat{T}$. It follows that the representations of level h are parametrized by $(\hat{T}/h.\hat{T})/S_n$, where S_n is the symmetric group, or in alternative language by the λs such that
$$0 \leq \lambda_1 \leq \lambda_2 \leq \ldots \leq \lambda_n < h.$$

In particular there is only *one* representation of level 1.

(iii) It may be worth giving an explicit example at this point of the complete set of weights of an irreducible representation of a loop group. For a representation of level h the weights lie in the hyperplane $\mathbb{Z} \times \hat{T} \times \{h\}$ of $\mathbb{Z} \times \hat{T} \times \mathbb{Z}$. Let us consider the universal central extension of LSU_2. Then $\hat{T} \cong \mathbb{Z}$ with $\mu \in \mathbb{Z}$ corresponding to
$$\text{diag}\{u, u^{-1}\} \mapsto u^\mu.$$

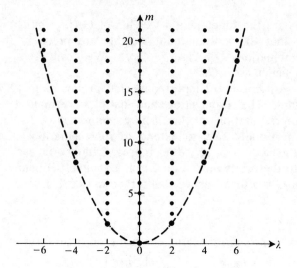

Fig. 3 The weights of the basic representation of $\tilde{L}SU_2$.

The fundamental antidominant weights are

$$\omega_0 = (0, 0, 1),$$

and

$$\omega_1 = (0, -1, 1).$$

The weights of the representation corresponding to ω_0—which is called the *basic* representation—consist of all $(m, \mu, 1)$ such that μ is even and $\mu^2 \leq 2m$; the representation corresponding to ω_1 has all $(m, \mu, 1)$ with μ odd and $\mu^2 \leq 2m + 1$. The multiplicity of $(m, \mu, 1)$, i.e. the dimension of $E_{(m,\mu,1)}$, is the number of partitions of $m - \frac{1}{2}\mu^2$ in the case of the basic representation, and of $m - \frac{1}{2}\mu^2 + \frac{1}{2}$ in the other case.

The justification for these assertions will appear later, when we construct the representations explicitly. The examples are typical in the following sense: in all cases the orbit of the lowest weight λ under the affine Weyl group consists of lattice points $\mathbf{\mu} = (m, \mu, h)$ on the paraboloid

$$\|\mathbf{\mu}\|^2 = \|\mu\|^2 - 2mh = \|\lambda\|^2 \tag{9.3.7}$$

in the hyperplane $\mathbb{Z} \times \hat{T} \times \{h\}$, and the remaining weights lie inside the paraboloid, i.e. they satisfy

$$\|\mathbf{\mu}\|^2 \leq \|\lambda\|^2. \tag{9.3.8}$$

(The inner product in (9.3.7) and (9.3.8) is induced by that of Section 4.9.) The proof of the inequality (9.3.8) is very simple. One may as well

9.3 THE BASIC REPRESENTATION

suppose that μ is antidominant as well as λ, for it can be made so by applying an element of W_{aff}, which does not change $\|\mu\|^2$. But then

$$\|\mu\|^2 - \|\lambda\|^2 = \langle \mu + \lambda, \mu - \lambda \rangle \leq 0,$$

because $\mu + \lambda$ is antidominant and $\mu - \lambda$ is a sum of positive roots (cf. (11.1.1)).

The basic representation

If the inner product on $\mathfrak{g}$ defining the extension $\tilde{L}G$ is positive semidefinite then the weight $\omega_0 = (0, 0, 1)$ is antidominant. The corresponding irreducible representation is called the *basic* representation of $\tilde{L}G$. The name is justified really only in the case of simply laced groups, for which we have

Proposition (9.3.9). *Suppose that G is simply connected and simply laced and that $\tilde{L}G$ is the universal central extension of LG described in Section 4.4. Then if E is the basic representation of $\tilde{L}G$ we have*

(i) *the irreducible representations of $\tilde{L}G$ of level one are precisely the conjugates $\alpha^* E$ of E by the outer automorphisms α of $\tilde{L}G$, and*

(ii) *every irreducible representation of $\tilde{L}G$ occurs as a summand in a representation of the form $i^* E$, where $i : \tilde{L}G \to \tilde{L}G$ is an endomorphism.*

Remarks.

(i) We recall that apart from automorphisms of G itself (which act trivially on ω_0 and hence on E) the outer automorphisms of $\tilde{L}G$ correspond to the elements of the centre Z of G (see Section 3.4). For each $g \in Z$ there is an automorphism α_g of $\tilde{L}G$ got by 'conjugation' by a path in G from the identity to g.

(ii) A more precise version of the second assertion above is that an irreducible representation of level h is a summand in $i_h^* F$, where F is of level 1 and $i_h : LG \to LG$ is induced by any map $S^1 \to S^1$ of degree h.

Proof of (9.3.9).

(i) The irreducible representations of level one correspond (by remark (iv) after Theorem (9.3.5)) to the orbits of W_{aff} on the lattice $\hat{T}$, when the translation sub-group $\check{T}$ of W_{aff} acts on $\hat{T}$ by the embedding $\check{T} \to \hat{T}$ given by the canonical inner product. The image of $\check{T}$ in $\hat{T}$ is the root lattice, and so $\hat{T}/\check{T}$ can be identified with $\hat{Z}$, the dual of the centre of G (see Section 2.4). The group W acts trivially on $\hat{T}/\check{T}$, so the representations are parametrized precisely by $\hat{Z}$. Now Z is contained in the centre of $\tilde{L}G$, and so acts by a character on any irreducible representation of $\tilde{L}G$. We have shown that there is precisely one irreducible representation of level one in which Z acts by a given character. The fact that the outer automorphisms of $\tilde{L}G$ permute the characters of Z transitively was shown in (4.6.5).

(ii) Let us consider the embedding $i_h: \tilde{L}G_{\mathbb{C}} \to \tilde{L}G_{\mathbb{C}}$ induced by the map $z \mapsto z^h$ of S^1. The Lie algebra of the subgroup $i_h(N^-)$—where $N^- = N^- G_{\mathbb{C}}$—is spanned by the root-vectors in $\tilde{L}\mathfrak{g}_{\mathbb{C}}$ corresponding to the negative roots (n, α) such that n is divisible by h. For each loop η belonging to the lattice $\check{T}$ in $LG_{\mathbb{C}}$ we have the operation of conjugation by η

$$c_\eta: \tilde{L}G_{\mathbb{C}} \to \tilde{L}G_{\mathbb{C}}.$$

But if we regard $\check{T}$ as a subgroup of $\hat{T}$ as above then we can think of elements of $\hat{T}$ as paths in T from the identity to points of Z, and so the automorphism c_η is defined for all $\eta \in \hat{T}$. It transforms the root-vector corresponding to the root (n, α) to that for $(n + \langle \alpha, \eta \rangle, \alpha)$, and consequently maps $i_h(N^-)$ into $L^- G_{\mathbb{C}}$ providing

$$-h \leq \langle \alpha, \eta \rangle \leq 0 \tag{9.3.10}$$

for all positive roots α of G. If η satisfies this condition it may still not be true that $c_\eta(i_h(N^-)) \subset N^-$, but because N^- is nilpotent we can certainly choose an element w_η of the finite Weyl group W such that

$$c_{w_\eta} c_\eta i_h(N^-) \subset N^-.$$

The lowest weight vector of the basic representation E will then also be a lowest weight vector in $i_h^* c_\eta^* c_{w_\eta}^* E$; and as an element of the latter it has weight $(\frac{1}{2}h\|\eta\|^2, \eta, h)$. An arbitrary representation of level h has its lowest weight of this form for some η satisfying (9.3.10)—the action of the rotations of S^1 is irrelevant—and so the proof is complete.

Let us record explicitly the conclusion of the first half of the preceding proof.

Proposition (9.3.11). *If G is simply connected and simply laced then the representations of level 1 of $\tilde{L}G$ are completely characterized by the action of the centre Z of G, which can be an arbitrary character.*

We recall from Section 4.4 that G, and hence Z, is naturally a subgroup of $\tilde{L}G$.

9.4 The Casimir operator and the infinitesimal action of the diffeomorphism group

In the theory of finite dimensional semisimple Lie algebras an important role is played by the Casimir operator. This is associated to an algebra $\mathfrak{g}_{\mathbb{C}}$ equipped with a non-degenerate invariant $\mathbb{C}$-bilinear form $\langle \, , \, \rangle$. If V is a vector space on which $\mathfrak{g}_{\mathbb{C}}$ acts, then the Casimir operator in V is the endomorphism Δ of V defined by

$$\Delta = -\tfrac{1}{2} \sum_j e_j^* e_j, \tag{9.4.1}$$

9.4 THE CASIMIR OPERATOR

where $\{e_j\}$ is a basis for $\mathfrak{g}_\mathbb{C}$ and $\{e_j^*\}$ is the dual basis with respect to $\langle\ ,\ \rangle$—i.e. $\langle e_j^*, e_k \rangle = \delta_{jk}$. It is easy to check that Δ does not depend on the choice of the basis $\{e_j\}$. The crucial property of Δ is that it commutes with the action of the elements of $\mathfrak{g}$. This follows immediately from the invariance of $\langle\ ,\ \rangle$: if $\xi \in \mathfrak{g}$ and

$$[\xi, e_j] = \sum x_{jk} e_k$$

then

$$[\xi, e_j^*] = -\sum x_{kj} e_k^*,$$

from which it follows that $[\xi, \Delta] = 0$.

Because of this intertwining property Δ acts as a scalar on any irreducible representation V of $\mathfrak{g}$. For use in Chapter 14 let us recall

Proposition (9.4.2.). *If V is an irreducible representation of $\mathfrak{g}$ with lowest weight $\lambda \in \mathfrak{t}^*$ then we have*

$$\Delta = c_\lambda = \tfrac{1}{2}\{\|\lambda - \rho\|^2 - \|\rho\|^2\},$$

on V, where ρ is half the sum of the positive roots of $\mathfrak{g}$, and the norm on $\mathfrak{t}^*$ is the one induced by the inner product of $\mathfrak{g}$.

Proof. Let e_α be the standard root vector in $\mathfrak{g}_\mathbb{C}$ corresponding to the root α. We take the e_α together with an orthonormal basis $\{h_j\}$ for $\mathfrak{t}_\mathbb{C}$ as our basis of $\mathfrak{g}_\mathbb{C}$. Then $e_\alpha^* = k_\alpha e_{-\alpha}$, where $k_\alpha = \langle e_\alpha, e_{-\alpha} \rangle^{-1} = 2/\|h_\alpha\|^2$. Because $[e_\alpha, e_{-\alpha}] = ih_\alpha$ we have

$$\tfrac{1}{2}(e_\alpha e_{-\alpha} + e_{-\alpha} e_\alpha) = e_\alpha e_{-\alpha} - \tfrac{1}{2} ih_\alpha,$$

and

$$\Delta = -\sum_{\alpha>0} k_\alpha e_\alpha e_{-\alpha} - \tfrac{1}{2} \sum h_j^2 + i \sum_{\alpha>0} h_\alpha \Big/ \|h_\alpha\|^2.$$

Applying Δ to the lowest weight vector ξ in V, which is such that $e_{-\alpha}\xi = 0$ and $h\xi = i\lambda(h)\xi$, we obtain

$$c_\lambda = \tfrac{1}{2}\sum \lambda(h_j)^2 - \sum_{\alpha>0} \lambda(h_\alpha/\|h_\alpha\|^2).$$

But when we identify $\mathfrak{t}$ with $\mathfrak{t}^*$ by the inner product we find that $2h_\alpha/\|h_\alpha\|^2$ corresponds to α, and so

$$c_\lambda = \tfrac{1}{2}\sum \lambda(h_j)^2 - \tfrac{1}{2}\sum_{\alpha>0} \langle \lambda, \alpha \rangle$$
$$= \tfrac{1}{2}\|\lambda\|^2 - \langle \lambda, \rho \rangle$$
$$= \tfrac{1}{2}(\|\lambda - \rho\|^2 - \|\rho\|^2).$$

9 REPRESENTATION THEORY

In the case of loop groups there is an invariant bilinear form on the Lie algebra $L\mathfrak{g}_\mathbb{C}$—cf. Section 4.9. On the other hand the expression (9.4.1) becomes an infinite sum, and does not immediately make sense. Having fixed a non-degenerate invariant inner product $\langle\ ,\ \rangle$ on $\mathfrak{g}$ let us choose an orthonormal basis $\{e_a\}$ for $\mathfrak{g}$, and let us write e_a^n for $e_a z^n$ in $L\mathfrak{g}_\mathbb{C}$. The dual basis element to e_a^n is then e_a^{-n}. If now a vector space V has an action of the central extension $\tilde{L}\mathfrak{g}_\mathbb{C}$ defined by $\langle\ ,\ \rangle$ then we find that, as operators on V,

$$\tfrac{1}{2}\sum_a (e_a^n e_a^{-n} + e_a^{-n} e_a^n) = \sum_a (e_a^n e_a^{-n} + \tfrac{1}{2}[e_a^{-n}, e_a^n])$$

$$= \sum_a e_a^n e_a^{-n} + \frac{i}{2} NnI,$$

where N is the dimension of $\mathfrak{g}$, and I is the generator of the central extension. If V is of positive energy then the operator

$$\Delta_0 = -\sum_{a;n>0} e_a^n e_a^{-n} - \tfrac{1}{2}\sum_a (e_a^0)^2 \qquad (9.4.3)$$

is well-defined at least on the part $\check{V}$ of finite energy in V, for any vector of finite energy is annihilated by $e_a^n e_a^{-n}$ for almost all n. This Δ_0 differs from the formal Casimir operator by 'subtracting the infinite constant' $(i/2)N\sum_{n>0} n$—an idea familiar in quantum field theory. On the other hand Δ_0 does not quite commute with the elements of $\tilde{L}\mathfrak{g}_\mathbb{C}$.

Let us calculate the commutator $[\xi z^m, \Delta_0]$ for some $\xi \in \mathfrak{g}_\mathbb{C}$. We write

$$[\xi, e_a] = \sum_b x_{ab} e_b,$$

where (x_{ab}) is a skew matrix. Then

$$\left[\xi z^m, \sum_a e_a^n e_a^{-n}\right] = X_{n+m} - X_n \quad \text{if}\quad m \neq \pm n,$$

$$= X_{n+m} - X_n - im\xi z^m I \quad \text{if}\quad m = \pm n, \qquad (9.4.4)$$

where

$$X_n = \sum_{a,b} x_{ab} e_b^n e_a^{-n+m}.$$

From this

$$[\xi z^m, \Delta_0] = \{\tfrac{1}{2}X_0 + X_1 + \ldots + X_{m-1} + \tfrac{1}{2}X_m\} + im\xi z^m I.$$

But if $0 \leq k \leq m$ we have

$$X_k + X_{m-k} = \sum_{a,b} x_{ab} [e_b^k, e_a^{m-k}]$$

$$= \sum_a [[\xi, e_a], e_a] z^m$$

$$= -2\Delta_\mathfrak{g} \xi \cdot z^m,$$

9.4 THE CASIMIR OPERATOR

where $\Delta_{\mathfrak{g}} : \mathfrak{g}_\mathbb{C} \to \mathfrak{g}_\mathbb{C}$ is the Casimir operator of the algebra $\mathfrak{g}_\mathbb{C}$ acting in its adjoint representation. Thus

$$[\xi z^m, \Delta_0] = m(\mathrm{i}\xi I - \Delta_{\mathfrak{g}}\xi)z^m. \tag{9.4.5}$$

At this point the theory becomes a little easier if the algebra $\mathfrak{g}$ is simple. For then $\Delta_{\mathfrak{g}}$ acts as a scalar c on $\mathfrak{g}_\mathbb{C}$, and we have

$$[\xi z^m, \Delta_0] = (I + \mathrm{i}c)\frac{d}{d\theta}(\xi z^m). \tag{9.4.6}$$

This leads us to

Definition (9.4.7). *If $\mathfrak{g}$ is simple then the Casimir operator Δ of $\tilde{L}\mathfrak{g}_\mathbb{C}$ is defined by*

$$\Delta = \Delta_0 + (I + \mathrm{i}c)\frac{d}{d\theta},$$

where Δ_0 is defined by (9.4.3).

Here $(d/d\theta)$ is regarded as an element of the Lie algebra of $\mathbb{T} \tilde{\times} \tilde{L}G$, and it therefore acts on any representation of positive energy. Apart from the reordering involved in Δ_0 the expression Δ differs from the naive Casimir operator one would have associated to the invariant inner product (4.9.3) on the Lie algebra of $\mathbb{T} \tilde{\times} \tilde{L}G$ by the addition of the term $\mathrm{i}c(d/d\theta)$.

If $\mathfrak{g}$ is not simple then we can write $\mathfrak{g} = \mathfrak{g}_1 \oplus \ldots \oplus \mathfrak{g}_k$, where $\mathfrak{g}_1$ is abelian and the other $\mathfrak{g}_j$ are simple. We know that there is an action of $\mathbb{T}^k$ on $\tilde{L}G$ which rotates loops in the different factors independently. Let us denote the corresponding elements of the Lie algebra of $\mathbb{T}^k \tilde{\times} \tilde{L}G$ by $\partial/\partial\theta_1, \ldots, \partial/\partial\theta_k$, and let the scalar action of the Casimir operator $\Delta_{\mathfrak{g}}$ on $\mathfrak{g}_{j,\mathbb{C}}$ be c_j. (Thus $c_1 = 0$.)

Definition (9.4.8). *The Casimir operator of $\tilde{L}\mathfrak{g}_\mathbb{C}$ is*

$$\Delta = \Delta_0 + \sum_{j=1}^{k}(I + \mathrm{i}c_j)\frac{\partial}{\partial\theta_j}.$$

This operator makes sense for any representation of positive energy, and commutes with the action of $\tilde{L}\mathfrak{g}_\mathbb{C}$.

We shall use the Casimir operator to prove the unitarity and complete reducibility of representations of loop groups, and also to give an algebraic proof of the Kac character formula. The main point is that in a cyclic representation of $\tilde{L}\mathfrak{g}_\mathbb{C}$ generated by a lowest weight vector the Casimir operator acts as the scalar given by

Proposition (9.4.9). *Suppose that V is a cyclic representation of $\tilde{L}\mathfrak{g}_\mathbb{C}$ generated by a lowest weight vector, and that it admits an intertwining*

action of $\mathbb{T}^k$ as above. Then the Casimir operator acts on V as the scalar

$$c_\lambda = \tfrac{1}{2}\|\lambda\|^2 - \langle \lambda, \rho\rangle - \sum(h+c_j)n_j,$$

where the lowest weight, as a linear function on $\mathbb{R}^k \times \mathfrak{t} \times \mathbb{R}$, is $\lambda = (\{n_j\}, \lambda, h)$.

The preceding formula is easiest to understand when $\mathfrak{g}$ is simple. If we then write $\lambda = (n, \lambda, h)$ and $\rho = (0, \rho, -c)$ the formula of (9.4.9) becomes

$$c_\lambda = \tfrac{1}{2}\{\|\lambda - \rho\|^2 - \|\rho\|^2\}, \tag{9.4.10}$$

in exact analogy with the finite dimensional case. In fact (9.4.10) makes sense and is valid in general provided we think of V as a representation of the algebra $\mathbb{R}^k \oplus \tilde{L}\mathfrak{g}$, where $\tilde{L}\mathfrak{g}$ is the universal central extension of $L\mathfrak{g}$ by $\mathbb{R}^k$ described in Section 4.2. Then $\lambda = (\{n_j\}, \lambda, \{h_j\})$—with $h_1 = \ldots = h_k = h$—and $\dot{\rho} = (0, \rho, \{-c_j\})$ belong to the dual of $\mathbb{R}^k \oplus \mathfrak{t} \oplus \mathbb{R}^k$, on which the inner product is defined by

$$\langle(\{n_j\}, \lambda, \{h_j\}), (\{n_j'\}, \lambda', \{h_j'\})\rangle = \langle \lambda, \lambda'\rangle - \sum n_j h_j' - \sum n_j' h_j. \tag{9.4.11}$$

Proof of (9.4.9). We have

$$\Delta = \Delta_\mathfrak{g} - \sum_{a; n>0} e_a^n e_a^{-n} + \sum(I + ic_j)\frac{\partial}{\partial \theta_j},$$

where $\Delta_\mathfrak{g}$ is the Casimir operator of the algebra $\mathfrak{g}$. We must calculate $\Delta\xi$, where ξ is the lowest weight vector of V. But $\sum e_a^n e_a^{-n}$ annihilates ξ, and the action of $\Delta_\mathfrak{g}$ on ξ is given by (9.4.2). Finally we have $I\xi = ih\xi$ and $\partial \xi/\partial \theta_j = in_j \xi$.

The infinitesimal action of Diff(S^1)

With the Casimir operator at our disposal we can see very easily that the Lie algebra Vect(S^1) of the group of diffeomorphisms of the circle acts on any positive energy representation E of a loop group LG. In the following discussion we shall assume that the Lie algebra $\mathfrak{g}$ is simple. If that were not true, and we decomposed $\mathfrak{g}$ as $\mathfrak{g}_1 \oplus \ldots \oplus \mathfrak{g}_p$ as before, then our argument would produce p different commuting actions of Vect(S^1) on E corresponding to the natural action of Diff(S^1)^p on LG which twists the factors independently. We may as well assume also that the representation E is of a definite level $h > 0$.

The complexified Lie algebra of vector fields has a basis consisting of the elements $e^{in\theta}(d/d\theta)$ for $n \in \mathbb{Z}$. What we must show is that for each n there is a densely-defined operator $L_n: E \to E$ such that

$$[L_n, \xi] = \xi_{(n)} \tag{9.4.12}$$

9.4 THE INFINITESIMAL ACTION OF DIFF(S^1)

(as operators on E), where $\xi \in L\mathfrak{g}_\mathbb{C}$, and $\xi_{(n)} = e^{in\theta} \cdot d\xi/d\theta$. In fact L_n will be defined on the space of smooth vectors of finite energy in E. If $n \neq 0$ we set

$$\Delta_n = -\tfrac{1}{2} \sum_{\substack{k \in \mathbb{Z} \\ 1 \leq a \leq N}} e_a^{n-k} e_a^k$$

in the notation we have been using. This makes sense as an operator on vectors of finite energy because for any k we have

$$\Delta_{n,k} = -\tfrac{1}{2} \sum_a e_a^{n-k} e_a^k = -\tfrac{1}{2} \sum_a e_a^k e_a^{n-k},$$

and so $\Delta_{n,k}$ annihilates a vector of energy m unless $-m \leq k \leq m + n$. A calculation analogous to (9.4.5) gives

$$[\Delta_n, e_a^k] = k(h+c) e_a^{k+n}, \tag{9.4.13}$$

just as when $n = 0$. The operators L_n defined by

$$L_n = \frac{i}{h+c} \Delta_n$$

accordingly have the desired commutation relations with the e_a^k.

From (9.4.13) we deduce easily that

$$[L_n, L_m] = i(m-n) L_{n+m}$$

if $n + m \neq 0$, corresponding to the commutation relation

$$\left[e^{in\theta} \frac{d}{d\theta}, e^{im\theta} \frac{d}{d\theta} \right] = i(m-n) e^{i(m+n)\theta} \frac{d}{d\theta}$$

of the vector fields. It is more complicated to calculate $[L_n, L_{-n}]$. The result is

$$[L_n, L_{-n}] = -2in L_0 + \frac{Nh}{12(h+c)} n(n^2 - 1). \tag{9.4.14}$$

The presence of the scalar term on the right-hand side of this equation tells us that the Lie algebra Vect(S^1) acts projectively on the representation space. The form of (9.4.14) is not very surprising. In (4.2.11) we saw that the most general central extension of Vect(S^1) by $\mathbb{R}$ is defined by the relation

$$[L_n, L_{-n}] = -2in L_0 + \lambda n(n^2 - 1)$$

for some $\lambda \in \mathbb{R}$.

The fact that the vector fields on the circle act projectively on representations of loop groups suggests strongly that the group Diff$^+(S^1)$ itself ought to act. We shall approach that question in Section 13.4 by a different method. (Goodman and Wallach [64] treat it by exponentiating the infinitesimal action.)

9.5 Heisenberg groups and their standard representations

Let V be a real topological vector space with a skew bilinear form $S: V \times V \to \mathbb{R}$. We shall suppose that S is continuous and nondegenerate, i.e. for each $\xi \in V$ there is some $\eta \in V$ such that $S(\xi, \eta) \neq 0$. The Heisenberg group $\tilde{V}$ is the central extension of V by the circle defined by S. In other words:

Definition (9.5.1). *The Heisenberg group $\tilde{V}$ associated to (V, S) is the set $V \times \mathbb{T}$ with the multiplication defined by*

$$(v, \lambda) \cdot (v', \lambda') = (v + v', \lambda \lambda' e^{iS(v, v')}).$$

We think of V and $\mathbb{T}$ as the subsets $V \times 1$ and $0 \times \mathbb{T}$ of $\tilde{V}$. The centre of $\tilde{V}$ is $\mathbb{T}$. There is an obvious complexification $\tilde{V}_\mathbb{C}$, which is an extension of $V_\mathbb{C}$ by $\mathbb{C}^\times$.

The purpose of this section is to describe a standard irreducible unitary representation of $\tilde{V}$. The construction depends on choosing a complex structure $J: V \to V$ (such that $J^2 = -1$) which is compatible with S and positive, i.e.
  (i) $S(Jv, Jv') = S(v, v')$ for all $v, v' \in V$, and
  (ii) $S(Jv, v) > 0$ for all non-zero $v \in V$.
Using J we can write $V_\mathbb{C} = A \oplus \bar{A}$, where A and $\bar{A}$ are the $+i$ and $-i$ eigenspaces of J. Both A and $\bar{A}$ are isotropic for S. Conversely a decomposition $V_\mathbb{C} = A \oplus \bar{A}$ into complex conjugate isotropic subspaces determines a suitable J providing the hermitian form $\langle \, , \, \rangle$ on A defined by

$$\langle \xi, \eta \rangle = -2iS(\bar{\xi}, \eta) \qquad (9.5.2)$$

is positive definite.

The standard representation of $\tilde{V}$ associated to J is on the Hilbert space $\hat{S}(A)$ obtained by completing the symmetric algebra $S(A)$ with respect to the inner product (9.5.2), which is extended from A to $S(A)$ by the formula

$$\langle a_1 a_2 \ldots a_n, a'_1 a'_2 \ldots a'_n \rangle = \sum \langle a_1, a'_{i_1} \rangle \langle a_2, a'_{i_2} \rangle \ldots \langle a_n, a'_{i_n} \rangle \qquad (9.5.3)$$

where the sum is over all permutations $\{i_1, \ldots, i_n\}$ of $\{1, \ldots, n\}$. To describe the representation we first describe an action of the complexified group $\tilde{V}_\mathbb{C}$ on the space $\text{Hol}(\bar{A})$ of holomorphic functions on $\bar{A}$. ($\text{Hol}(\bar{A})$ is given the topology of uniform convergence on compact sets.) The

9.5 HEISENBERG GROUPS

symmetric algebra $S(A)$ is regarded as a subalgebra of $\text{Hol}(A)$ by identifying $a \in A$ with the holomorphic function $x \mapsto \langle x, a \rangle$ on $\bar{A}$.

The vector spaces A and $\bar{A}$ can be regarded as subgroups of $\tilde{V}_C$ by identifying a with $(a, 1)$ and $\bar{a}$ with $(\bar{a}, 1)$. Together they generate $\tilde{V}_C$ subject to the relations

$$\bar{a}_1 a_2 = e^{-\langle a_1, a_2 \rangle} a_2 \bar{a}_1,$$

so we can define a representation of $\tilde{V}_C$ on $\text{Hol}(\bar{A})$ by making $\bar{A}$ act by translation, i.e.

$$(\bar{a}_1 . f)(\bar{a}_2) = f(\bar{a}_2 - \bar{a}_1), \tag{9.5.4a}$$

and A act by multiplication, i.e.

$$(a_1 . f)(\bar{a}_2) = e^{\langle a_2, a_1 \rangle} f(\bar{a}_2). \tag{9.5.4b}$$

The action of $\tilde{V}_C$ on $\text{Hol}(\bar{A})$ is defined by a smooth map $\tilde{V}_C \times \text{Hol}(\bar{A}) \to \text{Hol}(\bar{A})$: there is thus an action of the Lie algebra of $\tilde{V}_C$ on $\text{Hol}(\bar{A})$ which can be exponentiated to give the formulae (9.5.4). We shall write the action of the group element $v \in V_C \subset \tilde{V}_C$ as $\exp i\mathfrak{a}(v)$. The operators

$$\mathfrak{a}(v) : \text{Hol}(\bar{A}) \to \text{Hol}(\bar{A})$$

satisfy the commutation relations

$$[\mathfrak{a}(v_1), \mathfrak{a}(v_2)] = -2iS(v_1, v_2). \tag{9.5.5}$$

We shall now show that the action of $\tilde{V}_C$ on $\text{Hol}(\bar{A})$ induces a unitary action of $\tilde{V}$ on the Hilbert space completion $\hat{S}(A)$ of $S(A)$. We begin by noticing that for any $\xi \in A$ the function $e^\xi \in \text{Hol}(\bar{A})$ belongs to $\hat{S}(A)$: we easily check that

$$\langle e^\xi, e^\eta \rangle = e^{\langle \xi, \eta \rangle}.$$

We can also calculate that the action of $v \in V \subset \tilde{V}$ takes e^ξ to the element

$$v . e^\xi = e^{-\frac{1}{2}\langle a, a \rangle - \langle a, \xi \rangle} e^{\xi + a} \tag{9.5.6}$$

of $\hat{S}(A)$, where $v = a + \bar{a}$ with $a \in A$. We see that

$$\langle v . e^\xi, v . e^\eta \rangle = \langle e^\xi, e^\eta \rangle.$$

The following lemma implies that $\tilde{V}$ acts unitarily on the subspace $\hat{S}(A)$ of $\text{Hol}(\bar{A})$.

Lemma (9.5.7). *Let E denote the free complex vector space whose basis is a set of symbols ε_ξ in one-to-one correspondence with the elements ξ of A. Define a hermitian inner product in E by prescribing*

$$\langle \varepsilon_\xi, \varepsilon_\eta \rangle = e^{\langle \xi, \eta \rangle}.$$

Then the inner product is positive definite, and the completion of E with respect to the corresponding norm is canonically isomorphic to $\hat{S}(A)$.

Proof. We map E to $\hat{S}(A)$ by $\varepsilon_\xi \mapsto e^\xi$. This preserves the inner product, so we have only to show that the elements e^ξ span a dense subspace of $\hat{S}(A)$. Let F be the closure of the subspace they span. By differentiating $e^{t\xi}$ (for $t \in \mathbb{R}$) repeatedly with respect to t at $t = 0$ we find that $\xi^n \in F$ for all n. But then $\xi_1 \xi_2 \ldots \xi_n \in F$ for all $\xi_1, \ldots, \xi_n \in A$, because

$$\xi_1 \xi_2 \ldots \xi_n = \frac{1}{n!} \sum_\sigma (-1)^{n-|\sigma|} \left(\sum_{i \in \sigma} \xi_i \right)^n,$$

where σ runs through the subsets of $\{1, 2, \ldots, n\}$, and $|\sigma|$ denotes the number of elements in σ. Thus F must be the whole of $\hat{S}(A)$.

Remark. The Hilbert space $\hat{S}(A)$ is essentially the space of holomorphic functions on $\bar{A}$ which are square-summable for the natural Gaussian measure defined by the norm of $\bar{A}$. The problem with this description, however, is that unless $\bar{A}$ is finite dimensional the Gaussian measure is supported only on an enlargement of $\bar{A}$. Cf. [60] Section IV.3 and [98].

Lemma (9.5.7) implies that $\tilde{V}$ acts unitarily on $\hat{S}(A)$, for we can define an action of $\tilde{V}$ on the free vector space E by

$$v \cdot \varepsilon_\xi = e^{-\frac{1}{2}\langle a,a \rangle - \langle a,\xi \rangle} \varepsilon_{\xi+a}$$

when $v = a + \bar{a}$ (cf. (9.5.6)), and this preserves the inner product.

The infinitesimal generators $\mathfrak{a}(v)$ of the action of $\tilde{V}$ do not preserve the subspace $\hat{S}(A)$ of $\text{Hol}(\bar{A})$. They can, however, be regarded as densely-defined unbounded self-adjoint operators in $\hat{S}(A)$.

Proposition (9.5.8). *The unitary representation of the Heisenberg group $\tilde{V}$ on $\hat{S}(A)$ is irreducible.*

Proof. One may as well assume that V is separable. Furthermore one may as well assume that A is complete with respect to the form $\langle \ , \ \rangle$. In that case one can suppose that V is the space of real-valued L^2 functions on the circle with integral zero, and that A is spanned by z^k for $k > 0$. Then the group $\mathbb{T}$ of rotations of the circle acts as a group of automorphisms on $\tilde{V}$, and acts compatibly on $\hat{S}(A)$, which is accordingly a representation of $\tilde{V}$ of *positive energy* in the sense of Section 9.2. In this situation we know from Proposition (9.2.3) that any decomposition $\hat{S}(A) = P \oplus Q$ which is stable under $\tilde{V}$ is also stable under rotation, so that $\hat{S}(A)(k) = P(k) \oplus Q(k)$. Then the unit element $1 \in \hat{S}(A)$, which is (up to a multiple) the unique vector of energy zero, must belong to either P or Q. On the other hand the element 1 is a cyclic vector for the action of $\tilde{V}$ on $\hat{S}(A)$ because—as we see from (9.5.6)—the group element $v = a + \bar{a}$ transforms 1 to a multiple of e^a, and the vectors e^a generate

9.5 HEISENBERG GROUPS

$\hat{S}(A)$. So no non-trivial decomposition is possible, and the representation is irreducible.

If the vector space V is finite dimensional then the representations of $\tilde{V}$ corresponding to different positive complex structures J are all equivalent. In that case the symplectic group $\text{Sp}(V)$—the group of all automorphisms of V which preserve its skew form—acts projectively on $\hat{S}(A)$ compatibly with its action on $\tilde{V}$. The resulting projective representation of $\text{Sp}(V)$ is called the *metaplectic* representation. If V is not finite dimensional we have the following result, which is due to Shale [136].

Proposition (9.5.9).

(i) *The representations of the Heisenberg group $\tilde{V}$ corresponding to two complex structures J_0 and J_1 on V are equivalent if and only if $J_1 - J_0$ is a Hilbert–Schmidt operator.*

(ii) *For a given complex structure J on V the subgroup $\text{Sp}_{\text{res}}(V)$ of $\text{Sp}(V)$ consisting of automorphisms T such that $[T, J]$ is Hilbert–Schmidt acts projectively on the Hilbert space $\hat{S}(A)$ corresponding to J, and intertwines with the action of $\tilde{V}$.*

We shall not prove this here, because it exactly resembles—but is simpler than—the corresponding theory for the spin representation, which we shall treat in detail in Chapter 12. A proof in the spirit of this book is given in [131]; but there are many proofs in the literature. Cf. Shale [136], Vergne [150].

Let us mention, however, that the 'only if' part of statement (i) is very easy to prove. For if J_0 and J_1 correspond to decompositions

$$V_{\mathbb{C}} = A_0 \oplus \bar{A}_0 = A_1 \oplus \bar{A}_1$$

we can express $\bar{A}_1$ as the graph of a symmetric operator $\alpha : \bar{A}_0 \to A_0$. If $\hat{S}(A_0)$ and $\hat{S}(A_1)$ are equivalent than there must be a vector in $\hat{S}(A_0)$ which is invariant under the subgroup $\bar{A}_1$ of $\tilde{V}_{\mathbb{C}}$. A simple calculation shows that the only possible vector is $e^{\frac{1}{2}\alpha}$, where α is regarded as an element of $\hat{S}^2(A_0)$. But α belongs to $\hat{S}^2(A_0)$ only if it is Hilbert–Schmidt, which is the case only if $J_1 - J_0$ is Hilbert–Schmidt.

The Heisenberg groups that are relevant in this book arise from the central extensions of the loop groups of tori. If T is a torus with Lie algebra $\mathfrak{t}$ we can write $LT \cong \check{T} \times T \times V$, where $\check{T} = \text{Hom}(\mathbb{T}; T)$, T is the subgroup of constant loops, and V is the vector space of maps $f : S^1 \to \mathfrak{t}$ with integral 0, which is regarded as a subgroup of LT by the exponential map. The extensions $\tilde{L}T$ we are concerned with all have the property that the identity component $(\tilde{L}T)^0$ is canonically a product $T \times \tilde{V}$, where V is the Heisenberg group defined by some skew form S on V. There is a homomorphism

$$s: \check{T} \to \hat{T}$$

associated to $\tilde{L}T$: it is defined by the fact that the conjugation action of $\xi \in \check{T}$ on the centre $T \times \mathbb{T}$ of $(\tilde{L}T)^0$ is necessarily of the form

$$(t, u) \mapsto (t, u \cdot s(\xi)(t)).$$

In our examples s is induced by the inner product on the Lie algebra $\mathfrak{t}$, and is therefore injective.

Let us now consider the classification of the irreducible positive energy representations of $\tilde{L}T$. There is a canonical decomposition $V_\mathbb{C} = A \oplus \bar{A}$ into the positive and negative energy parts of the rotation action of $\mathbb{T}$, and the hermitian form (9.5.2) on A is positive-definite. Thus $\hat{S}(A)$ is a representation of $\tilde{V}$ of positive energy. The crucial fact for the classification is

Proposition (9.5.10). *$\hat{S}(A)$ is the only irreducible unitary representation of $\tilde{V}$ of positive energy which is faithful on the centre. More generally, for each $h > 0$ the only irreducible positive energy unitary representation of $\tilde{V}$ of level h is the canonical representation (also on $\hat{S}(A)$) of the Heisenberg group associated to the skew form hS on V.*

Granting this for the moment, let us proceed with the classification. The possible irreducible representations of the identity component $(\tilde{L}T)^0 \cong T \times \tilde{V}$ correspond to pairs (λ, h), where $\lambda \in \hat{T}$ is the character by which T acts, and $h > 0$ is the level of the representation of $\tilde{V}$. For each such representation $\mathcal{H}_{\lambda, h}$ of $(\tilde{L}T)^0$ there is the corresponding induced representation of $\tilde{L}T$. Conjugation by $\xi \in \check{T}$ transforms $\mathcal{H}_{\lambda, h}$ to $\mathcal{H}_{\lambda + h \cdot s(\xi), h}$, so the restriction to $(\tilde{L}T)^0$ of the representation $\mathcal{H}_{[\lambda], h}$ of $\tilde{L}T$ induced from $\mathcal{H}_{\lambda, h}$ is

$$\bigoplus_{\xi \in \check{T}} \mathcal{H}_{\lambda + h \cdot s(\xi), h}.$$

It follows that if $h > 0$ and $s: \check{T} \to \hat{T}$ is injective the representation $\mathcal{H}_{[\lambda], h}$ is irreducible; and every representation of positive level must be of this form.

We have therefore proved

Proposition (9.5.11). *If the homomorphism $s: \check{T} \to \hat{T}$ associated to the extension $\tilde{L}T$ is injective then the only irreducible unitary representations of $\tilde{L}T$ of positive energy are the representations $\mathcal{H}_{[\lambda], h}$ just constructed. Those of level h correspond to the elements of $\hat{T}/h \cdot \check{T}$.*

This means that the representations correspond to the orbits of the action of $W_{\text{aff}} = \check{T}$ on the dual of the maximal torus of $\tilde{L}T$.

9.5 HEISENBERG GROUPS

We shall now return to (9.5.10). This is proved by considering the action of the Lie algebra of $\tilde{V}$. If $\mathcal{H}$ is any unitary representation of $\tilde{V}$ then for every $v \in V$ there is a densely-defined unbounded self-adjoint operator $\mathfrak{a}(v)$ in $\mathcal{H}$ such that $v \in V \subset \tilde{V}$ acts on $\mathcal{H}$ by $\exp i\mathfrak{a}(v)$. We define $\mathfrak{a}(v)$ also for $v \in V_C$ by complex-linearity.

Proposition (9.5.12). *If $\mathcal{H}$ is of positive energy, and v belongs to the finite energy part $\check{V}_C$ of V_C, then the domain of $\mathfrak{a}(v)$ contains $\check{\mathcal{H}}$. Furthermore $\mathfrak{a}(v).\check{\mathcal{H}} \subset \check{\mathcal{H}}$, and the relations*

$$[\mathfrak{a}(v_1), \mathfrak{a}(v_2)] = -2iS(v_1, v_2)$$

hold in $\check{\mathcal{H}}$ for $v_1, v_2 \in \check{V}_C$.

Proof. We may as well assume that the centre of $\tilde{V}$ acts faithfully on $\mathcal{H}$.

Suppose $v \in \check{V} = V \cap \check{V}_C$. Let $v = a + \bar{a}$ with $a \in \check{A}$. Then v belongs to the three-dimensional Heisenberg group Γ generated by v and $Jv = i(a - \bar{a})$. We can assume that a has a definite energy, i.e. is an eigenvector of $\mathbb{T}$. Then Γ is stable under rotations. We shall assume known that Γ has a unique faithful irreducible representation, which can be realized on the completion of the polynomial algebra $\mathbb{C}[a]$. This means that $\mathcal{H} \cong \mathbb{C}[a]^\wedge \hat{\otimes} \mathcal{K}$, where $\mathcal{K}$ is a Hilbert space on which Γ acts trivially, and which has an action of the rotation group. Evidently

$$\check{\mathcal{H}} = \mathbb{C}[a] \otimes \check{\mathcal{K}}.$$

But $\mathfrak{a}(v)$ acts on $\mathbb{C}[a]^\wedge$ as $-ia + id/da$, so the domain of $\mathfrak{a}(v)$ contains $\mathbb{C}[a] \otimes \mathcal{K}$, which contains $\check{\mathcal{H}}$. It is also clear that $\mathfrak{a}(v)$ preserves $\check{\mathcal{H}}$.

The proof of the commutation relation is now elementary, for it is enough to consider the three cases (i) $v_2 = v_1$, (ii) $v_2 = Jv_1$, and (iii) $S(v_1, v_2) = S(Jv_1, v_2) = 0$.

Proof of (9.5.10). We shall only consider the case where the centre acts faithfully: the general case is essentially the same.

Suppose that $\mathcal{H}$ is an irreducible representation of positive energy, and let Ω be a unit vector of minimum energy in $\mathcal{H}$. We define a map

$$F: S(\check{A}) \to \mathcal{H}$$

by

$$a_1 a_2 \ldots a_k \mapsto \mathfrak{a}(a_1)\mathfrak{a}(a_2) \ldots \mathfrak{a}(a_k) . \Omega.$$

It is very easy to check that F preserves inner products: for example

$$\begin{aligned}
\langle \mathfrak{a}(a_1)\Omega, \mathfrak{a}(a_2)\Omega \rangle &= \langle \Omega, \mathfrak{a}(a_1)^*\mathfrak{a}(a_2)\Omega \rangle \\
&= \langle \Omega, \mathfrak{a}(\overline{a_1})\mathfrak{a}(a_2)\Omega \rangle \\
&= \langle \Omega, [\mathfrak{a}(\overline{a_1}), \mathfrak{a}(a_2)]\Omega \rangle \\
&= -2iS(\overline{a_1}, a_2) \\
&= \langle a_1, a_2 \rangle. \tag{9.5.13}
\end{aligned}$$

(Here, in obtaining (9.5.13), we have used the fact that $\mathfrak{a}(\bar{a}_1)\Omega = 0$ because $\mathfrak{a}(\bar{a}_1)$ lowers energy.) It follows that F extends to an isometric isomorphism of $\hat{S}(A)$ with a closed subspace of $\mathcal{H}$. It is also easy to check that F is compatible with the action of $\mathfrak{a}(v)$ for any $v \in V$, and therefore $\hat{S}(A)$ is identified with a subrepresentation of $\mathcal{H}$. As $\mathcal{H}$ is irreducible, that completes the proof.

Because we now know the form of all positive energy representations of $\tilde{L}T$ we can assert

Proposition (9.5.14). *Any positive energy representation of $\tilde{L}T$ admits a projective intertwining action of* $\text{Diff}^+(S^1)$.

Proof. It is enough to show that $\text{Diff}^+(S^1)$ acts on the representation $\hat{S}(A)$ of $(\tilde{L}T)^0$. The action of $\text{Diff}^+(S^1)$ on $(LT)^0$ induces an action on $V = (LT)^0/T$, and by (6.8.2) we know that $\text{Diff}^+(S^1) \subset Sp_{\text{res}}(V)$. So $\text{Diff}^+(S^1)$ acts projectively on $\hat{S}(A)$ by the metaplectic representation of (9.5.9).

In Section 13.4 we shall use (9.5.14) to derive the corresponding result for a general loop group.

To conclude this section we record the following result for use in Chapter 13.

Proposition (9.5.15). *If $\mathcal{H}$ is a representation of $\tilde{L}T$ of positive energy, then the vectors of finite energy in $\mathcal{H}$ are smooth. (I.e. if $\xi \in \mathcal{H}$ then $\gamma \mapsto \gamma \cdot \xi$ is a smooth map $\tilde{L}T \to \mathcal{H}$.)*

Proof. As usual it is enough to show that vectors of finite energy in $\hat{S}(A)$ are smooth for the action of $\tilde{V}$. But $1 \in \hat{S}(A)$ is a smooth vector because of the explicit formula

$$v \cdot 1 = e^{-\frac{1}{2}\langle a,a\rangle} e^a,$$

where $v = a + \bar{a}$. Any other vector of finite energy is a linear combination of vectors of the form

$$\mathfrak{a}(a_1)\mathfrak{a}(a_2)\ldots \mathfrak{a}(a_k) \cdot 1,$$

with $a_i \in A$, and is therefore smooth because

$$v \cdot \mathfrak{a}(a) \cdot v^{-1} = \mathfrak{a}(a) + 2S(v, a).$$

10
THE FUNDAMENTAL REPRESENTATION

For a polarized Hilbert space $H = H_+ \oplus H_-$ we saw in Chapter 7 that the holomorphic line bundle Det on the Grassmannian $\mathrm{Gr} = \mathrm{Gr}(H)$ is acted on by a central extension $\widetilde{GL}_{\mathrm{res}}(H)$ of $GL_{\mathrm{res}}(H)$. The bundle Det has no non-zero holomorphic cross-sections, but its dual Det* (whose fibre at a point of Gr is the dual of the fibre of Det) has an infinite dimensional space Γ of holomorphic sections. This space Γ is the *fundamental representation* of $\widetilde{GL}_{\mathrm{res}}(H)$. We give it the topology of uniform convergence on compact subsets of Gr. It is then a complete locally convex topological vector space (cf. [75]) on which the group $\widetilde{GL}_{\mathrm{res}}(H)$ acts continuously. (The space Gr in this chapter means the Hilbert manifold defined in Chapter 7; but in fact almost nothing would be changed if we considered the smooth Grassmannian instead.)

We shall give two different explicit descriptions of the space Γ in this chapter. The first is as the (completed) exterior algebra—or 'fermionic Fock space'—on $H_+ \oplus \bar{H}_-$.† The second is as a sum of symmetric algebras or 'bosonic Fock spaces'. These symmetric algebras can be identified with the classical ring of universal symmetric polynomials. The second description amounts to interpreting Γ as the standard representation of the Heisenberg group $\tilde{L}\mathbb{T}$. It makes clear the irreducibility of Γ. We shall also show that Γ contains a Hilbert space $\mathcal{H}$ as a dense subspace on which the group $\tilde{U}_{\mathrm{res}}(H)$ acts unitarily: $\mathcal{H}$ should be thought of as the space of 'square-summable' holomorphic sections of Det*. (The terminology 'square-summable' has been justified by Pickrell [122].)

We shall see not only that Γ is an irreducible representation of $\widetilde{GL}_{\mathrm{res}}(H)$, but that it is an irreducible representation of positive energy of each of the loop groups $\tilde{L}U_n$. In Section 10.6 we shall describe its decomposition under $\tilde{L}SU_n$.

The representation Γ is properly described as the spin representation of the restricted orthogonal group of the real Hilbert space underlying H. That point of view is explained in the next chapter.

In Section 10.7 we shall describe how the equivalence between bosonic

---

† We should emphasize here that $\bar{H}_-$ denotes the abstract complex conjugate space to H_-. It is a replica of H_- with the scalars acting in a conjugate way: $\lambda \cdot \bar{\xi} = (\bar{\lambda} \cdot \xi)^-$. In particular it is disjoint from H; we do not suppose that there is an operation of conjugation defined inside the Hilbert space H.

and fermionic Fock spaces presents itself in two-dimensional quantum field theory.

10.1 Γ as an exterior algebra: the fermionic Fock space

In this section we shall explain how Γ can be identified with a completion of the exterior algebra $\Lambda(H_+ \oplus \bar{H}_-)$. Before doing so, let us point out that the passage from H to $\Lambda(H_+ \oplus \bar{H}_-)$ is very familiar in quantum field theory as the process of 'second quantization'. If H is the space of solutions of a relativistic field equation, and one writes $H = H_+ \oplus H_-$, where H_+ and H_- are the states of positive and negative energy, then $\Lambda(H_+ \oplus \bar{H}_-)$ is the state space of an assembly consisting of an indeterminate number of free fermionic particles and antiparticles of type H. Notice that

$$\Lambda(H_+ \oplus \bar{H}_-) = \Lambda(H_+) \otimes \Lambda(\bar{H}_-) = \bigoplus_{p,q} \Lambda^p(H_+) \otimes \Lambda^q(\bar{H}_-).$$

The subspace $\Lambda^p(H_+) \otimes \Lambda^q(\bar{H}_-)$ consists of the states with p particles and q antiparticles, all of positive energy. We shall return to this picture in Remark (iv) on p. 199.

To identify Γ with an exterior algebra we begin by recalling from Section 2.9 that if V is a finite dimensional vector space then the space of holomorphic sections of the bundle Det* on the Grassmannian $\text{Gr}_k(V)$ of k-dimensional subspaces of V is precisely the exterior power $\Lambda^k(V^*)$. Taking all the connected components of $\text{Gr}(V)$ together, the space of sections of Det* on $\text{Gr}(V)$ is the whole exterior algebra $\Lambda(V^*)$.

As in Chapter 7, let us identify the Hilbert space H with $L^2(S^1; \mathbb{C})$ with its orthonormal basis $\{z^k\}_{k \in \mathbb{Z}}$, so that H_+ is spanned by $\{z^k\}_{k \geq 0}$. We have a filtration

$$\ldots \subset H_2 \subset H_1 \subset H_0 \subset H_{-1} \subset H_{-2} \subset \ldots \subset H,$$

where $H_k = z^k H_+$. Subspaces of H which are sandwiched between H_n and H_{-n} correspond to subspaces of $H_{-n,n} = H_{-n}/H_n$, and so the finite dimensional Grassmannians $\text{Gr}(H_{-n,n})$ form a nested sequence of subspaces of Gr. We know from Section 7.2 that their union is dense. The determinant bundle on Gr restricts to the determinant bundle on each $\text{Gr}(H_{-n,n})$. Restriction of sections gives us a map

$$\Gamma \to \varprojlim_n \Lambda(H^*_{-n,n}) \qquad (10.1.1)$$

which is injective because the union of the $\text{Gr}(H_{-n,n})$ is dense. We shall show that the right-hand-side of (10.1.1) can be naturally identified with a completion of the exterior algebra not of H^* but of $H^*_- \oplus H_+$, or equivalently $\bar{H}_- \oplus H_+$.

10.1 Γ AS AN EXTERIOR ALGEBRA

The inclusion $\mathrm{Gr}(H_{-n,n}) \to \mathrm{Gr}(H_{-n-1,n+1})$ takes the pth connected component into the $(p+1)^{\mathrm{th}}$, so the corresponding restriction map

$$\Lambda(H^*_{-n-1,n+1}) \to \Lambda(H^*_{-n,n})$$

lowers degree by 1. In fact it is the derivation D_{z^n} associated to the basis element z^n of H_n/H_{n+1}. Now $H_{-n,n} = H_{-n,0} \oplus H_{0,n}$, so

$$\Lambda(H^*_{-n,n}) \cong \Lambda(H^*_{-n,0}) \otimes \Lambda(H^*_{0,n}).$$

If we identify $\Lambda(H_{0,n})$ with $\Lambda(H^*_{0,n})$ by

$$\xi_1 \wedge \ldots \wedge \xi_p \mapsto D_{\xi_1} D_{\xi_2} \ldots D_{\xi_p}(1 \wedge z \wedge \ldots \wedge z^{n-1})$$

(which takes Λ^p to Λ^{n-p}), then we have a commutative diagram

$$\begin{array}{ccc} \Lambda(H^*_{-n-1,0}) \otimes \Lambda(H_{0,n+1}) & \xrightarrow{\cong} & \Lambda(H^*_{-n-1,n+1}) \\ \downarrow & & \downarrow \\ \Lambda(H^*_{-n,0}) \otimes \Lambda(H_{0,n}) & \xrightarrow{\cong} & \Lambda(H^*_{-n,n}), \end{array} \qquad (10.1.2)$$

where the right-hand vertical map is the one in the inverse system (10.1.1), and the left-hand vertical map is the natural map induced by the inclusion $H_{-n,0} \to H_{-n-1,0}$ and the projection $H_{0,n+1} \to H_{0,n}$. This means that the inverse limit can essentially be identified with

$$\Lambda(H^*_-) \otimes \Lambda(H_+) = \Lambda(H^*_- \oplus H_+).$$

To explain the word 'essentially', let us decompose all the vector spaces under consideration according to the natural action of the circle $\mathbb{T}$ on $H = L^2(S^1; \mathbb{C})$ defined by rotation. If V is a complex vector space with $\mathbb{T}$-action we recall (see Section 9.2) that $V(k)$ denotes the subspace where $e^{i\theta} \in \mathbb{T}$ acts as $e^{-ik\theta}$, and is called the part of *energy* k. Thus the energies which occur in $H_{0,n}$ are $0, 1, \ldots, n-1$, and those which occur in $H^*_{-n,0}$ are $1, 2, \ldots, n$. The left-hand vertical map in (10.1.2) induces an isomorphism of the parts of energy k when $k < n$. It follows that the inverse limit is simply

$$\hat{\Lambda}(H^*_- \oplus H_+) = \prod_{k \geq 0} \Lambda(H^*_- \oplus H_+)(k), \qquad (10.1.3)$$

as a topological vector space. Notice that only positive energies occur, and that the part with a given energy is finite dimensional.

Proposition (10.1.4).
(i) $\varprojlim \Lambda(H^*_{-n,n}) \cong \hat{\Lambda}(H^*_- \oplus H_+)$.
(ii) $\Gamma \to \hat{\Lambda}(H^*_- \oplus H_+)$ *is injective with a dense image.*
(iii) $\Gamma(k) \cong \Lambda(H^*_- \oplus H_+)(k)$ *for each k.*

Proof. The assertion (i) has already been proved, and (ii) is a consequence of (iii). To prove (iii) it is enough to show that

$$\Gamma \to \Lambda(H^*_{-n,n})$$

is surjective for each n. If we regard $\Lambda(H^*_{-n,n})$ as the dual of $\Lambda(H_{-n,n})$ then it has a standard basis $\{\omega_\sigma\}$, indexed by the subsets σ of $\{-n, -n+1, \ldots, n-1\}$, which is dual to the basis $\{z^\sigma\}$ of $\Lambda(H_{-n,n})$: here

$$z^\sigma = z^{a_1} \wedge \ldots \wedge z^{a_p}$$

if $\sigma = \{a_1, \ldots, a_p\}$. The subsets σ correspond to subsets S of $\mathbb{Z}$ such that $[n, \infty) \subset S \subset [-n, \infty)$. For each such S we have the Plücker coordinate π_S which was defined in Section 7.5. This is an element of Γ which maps to ω_σ.

Our argument has also proved the following.

Proposition (10.1.5). *The Plücker coordinates π_S form an algebraic basis for the dense subspace $\check{\Gamma}$ of Γ consisting of the elements of finite energy.*

Remarks.

(i) Notice that if $S \subset \mathbb{Z}$ is an indexing set of our usual type, obtained from $\mathbb{N}$ by deleting some positive integers $a_1, \ldots, a_p$ and adjoining some negative integers $b_1, \ldots, b_q$, then $\pi_S \in \Gamma$ corresponds to the basis element

$$\bar{z}^{b_1} \wedge \ldots \wedge \bar{z}^{b_q} \wedge z^{a_1} \wedge \ldots \wedge z^{a_p}$$

of $\Lambda(\bar{H}_- \oplus H_+)$ of energy $\ell^*(S) = \sum a_i - \sum b_i$.† In the language of quantum field theory this is a state with p particles and q antiparticles; the indexing by the sets S corresponds to a Dirac 'hole' description, where only antiparticles exist, and the vacuum state is the one with all the negative energy states filled by antiparticles.

(ii) The Grassmannian Gr is the union of connected components $_d\text{Gr}$, where $_d\text{Gr}$ is the set of subspaces of virtual dimension d. The space of sections Γ breaks up accordingly as a direct product $\Gamma = \prod \Gamma_d$. In terms of $\Lambda(\bar{H}_- \oplus H_+)$ the factor Γ_d corresponds to

$$\bigoplus_{q-p=d} \Lambda^p(H_+) \otimes \Lambda^q(\bar{H}_-), \tag{10.1.6}$$

i.e. to the 'states of charge $-d$'. One can characterize Γ_d as the part of Γ where the scalars $u \in \mathbb{T} \subset U_{\text{res}}(H)$ act by multiplication by u^{-d}.

(iii) Our use of the standard basis $\{z^k\}$ and the associated circle-action may have obscured the naturality of the relation between Γ and

---

† In terms of the length $\ell(S)$ of S defined in Section 7.3 we have $\ell^*(S) = \ell(S) + \frac{1}{2}d(d+1)$, where $d = q - p = \text{card}(S)$. Cf (7.8.5).

10.1 Γ AS AN EXTERIOR ALGEBRA

$\Lambda(H_-^* \oplus H_+)$, which depends only on the splitting $H = H_+ \oplus H_-$. Let us therefore review the situation in a purely algebraic light.

Suppose that H is any vector space with a given subspace H_+. Let Gr denote for the moment the Grassmannian of all subspaces W of H which are commensurable with H_+ (see Section 7.1); and let Γ denote the space of algebraic sections of the dual of the determinant bundle on Gr. Let A and B be subspaces of H such that

$$B \subset H_+ \subset A,$$

with $\dim(A/B) < \infty$. If W is any other subspace sandwiched between A and B then the fibre of the determinant bundle at W is canonically isomorphic to

$$\det(W/B) \otimes \det(H_+/B)^*,$$

and so the restriction of Det* to Gr(A/B) is

$$\text{Det}^*_{A/B} \otimes \det(H_+/B), \qquad (10.1.7)$$

where $\text{Det}_{A/B}$ is the determinant bundle of Gr(A/B), and the one-dimensional vector space $\det(H_+/B)$ is regarded as a trivial line-bundle. The space of sections of (10.1.7) is

$$\Lambda((A/B)^*) \otimes \det(H_+/B)$$
$$\cong \Lambda((A/H_+)^* \oplus (H_+/B)^*) \otimes \det(H_+/B)$$
$$\cong \Lambda((A/H_+)^*) \otimes \Lambda((H_+/B)^*) \otimes \det(H_+/B)$$
$$\cong \Lambda((A/H_+)^*) \otimes \Lambda(H_+/B)$$
$$\cong \Lambda((A/H_+)^* \oplus (H_+/B)), \qquad (10.1.8)$$

all these isomorphisms being canonical.† Thus Γ maps canonically to the inverse limit of the exterior algebras (10.1.8) which form an inverse system as A expands and B contracts. The limit is simply the completion of the vector space $\Lambda((H/H_+)^* \oplus H_+)$ for the topology for which the neighbourhoods of 0 are the subspaces of finite codimension. It should be noticed that we have not made any use of an inner product in H in the discussion.

(iv) In quantum field theory it is extremely important that (if H is a Hilbert space) the projective space of rays in $\mathcal{H} = \Lambda(\bar{H}_- \oplus H_+)$ depends on the decomposition $H = H_+ \oplus H_-$ only up to commensurability—i.e. $P(\mathcal{H})$ does not change if a finite dimensional piece is transferred from H_+ to H_-. To see this, suppose that

$$H = H_+ \oplus H_- = H'_+ \oplus H'_-,$$

where

$$H'_+ = H_+ \oplus N \quad \text{and} \quad H_- = H'_- \oplus N,$$

---

† The first isomorphism of (10.1.8) depends on the splitting $H = H_+ \oplus H_-$, so although both Γ and $\Lambda((H/H_+)^* \oplus H_+)$ depend only on H_+, the map between them depends on the splitting.

with $\dim(N) = n < \infty$. Then
$$\Lambda(\bar{H}_- \oplus H_+) = \Lambda(\bar{H}'_-) \otimes \Lambda(\bar{N}) \otimes \Lambda(H_+),$$
and
$$\Lambda(\bar{H}'_- \oplus H'_+) = \Lambda(\bar{H}'_-) \otimes \Lambda(N) \otimes \Lambda(H_+).$$
But $\Lambda(\bar{N})$ is *canonically* isomorphic to $\Lambda(N)$ up to a scalar multiplier: choosing an element of the one-dimensional space $\Lambda^n(N)$ defines an isomorphism $\Lambda^p(\bar{N}) \to \Lambda^{n-p}(N)$ for each p.

The effect of this is that $P(\mathcal{H})$ can be constructed from H and a (possibly unbounded) self-adjoint Fredholm operator $\Delta : H \to H$, even though Δ has a finite number of zero eigenvalues. We choose any real number λ which does not belong to the spectrum of Δ, and define H_+ as the largest subspace on which $\Delta - \lambda$ is positive. If H and Δ depend continuously on a parameter belonging to a space $\mathcal{F}$ the associated projective spaces $P(\mathcal{H})$ form a fibre bundle on $\mathcal{F}$. The interesting case is when $\mathcal{F}$ is the space of gauge-equivalence classes of connections in a vector bundle on an odd dimensional manifold X, and Δ is the Dirac operator associated to a connection. The topological type of a bundle of projective spaces on a base space $\mathcal{F}$ is completely described by a cohomology class in $H^3(\mathcal{F}; \mathbb{Z})$. This vanishes if and only if the bundle comes from a bundle of Hilbert spaces on $\mathcal{F}$. The class is called the 'non-abelian anomaly' in quantum field theory. (Cf. Mickelsson [112], Faddeev [41], and Zumino [157]. The corresponding situation for even dimensional manifolds is described in Singer [138].)

10.2 The Hilbert space structure

We shall now show that there is a Hilbert space $\mathcal{H}$ contained as a dense subspace (with a finer topology) in Γ, and that the subgroup $U_{\text{res}}^\sim(H)$ preserves $\mathcal{H}$ and acts unitarily on it. Of course $\Lambda(\bar{H}_- \oplus H_+)$ is a pre-Hilbert space: if V is any vector space with an inner product we define an inner product on $\Lambda(V)$ by
$$\langle v_1 \wedge \ldots \wedge v_k, v'_1 \wedge \ldots \wedge v'_m \rangle = \det(\langle v_i, v'_j \rangle) \quad \text{if} \quad k = m,$$
$$= 0 \quad \text{otherwise}.$$

We shall see that $\mathcal{H}$ is the Hilbert space completion $\hat{\Lambda}(\bar{H}_- \oplus H_+)$. But as yet we do not know that $U_{\text{res}}^\sim$ acts on this.

Our method is as follows. Let Γ^* denote the dual space of Γ, with the topology of uniform convergence on compact sets. We shall define a continuous linear map
$$\beta : \bar{\Gamma}^* \to \Gamma$$
which is equivariant with respect to $U_{\text{res}}^\sim$. This gives us a pairing
$$\langle \ , \ \rangle : \Gamma^* \times \bar{\Gamma}^* \to \mathbb{C} \tag{10.2.1}$$

10.2 THE HILBERT SPACE STRUCTURE

defined by
$$\langle \xi, \eta \rangle = \xi(\beta(\eta))$$
which is continuous in each variable. We shall show that it is hermitian and positive-definite, and so makes $\bar{\Gamma}^*$ into a pre-Hilbert space. The equivariance of β implies that $\langle \, , \, \rangle$ is preserved by $U_{\text{res}}^\sim$. The Hilbert space completion of $\bar{\Gamma}^*$ is denoted by $\mathcal{H}$. It can automatically be identified with a subspace of the antidual of $\bar{\Gamma}^*$, i.e. with a subspace of Γ. (Recall that if the dual of any complete locally convex topological vector space Γ is given the topology of uniform convergence on compact sets, then $\Gamma^{**} \cong \Gamma$. (Cf. [19] Chapter 4 Section 2.3.)) Thus we have the situation
$$\bar{\Gamma}^* \hookrightarrow \mathcal{H} \hookrightarrow \Gamma,$$
where both injections are continuous, and the composite is β.

Because an element of Γ is a holomorphic map $\text{Det} \to \mathbb{C}$ which is linear on each fibre, to give a map $\bar{\Gamma}^* \to \Gamma$ it is enough to give a holomorphic map $\text{Det} \to \bar{\Gamma}$ which is linear on each fibre. This in turn amounts to giving an equivariant map
$$\beta_0 : \text{Det} \times \text{Det} \to \mathbb{C}$$
which is holomorphic in the second variable and antiholomorphic in the first, and linear and antilinear on the respective fibres. The definition of β_0 is:
$$\beta_0(w, \bar{w}) = \det(\langle w_i, \bar{w}_j \rangle), \qquad (10.2.2)$$
when w and $\bar{w}$ are admissible bases (cf. Section 7.5) for spaces in Gr of the same virtual dimension, and
$$\beta_0(w, \bar{w}) = 0$$
otherwise. (We observe that (10.2.2) makes sense because for two admissible bases the matrix $(\langle w_i, \bar{w}_j \rangle)$ differs from the identity by a matrix of trace class.) Because
$$\beta_0(\bar{w}, w) = \beta_0(w, \bar{w})^-$$
the pairing (10.2.1) is hermitian.

For each indexing set $S \in \mathcal{S}$ we have the subspace $H_S \in \text{Gr}$ with its canonical basis z^S. Evaluation of sections of Det^* at z^S gives us an element of Γ^* or $\bar{\Gamma}^*$, again denoted z^S. From the definition of $\beta : \bar{\Gamma}^* \to \Gamma$ we see at once that $\beta(z^S)$ is the Plücker coordinate π_S. On the other hand
$$\pi_S(z^{S'}) = 1 \quad \text{if} \quad S = S', \text{ and}$$
$$= 0 \quad \text{if} \quad S \neq S'.$$
Thus $\{z^S\}_{S \in \mathcal{S}}$ is an orthonormal family in $\bar{\Gamma}^*$. As we already know that the Plücker coordinates are an algebraic basis for the dense subspace $\check{\Gamma}$ of

Γ it follows that β induces an isomorphism $(\bar{\Gamma}^*)^{\vee} \to \check{\Gamma}$, and that the inner product (10.2.1) is positive definite. We have now proved

Proposition (10.2.3). *The Plücker coordinates π_S form an orthonormal basis for a Hilbert space $\mathcal{H}$ which is a dense subspace of Γ.*

The space $\mathcal{H}$ is of course the one which was introduced formally in Section 7.5. Proposition (10.2.3) tells us at the same time that $\mathcal{H}$ is the Hilbert space completion $\hat{\Lambda}(\bar{H}_- \oplus H_+)$.

10.3 The ring of symmetric polynomials

Our next objective is to show that Γ can be expressed as a sum of polynomial algebras, each isomorphic to the ring of universal symmetric polynomials. That will be done in Section 10.4. In this section we shall briefly recall the main points of the classical theory of symmetric polynomials, following the excellent account in Macdonald [108].

If $u_1, \ldots, u_N$ are indeterminates, the subring R_N of the integral polynomial ring $\mathbb{Z}[u_1, \ldots, u_N]$ formed by the polynomials which are symmetric in $u_1, \ldots, u_N$ is itself a polynomial ring

$$R_N = \mathbb{Z}[\sigma_1, \ldots, \sigma_N]$$

generated by the 'elementary symmetric functions' $\sigma_1, \ldots, \sigma_N$ in $u_1, \ldots, u_N$. The polynomial σ_k is defined as the coefficient of t^k in $\prod_{i=1}^{N}(1 + u_i t)$. We notice that when $N \geq N'$ there is a natural restriction map

$$\mathbb{Z}[u_1, \ldots, u_N] \to \mathbb{Z}[u_1, \ldots, u_{N'}]$$

which takes u_i to 0 for $i > N'$; and the induced map $R_N \to R_{N'}$ takes σ_i to σ_i for $i \leq N'$ and σ_i to 0 for $i > N'$.

The ring of universal symmetric polynomials R is defined as the abstract polynomial ring $\mathbb{Z}[\sigma_1, \sigma_2, \ldots]$ in a sequence of indeterminates $\{\sigma_i\}$. It maps to each R_N, so one can think of its elements, a little loosely, as 'symmetric polynomials in a indefinitely large number of variables'. The most important theorem about symmetric functions is that the *Schur functions* Φ_ν, indexed by partitions ν, form a integral basis for R. Here a *partition* means a sequence $\nu = \{\nu_i\}_{i \geq 1}$ of integers, almost all zero, such that

$$\nu_1 \geq \nu_2 \geq \nu_3 \geq \ldots.$$

We shall give the definition of Φ_ν below; the usual notation for it is S_ν. For a proof of the theorem we refer the reader to Macdonald [108].

We should mention the main reason for the importance of Schur functions, though it is not directly relevant here. The functions Φ_ν corresponding to partitions ν with $\leq N$ parts, i.e. those such that $\nu_i = 0$ for $i > N$, form a basis of R_N. They can be thought of as functions on the

10.3 THE RING OF SYMMETRIC POLYNOMIALS

unitary group U_N: if $A \in U_N$, let $u_1, \ldots, u_N$ be the eigenvalues of A, and associate to A the corresponding symmetric function of $u_1, \ldots, u_N$. Then the functions so obtained are precisely the characters of all the irreducible representations of SU_N. More precisely, Φ_v is the character of the representation of U_N whose highest weight is

$$(u_1, \ldots, u_N) \mapsto u_1^{v_1} u_2^{v_2} \ldots u_N^{v_N},$$

where $u_1, \ldots, u_N$ are the entries of a diagonal matrix in U_N. (The general irreducible representation of U_N is got by multiplying this one by a negative power of $\det(A)$.) As a function of $u_1, \ldots, u_N$ we have $\Phi_v = \Delta(v)/\Delta(0)$, where $\Delta(v)$ is the determinant

$$\Delta(v) = \begin{vmatrix} u_1^{v_1+N-1} u_1^{v_2+N-2} & \ldots & u_1^{v_N} \\ \vdots & \vdots & \vdots \\ u_N^{v_1+N-1} & \ldots & u_N^{v_N} \end{vmatrix}.$$

Before giving the definition of Φ_v as a polynomial in the σ_k we must make two more remarks. First, the ring $R_N = \mathbb{Z}[\sigma_1, \ldots, \sigma_N]$ is equally well the polynomial ring $\mathbb{Z}[h_1, \ldots, h_N]$, where the h_k are the 'complete symmetric functions', i.e. h_k is the sum of all monomials of degree k in $u_1, \ldots, u_N$, each taken with coefficient 1. Thus h_k is the coefficient of t^k in $\prod (1 - u_i t)^{-1}$, and the h_k and σ_k are expressed in terms of each other by

$$\sum_{k \geq 0} (-1)^k h_k t^k = \left(\sum_{k \geq 0} \sigma_k t^k \right)^{-1}, \qquad (10.3.1)$$

where we are using the convention that $h_0 = \sigma_0 = 1$. We shall correspondingly identify the ring R with $\mathbb{Z}[h_1, h_2, \ldots]$ by using the relation (10.3.1).

The second remark is that partitions correspond one-to-one to indexing sets $S \in \mathcal{S}$ of virtual cardinal zero: the partition v corresponding to $S = \{s_k\}$ is defined by $v_k = k - s_k$. We shall denote the Schur function Φ_v also by Φ_S.

The definition of Φ_S is as follows. Consider the $\mathbb{Z} \times \mathbb{N}$ matrix

$$\Xi = \begin{pmatrix} \vdots & \vdots & \vdots & \vdots \\ \ldots & h_1 & 1 & 0 & 0 \\ \ldots & h_2 & h_1 & 1 & 0 \\ \ldots & h_3 & h_2 & h_1 & 1 \\ \hline \ldots & h_4 & h_3 & h_2 & h_1 \\ \ldots & h_5 & h_4 & h_3 & h_2 \\ \ldots & h_6 & h_5 & h_4 & h_3 \\ \vdots & \vdots & \vdots & \vdots \end{pmatrix} \qquad (10.3.2)$$

whose $(p, q)^{\text{th}}$ entry is h_{p-q}. (We write $h_0 = 1$, and $h_k = 0$ if $k < 0$.)

10 THE FUNDAMENTAL REPRESENTATION

Definition (10.3.3). *The Schur function Φ_S is the determinant formed from the rows of the matrix (10.3.2) which belong to S.*

Note. If S satisfies $s_k = k$ for $k > N$, i.e. if the associated partition v has $\leq N$ parts, then the matrix Ξ_S whose determinant is Φ_S is of the form

$$\Xi_S = \begin{pmatrix} A & 0 \\ C & D \end{pmatrix},$$

where A is lower triangular with ones on the diagonal, and D is a $N \times N$ matrix. We interpret $\det(\Xi_S)$ as the finite determinant $\det(D)$.

Examples. If $v = (n\,0\,0\ldots)$ then $\Phi_v = h_n$, and if $v = (\underset{\leftarrow n \rightarrow}{1\,1\ldots1}\,0\,0\ldots)$ then $\Phi_v = \sigma_n$.

The other symmetric functions which need to be mentioned are the *power sums* p_k. In R_N we define

$$p_k = \sum u_i^k;$$

in the universal ring R the p_k are defined as integral polynomials in the h_k by the relation

$$\log\left(\sum_{k \geq 0} h_k t^k\right) = \sum_{k \geq 1} \frac{1}{k} p_k t^k. \tag{10.3.4}$$

This relation also expresses the h_k in terms of the p_k, but with rational coefficients. We have therefore

$$R_\mathbb{Q} = \mathbb{Q}[\sigma_1, \sigma_2, \sigma_3, \ldots] = \mathbb{Q}[p_1, p_2, p_3, \ldots].$$

By using the power sums we can introduce an inner product in $R_\mathbb{Q}$. Let P denote the abstract vector space over the rationals with the symbols p_k as basis. Define an inner product on P by prescribing

$$\langle p_k, p_k \rangle = k,$$

and $\langle p_k, p_m \rangle = 0$ if $k \neq m$. The ring $R_\mathbb{Q}$ is the symmetric algebra $S(P)$ of P, and the inner product on P extends in the standard way to give an inner product on $S(P)$: we define

$$\langle x_1 x_2 \ldots x_n, y_1 y_2 \ldots y_n \rangle = \sum \langle x_1, y_{i_1} \rangle \langle x_2, y_{i_2} \rangle \ldots \langle x_n, y_{i_n} \rangle, \tag{10.3.5}$$

where the sum is over all permutations $\{i_1, \ldots, i_n\}$ of $\{1, \ldots, n\}$.

Proposition (10.3.6). *The Schur functions Φ_v form an orthonormal basis for $R_\mathbb{Q}$ with respect to the inner product just defined.*

We refer to [108] for the proof.

Let us mention in conclusion that if the p_k are regarded as the coordinate functions on $\mathbb{C}^\infty$, so that R_Q becomes a ring of complex-valued functions on $\mathbb{C}^\infty$, then the inner product is the usual L^2 inner product with respect to the product Gaussian measure

$$e^{-\Sigma |p_k|^2/k} \prod \frac{dp_k \, d\bar{p}_k}{2\pi i k}. \tag{10.3.7}$$

10.4 Γ as a sum of symmetric algebras

We have already observed (cf. (10.1.6)) that the space of sections Γ is a product of spaces Γ_d which correspond to the different connected components of Gr. Each Γ_d is a representation of the identity component of $GL_{\text{res}}^{\sim}(H)$, and elements of the kth component of $GL_{\text{res}}^{\sim}$ map Γ_d to Γ_{d+k}. In this section we shall show that Γ_0 can be identified with the completion of a symmetric algebra, and that it is the standard Heisenberg representation of the subgroup $(\tilde{L}\mathbb{T})^0$ of $U_{\text{res}}^{\sim}(H)$. The essential point is that the space Γ of holomorphic sections of Det* over Gr is exactly the same as the space of holomorphic sections of its restriction to the subspace $\text{Gr}^{(1)}$ of Section 8.3.

The group $L\mathbb{C}^\times$ acts on $H = L^2(S^1; \mathbb{C})$ by multiplication operators. Let N^- denote the subgroup of $L\mathbb{C}^\times$ consisting of elements of the form

$$1 + h_1 z^{-1} + h_2 z^{-2} + \ldots \tag{10.4.1}$$

with $h_i \in \mathbb{C}$.

Although we shall not use it it seems appropriate to mention the following result.

Proposition (10.4.2). *The group N^- acts freely on* Gr.

Proof. Suppose that $\eta \cdot W = W$, with $\eta \in N^-$ and $W \in \text{Gr}$. Let w be an element of W of minimal order (see Section 7.3). This is unique if it is normalized so that

$$w = z^s + \sum_{k<s} \lambda_k z^k;$$

and as $\eta \cdot w$ is of the same form we must have $\eta \cdot w = w$. This implies that $\eta = 1$.

Let us consider the orbit of the base-point H_+ of Gr under N^-. The points W of this orbit do not meet H_- (which is invariant under N^-), so they project isomorphically on to H_+. The restriction of the bundle Det to the orbit can be trivialized canonically be identifying $\text{Det}(W)$ with $\text{Det}(H_+)$ by projection. This means that holomorphic sections of Det* restrict to give ordinary holomorphic functions on N^-:

$$\Gamma_0 \to \text{Hol}(N^-).$$

If we write a general element of N^- in the form (10.4.1) then the polynomial ring $R_{\mathbb{C}} = \mathbb{C}[h_1, h_2, \ldots]$ is naturally contained in $\mathrm{Hol}(N^-)$ as a dense subset, in fact as the set of all elements of finite energy. We have

Proposition (10.4.3). *The Plücker coordinate π_S for $S \in \mathscr{S}$ restricts to the Schur function Φ_S in $R_{\mathbb{C}} \subset \mathrm{Hol}(N^-)$.*

Corollary (10.4.4). *The restriction $\Gamma_0 \to \mathrm{Hol}(N^-)$ is injective, and its image is dense. It induces an isomorphism $\check{\Gamma}_0 \to R_{\mathbb{C}}$ of the subspaces of elements of finite energy.*

Proof of (10.4.3). We must calculate the Plücker coordinate π_S of the space $W = \eta \cdot H_+$, where η is the element (10.4.1) of N^-. The columns of the $\mathbb{Z} \times \mathbb{N}$ matrix Ξ of (10.3.2) form a basis for W, but not its canonical basis. The canonical basis is $\Xi \cdot \Xi_+^{-1}$, where Ξ_+ is the top $\mathbb{N} \times \mathbb{N}$ block of Ξ. The coordinate π_S is therefore $\det((\Xi \cdot \Xi_+^{-1})_S)$, where the subscript S denotes as usual the square submatrix formed by the rows in S. Formally this is

$$\det(\Xi_S) \cdot \det(\Xi_+^{-1}),$$

which equals $\det(\Xi_S) = \Phi_S$ as we want. We shall leave it to the reader to justify the formal calculation, using the remark after the definition (10.3.3).

The orbit Ω of $H_+ \in \mathrm{Gr}$ under N^- is one connected component of the space which in Chapter 8 we denoted by $\mathrm{Gr}_\infty^{(1)}$, i.e. our Grassmannian model of the loop space $\Omega \mathbb{T}$. It is a homogeneous space of the identity component $(L\mathbb{C}^\times)^0$ of $L\mathbb{C}^\times$. Thus the space of sections of $\mathrm{Det}^* | \Omega$, which we have identified with $\mathrm{Hol}(N^-)$, is naturally a representation space of $(\tilde{L}\mathbb{C}^\times)^0$. In obvious notation we have

$$(L\mathbb{C}^\times)^0 = N^- \times \mathbb{C}^\times \times N^+.$$

The subgroup N^- acts on $\mathrm{Hol}(N^-)$ by translation. The subgroup N^+ acts trivially on the orbit Ω, and acts trivially on the fibre of Det at H_+; an element $\xi \in N^+$ therefore acts on the fibre of Det at $\eta \cdot H_+$ by multiplication by $c(\xi, \eta)$, where $c(\xi, \eta) \in \mathbb{C}^\times$ is the commutator in $\tilde{L}\mathbb{C}^\times$ of representatives of ξ and η. Thus $\xi \in N^+$ acts on $\mathrm{Hol}(N^-)$ by multiplication by the holomorphic function $c(\xi, .)^{-1}$.

We are now in the situation described in Section 9.5. The group $\tilde{L}\mathbb{T}$ is essentially a Heisenberg group. In fact we can write the identity component $(L\mathbb{T})^0$ of $L\mathbb{T}$ as $\mathbb{T} \times V$, where $\mathbb{T}$ is the constant loops and V is the vector space of smooth maps $S^1 \to \mathbb{R}$ with integral zero, identified with a subgroup of $L\mathbb{T}$ by the exponential map. The skew bilinear form S of (4.7.5) which defines the extension $\tilde{L}\mathbb{T}$ is nondegenerate on V, and the identity component $(\tilde{L}\mathbb{T})^0$ is $\mathbb{T} \times \tilde{V}$, where $\tilde{V}$ is the Heisenberg group associated to (V, S). We have $V_\mathbb{C} = \bar{A} \oplus A$, where $\bar{A}$ is the Lie algebra of

10.4 Γ AS A SUM OF SYMMETRIC ALGEBRAS

N^- and A that of N^+. The exponential map $\bar{A} \to N^-$ is a holomorphic isomorphism, and so $\mathrm{Hol}(N^-)$ can be identified with $\mathrm{Hol}(\bar{A})$. When that is done the action of $\tilde{V} \subset \tilde{L}\mathbb{T}$ is the standard representation described in Section 9.5. Thus we have

Proposition (10.4.5). *The action of $(\tilde{L}\mathbb{T})^0$ on Γ_0, and hence of $\tilde{L}\mathbb{T}$ on Γ, is irreducible.*

Corollary (10.4.6). *The action of $GL^{\sim}_{\mathrm{res}}(H)$ on Γ is irreducible.*

We observe that Γ is the representation of $\tilde{L}\mathbb{T}$ induced from the representation Γ_0 of $(\tilde{L}\mathbb{T})^0$. Each space Γ_d is identical as a representation of $\tilde{V}$, but the constant loops $\mathbb{T} \subset (\tilde{L}\mathbb{T})^0$ act on Γ_d by $u \mapsto u^{-d}$. Notice also that the V-equivariant map $\Gamma_0 \to \Gamma_d$ (unique up to a scalar multiple) got from the action of the loop $e^{id\theta} \in L\mathbb{T}$ *raises energy* by $\frac{1}{2}d(d+1)$, for it takes the Plücker coordinate π_S to π_{S-d} (cf. (7.7.5)). Thus we have

$$\Gamma_d(k) \cong \Gamma_0(k - \tfrac{1}{2}d(d+1)). \tag{10.4.7}$$

We shall now consider the unitary structure. We know from (10.4.3) that the Hilbert space $\mathcal{H}_0 \subset \Gamma_0$ is isomorphic to the completion of the polynomial ring $R_\mathbb{C}$ with respect to the inner product defined algebraically in Section 10.3. This is the same inner product as that of the symmetric algebra $S(A) \subset \mathrm{Hol}(\bar{A})$, where A has its natural inner product given by

$$\langle \xi, \eta \rangle = -2\mathrm{i}S(\bar{\xi}, \eta). \tag{10.4.8}$$

For if we define coordinate functions p_k on $\bar{A}$ by writing a general element $\eta \in \bar{A}$ in the form

$$\eta = \Sigma \frac{1}{k} p_k z^{-k}$$

(cf. (10.3.4)) then $\{k^{-\frac{1}{2}} p_k\}$ is the basis of $\bar{A}^*$ dual to the basis $\{k^{-\frac{1}{2}} z^{-k}\}$ of $\bar{A}$. The latter basis is orthonormal for the form (10.4.8) and so the former is too, and

$$\langle p_k, p_m \rangle = k \delta_{km}$$

as in Section 10.3. We have proved

Proposition (10.4.9). *The Hilbert space $\mathcal{H}_0$ contained in Γ_0 can be identified canonically with the Hilbert space completion of the symmetric algebra $S(A)$ with the inner product (10.4.8).*

Remark. We can if we like regard $\mathcal{H}_0$ as the space of holomorphic functions on a suitable completion of $\bar{A}$ which are square-summable with respect to the natural Gaussian measure. One suitable completion of $\bar{A}$ is the dual space A^* of A, which is the space of all holomorphic maps

$$\eta : \{z : |z| > 1\} \to \mathbb{C}$$

which have distributional boundary values on $|z| = 1$ and satisfy $\eta(\infty) = 0$.

To conclude this section we should mention the action of $\text{Diff}^+(S^1)$ on Γ. There is an obvious action of $\text{Diff}^+(S^1)$ on $H = L^2(S^1; \mathbb{C})$ given by $(\phi^{-1}, f) \mapsto \phi^* f$ for $\phi \in \text{Diff}^+(S^1)$ and $f \in H$, where $\phi^* f(\theta) = f(\phi(\theta))$. For any complex number λ there is also an action got by regarding the elements of H as 'λ-densities', i.e. the action

$$(\phi^{-1}, f) \mapsto \phi^* f \cdot (\phi')^\lambda.$$

Let us for the moment write H_λ for H with this action. The action is unitary if and only if $\text{Re}(\lambda) = \frac{1}{2}$. In any case $\text{Diff}^+(S^1)$ is contained in $GL_{\text{res}}(H_\lambda)$, and there is an induced projective representation of $\text{Diff}^+(S^1)$ on Γ which intertwines with the representation of $\tilde{L}\mathbb{C}^\times$.

The subspace $\text{Gr}^{(1)}$ of $\text{Gr}(H_\lambda)$ is not preserved by $\text{Diff}^+(S^1)$ for any λ, so we have no reason at first to expect that the description of Γ_0 as $S(A)$ should be compatible with an action of $\text{Diff}^+(S^1)$. But $\text{Diff}^+(S^1)$ is a group of automorphisms of $\tilde{L}\mathbb{T}$ and of the Heisenberg group $\tilde{V}$, and it does act on $S(A)$ by the metaplectic representation. We shall see in Chapter 13 that this action agrees with the one arising from $H_{\frac{1}{2}}$.

10.5 Jacobi's triple product identity

In this chapter we have established an isomorphism of Hilbert spaces

$$\bigoplus_{d \in \mathbb{Z}} \hat{S}(A)_{(d)} \cong \hat{\Lambda}(H_+ \oplus \bar{H}_-), \tag{10.5.1}$$

where $\hat{S}(A)_{(d)}$ is a copy of $\hat{S}(A)$ corresponding to the dth connected component of the Grassmannian. From its construction the isomorphism is compatible with the action of $\tilde{L}\mathbb{T}$ on the two spaces, and also with the action of $\mathbb{T}$ defined by rotating functions on the circle. We shall now consider the decomposition of the spaces (10.5.1) under the abelian subgroup $\mathbb{T} \times \mathbb{T}$ of $\mathbb{T} \tilde{\times} \tilde{L}\mathbb{T}$. (The left-hand $\mathbb{T}$ is the group of rotations of S^1; the right-hand $\mathbb{T}$ is the constant loops in $\mathbb{T}$, regarded as a subgroup of $\tilde{L}\mathbb{T}$.)

We have pointed out in Chapter 2 that if a compact group K acts continuously on a complete topological vector space E, and E_ρ denotes the subspace of E where K acts by an irreducible representation ρ, then the algebraic direct sum $\bigoplus_\rho E_\rho$ is a dense subspace of E. If E_ρ has finite dimension for every ρ then the formal sum $\Sigma_\rho n_\rho \rho$, where n_ρ is the multiplicity of ρ in E, is called the (formal) *character* of E. If K is abelian then the irreducible representations are one-dimensional, and $n_\rho = \dim E_\rho$.

10.5 JACOBI'S TRIPLE PRODUCT IDENTITY

Let us write down the formal character of each side of (10.5.1). We shall denote the representation $(R_\zeta, u) \mapsto \zeta^{-p}u^q$ of $\mathbb{T} \times \mathbb{T}$ simply by $z^p u^q$. Then the character of H_+ is clearly $u(1 + z + z^2 + \ldots)$, while that of $\bar{H}_-$ is $u^{-1}(z + z^2 + z^3 + \ldots)$. To calculate the characters of exterior and symmetric algebras we have the following obvious result.

Proposition (10.5.2). *If a vector space E is a finite sum of one-dimensional representations ρ_i of K then the character of $\Lambda(E)$ is $\prod(1 + \rho_i)$, while that of $S(E)$ is $\prod(1 - \rho_i)^{-1}$.*

This holds even when E is infinite dimensional provided that it makes sense. When K is a torus whose characters form a lattice $\hat{K}$ then it holds providing the characters ρ_i which occur in E all belong to a positive convex cone in $\hat{K}$. That is the case in our examples.

Thus the character of $\Lambda(H_+ \oplus \bar{H}_-)$ is

$$\prod_{k \geq 0}(1 + uz^k) \prod_{k > 0}(1 + u^{-1}z^k).$$

On the other side, the circle of constant loops acts on $S(A)_{(d)}$ by u^{-d}, and the circle of rotations acts on A by $z + z^2 + z^3 + \ldots$, and hence on $S(A)_{(0)}$ by $\prod_{k>0}(1 - z^k)^{-1}$ and on $S(A)_{(d)}$ by (see (10.4.7))

$$z^{\frac{1}{2}d(d+1)} \prod_{k>0}(1 - z^k)^{-1}.$$

The isomorphism of (10.5.1) gives therefore

$$\sum_{d \in \mathbb{Z}} u^{-d} z^{\frac{1}{2}d(d+1)} \prod_{k>0}(1 - z^k)^{-1} = \prod_{k \geq 0}(1 + uz^k) \prod_{k>0}(1 + u^{-1}z^k), \quad (10.5.3)$$

or equivalently, replacing u by $-u$,

$$\sum_{d \in \mathbb{Z}}(-1)^d u^{-d} z^{\frac{1}{2}d(d+1)} = (1 - u)\prod_{k>0}\{(1 - uz^k)(1 - z^k)(1 - u^{-1}z^k)\}, \quad (10.5.4)$$

or

$$\sum_{d \geq 0}(-1)^d \chi_d(u) z^{\frac{1}{2}d(d+1)} = \prod_{k>0}\{(1 - uz^k)(1 - z^k)(1 - u^{-1}z^k)\}, \quad (10.5.5)$$

where

$$\chi_d(u) = u^d + u^{d-1} + u^{d-2} + \ldots + u^{-d}$$

is the character of the $(2d + 1)$-dimensional representation of SO_3.

The identity (10.5.4) or (10.5.5) is called 'Jacobi's triple product' formula. It is not hard to prove it directly (see Hardy and Wright [71] 19.9). The formula obtained by setting $u = 1$ is worth noting explicitly:

$$\sum_{d \geq 0}(-1)^d (2d + 1) z^{\frac{1}{2}d(d+1)} = \prod_{k>0}(1 - z^k)^3. \quad (10.5.6)$$

The identity (10.5.5) is the simplest case of a more general formula called *Macdonald's identity* [107], which we shall discuss in Chapter 14. The general formula involves a compact semisimple group G, and (10.5.5) is the case $G = SO_3$. The generalization of (10.5.6) is of the form

$$\sum_\rho \varepsilon_\rho d_\rho z^{\lambda_\rho} = \prod_{k>0} (1-z^k)^n,$$

where the sum is over all irreducible representations ρ of G, and
n is the dimension of G,
d_ρ is the dimension of ρ,
λ_ρ is the value of the (suitably normalized) Casimir operator in the representation ρ, and
ε_ρ is ± 1 or 0.

From the present point of view the role of SO_3 in (10.5.5) is mysterious.

If we consider the subspace of (10.5.1) where the constant loops act trivially then we obtain an isomorphism

$$\hat{S}(A) \cong \bigoplus_{p \geq 0} \hat{\Lambda}^p(H_+) \otimes \hat{\Lambda}^p(\bar{H}_-). \tag{10.5.7}$$

The character of $\Lambda^p(H_+)$ as a representation of the group of rotations is

$$\sum z^{k_1+k_2+\ldots+k_p},$$

the sum being over all p-tuples such that $0 \leq k_1 < k_2 < \ldots < k_p$. This is easily seen, by induction on p, to be

$$z^{\frac{1}{2}p(p-1)}/(1-z)(1-z^2)\ldots(1-z^p).$$

Similarly the character of $\Lambda^p(\bar{H}_-)$ is

$$z^{\frac{1}{2}p(p+1)}/(1-z)(1-z^2)\ldots(1-z^p).$$

So from (10.5.7) we can read off another well-known but not completely self-evident identity (cf. [71] 19.7):

$$\prod_{k>0}(1-z^k)^{-1} = \sum_{p \geq 0} z^{p^2}/(1-z)^2(1-z^2)^2\ldots(1-z^p)^2. \tag{10.5.8}$$

The function on the left here is called the partition function, as it can be written $\sum p_n z^n$ where p_n is the number of partitions of n. The combinatorial interpretation of (10.5.8), as explained in [71], is simply the fact we have already noticed, that partitions correspond one-to-one to pairs of finite sequences

$$0 \leq a_1 < \ldots < a_p; \quad 0 < b_1 < \ldots < b_p.$$

10.6 The basic representations of LU_n and LSU_n

The loop group LU_n acts on the Hilbert space $H^{(n)} = L^2(S^1; \mathbb{C}^n)$ by multiplication operators, and we saw in Section 6.3 that it is contained in $U_{\text{res}}(H^{(n)})$. Furthermore (cf. (6.7.1)), the central extension $\tilde{U}_{\text{res}}$ induces the basic extension $\tilde{L}U_n$ of LU_n. We can therefore restrict the fundamental representation of $\tilde{U}_{\text{res}}$ to obtain a unitary representation of $\tilde{L}U_n$ on $\mathcal{H}$ which is of level 1. This is called the *basic* representation of $\tilde{L}U_n$. It is irreducible because we have seen that it is irreducible even as a representation of $\tilde{L}\mathbb{T}$, where $L\mathbb{T}$ denotes the maximal abelian subgroup of LU_n obtained, as in Section 6.5, from the identification of $H^{(n)}$ with $H^{(1)}$. The basic representation is the *only* representation of $\tilde{L}U_n$ of level 1—cf. Section 9.3.

In this section we shall denote $\mathcal{H}$ by $\mathcal{H}^{(n)}$ when we are regarding it as the basic representation of $\tilde{L}U_n$ in the standard way.

Let us now consider the decomposition of $\mathcal{H}^{(n)}$ under the subgroup $\tilde{L}SU_n$. We know that under the identity component of $\tilde{U}_{\text{res}}$ the space $\mathcal{H}^{(n)}$ breaks up as $\bigoplus_{d \in \mathbb{Z}} \mathcal{H}_d^{(n)}$, corresponding to the decomposition of the Grassmannian into its connected components. Loops of winding number k in U_n map $\mathcal{H}_d^{(n)}$ to $\mathcal{H}_{d+k}^{(n)}$. The n-fold covering map

$$\mathbb{T} \times SU_n \to U_n,$$

where $\mathbb{T}$ is the centre of U_n, induces an n-fold covering map of identity components

$$(L\mathbb{T})^0 \times LSU_n \to (LU_n)^0.$$

(Notice that this map $L\mathbb{T} \to LU_n$ is quite different from the inclusion of $L\mathbb{T}$ in LU_n as a maximal abelian subgroup mentioned above.)

Proposition (10.6.1). *Each space $\mathcal{H}_d^{(n)}$ is an irreducible representation of $(\tilde{L}\mathbb{T})^0 \times \tilde{L}SU_n$, and breaks up as a tensor product*

$$\mathcal{H}_d^{(n)} \cong \mathcal{F}_d^{(n)} \otimes \mathcal{K}_d^{(n)},$$

where $\mathcal{K}_d^{(n)}$ is an irreducible representation of $\tilde{L}SU_n$ of level 1 which depends only on the congruence class of d modulo n, and $\mathcal{F}_d^{(n)}$ is an irreducible representation of $(\tilde{L}\mathbb{T})^0$ of level n in which the constants $\mathbb{T} \subset \tilde{L}\mathbb{T}$ act by $u \mapsto u^{-d}$. The centre of $SU_n \subset \tilde{L}SU_n$ also acts on $\mathcal{K}_d^{(n)}$ by $u \mapsto u^{-d}$.

We shall postpone the proof of this for a little, and shall discuss its significance instead. The group $(L\mathbb{T})^0$ can be written $\mathbb{T} \times V$ as in Section 10.4, where V is a vector space. The extension $(\tilde{L}\mathbb{T})^0$ is $\mathbb{T} \times \tilde{V}$. The Heisenberg group $\tilde{V}$ has a unique positive-energy representation $\mathcal{F}^{(n)}$ of each level n (see (9.5.10)), and $\mathcal{F}_d^{(n)}$ is $\mathcal{F}^{(n)}$ with $\mathbb{T}$ acting by $u \mapsto u^{-d}$. The representations of the complete group $\tilde{L}\mathbb{T}$ are all induced from representations of $(\tilde{L}\mathbb{T})^0$: there are n representations $\mathcal{F}^{(n)}_{\{d\}}$ of level n, where

$$\mathcal{F}^{(n)}_{\{d\}} = \bigoplus_{k \equiv d \bmod n} \mathcal{F}^{(n)}_k.$$

We can therefore restate (10.6.1) as follows.

Proposition (10.6.2). *Under the action of $\tilde{L}\mathbb{T} \times \tilde{L}SU_n$ the basic representation $\mathcal{H}^{(n)}$ breaks up into n pieces*

$$\mathcal{F}^{(n)}_{\{d\}} \otimes \mathcal{K}^{(n)}_d,$$

where $\mathcal{F}^{(n)}_{\{d\}}$ are the irreducible representations of $\tilde{L}\mathbb{T}$ of level n and the $\mathcal{K}^{(n)}_d$ the irreducible representations of $\tilde{L}SU_n$ of level 1.

In this formulation the result has a very beautiful generalization which was told to us by I. B. Frenkel. We shall state it here without proof. Let us first recall from Section 9.3 that LSU_n has $\binom{n+m-1}{m}$ representations $\mathcal{K}_\lambda$ of level m which are indexed by the sequences $\lambda = (\lambda_1, \ldots, \lambda_n)$ such that

$$0 = \lambda_1 \leq \lambda_2 \leq \ldots \leq \lambda_n \leq m.$$

The group $\tilde{L}U_m$ has the same number $\binom{n+m-1}{m}$ of irreducible representations $\mathcal{F}_\mu$ of level n, indexed by $\mu = (\mu_1, \ldots, \mu_m)$ such that

$$0 \leq \mu_1 \leq \mu_2 \leq \ldots \leq \mu_m < n. \tag{10.6.3}$$

The two indexing sets correspond by the usual operation $\lambda \mapsto \lambda'$ of conjugating partitions [108].

Let us now consider the homomorphism

$$\tilde{L}U_m \times \tilde{L}SU_n \to \tilde{L}U_{mn}$$

defined by the pointwise tensor product of matrices. We have

Proposition (10.6.4). *Under the action of $\tilde{L}U_m \times \tilde{L}SU_n$ the basic representation of $\tilde{L}U_{mn}$ breaks up*

$$\mathcal{H}^{(mn)} = \bigoplus_\mu \mathcal{F}_\mu \otimes \mathcal{K}_{\mu'},$$

where the $\mathcal{F}_\mu$ are the representations of $\tilde{L}U_m$ of level n and the $\mathcal{K}_{\mu'}$ are the representations of $\tilde{L}SU_n$ of level m.

This theorem, which Frenkel ([46], [48]) proved by using the character formula, is made more interesting by the following elementary observation. We shall write $p_n : \tilde{L}U_m \to \tilde{L}U_{mn}$ for the homomorphism defined by $A \mapsto A \otimes 1_n$, and

$$q_n : \tilde{L}U_m \to \tilde{L}U_m$$

for the homomorphism defined by 'n-fold iteration of loops', i.e. composition with the map $S^1 \to S^1$ given by $z \mapsto z^n$.

10.6 BASIC REPRESENTATIONS OF LU_n AND LSU_n

Proposition (10.6.5). *We have*

$$(\mathcal{H}^{(m)})^{\otimes n} \cong p_n^* \mathcal{H}^{(mn)} \cong q_n^* \mathcal{H}^{(m)},$$

where $(\mathcal{H}^{(m)})^{\otimes n}$ denotes the n-fold tensor product of $\mathcal{H}^{(m)}$ with itself.

Proof. This amounts to little more than the observation that

$$H^{(m)} \oplus \underset{\leftarrow n \rightarrow}{\ldots} \oplus H^{(m)} \cong p_n^* H^{(mn)} \cong q_n^* H^{(m)}$$

as representations of LU_m.

Putting (10.6.4) and (10.6.5) together tells us how to decompose the tensor powers of the basic representation of $\tilde{L}U_m$. The result is strikingly reminiscent of Weyl's decomposition of $(\mathbb{C}^m)^{\otimes n}$ under $U_m \times S_n$; but the role of the symmetric group S_n is taken over by $\tilde{L}SU_n$.

We shall now return to (10.6.1).

Proof of (10.6.1). Let us write $\mathbb{T}_Z$ for $\mathbb{T}$ when we wish to emphasize that we mean the centre of U_n.

It is elementary that $\mathcal{H}_d^{(n)}$ is irreducible under the identity component $(\tilde{L}U_n)^0$, and hence under $(\tilde{L}\mathbb{T}_Z)^0 \times \tilde{L}SU_n$.

Now let

$$i : L\mathbb{T} \to LU_n$$

be the inclusion of $L\mathbb{T}$ as a maximal abelian subgroup of LU_n, as in Section 6.5. We notice that if $\gamma \in L\mathbb{T}$ then

$$\det i(\gamma)(z) = \prod_\zeta \gamma(\zeta), \qquad (10.6.6)$$

where ζ runs through the nth roots of z. Let us write $(L\mathbb{T})^0 = \mathbb{T} \times V$ as in Section 10.4, where V is the space of functions $f : S^1 \to \mathbb{R}$ with integral 0. We have

$$V = V_1 \times V_2,$$

where V_1 is the subspace of functions which are periodic with period $2\pi/n$, and V_2 consists of the functions f which satisfy

$$\sum_{k=0}^{n-1} f(z \cdot e^{2\pi i k/n}) = 0.$$

From (10.6.6) we find that $i(V_2) \subset LSU_n$, and the formulae of Section 6.5 show that $i(\mathbb{T} \times V_1) \subset L\mathbb{T}_Z$.

Because $\mathcal{H}^{(n)}$ is irreducible under $i(\tilde{L}\mathbb{T})$ we know that $\mathcal{H}_d^{(n)}$ is irreducible under $i((\tilde{L}\mathbb{T})^0)$, in fact under $i(\tilde{V})$. We even know from Section 10.4 that $\mathcal{H}_d^{(n)} \cong \hat{S}(A)$ under $i(\tilde{V})$, where $V_\mathbb{C} = A \oplus \bar{A}$. The decomposition $V = V_1 \times V_2$ induces a decomposition $A = A_1 \oplus A_2$, and so

$$\mathcal{H}_d^{(n)} = \hat{S}(A_1) \otimes \hat{S}(A_2)$$

compatibly with the action of $i(\tilde{V}_1) \times i(\tilde{V}_2)$. This means that $(\tilde{L}T_Z)^0 \subset (\tilde{L}U_n)^0$ acts irreducibly on $\hat{S}(A_1)$ and trivially on $\hat{S}(A_2)$. It follows that any endomorphism of $\hat{S}(A_1) \otimes \hat{S}(A_2)$ which commutes with the action of $(\tilde{L}T_Z)^0$ is of the form (identity) $\otimes T$. As $\tilde{L}SU_n$ commutes with $(\tilde{L}T_Z)^0$ in $(\tilde{L}U_n)^0$ we conclude that $\tilde{L}SU_n$ acts on $\hat{S}(A_2)$, and acts trivially on $\hat{S}(A_1)$. It acts irreducibly on $\hat{S}(A_2)$ because even the subgroup $i(\tilde{V}_2)$ acts irreducibly.

That completes the proof of (10.6.1).

10.7 Quantum field theory in two dimensions

If H^Δ is, as in Section 6.9, the space of solutions of the Dirac equation†

$$\gamma_0 \frac{\partial \psi}{\partial t} + \gamma_1 \frac{\partial \psi}{\partial x} = im\psi \tag{10.7.1}$$

for functions $\psi: \mathbb{R}^2 \to \mathbb{C}^2$, polarized by the energy operator $-i\partial/\partial t$, then the Hilbert space

$$\mathcal{H}^\Delta = \hat{\Lambda}(H_+^\Delta \oplus \bar{H}_-^\Delta)$$

is the space whose rays represent the states of an assembly of fermionic particles and antiparticles governed by (10.7.1).

The Lie algebra $\mathfrak{gl}_{\widetilde{res}}(H^\Delta)$ acts on $\mathcal{H}^\Delta$. In (6.9.9) we saw that one can find elements in the Lie algebra which satisfy the 'canonical commutation relations', i.e. elements $\Phi(f), \dot{\Phi}(f)$ for each smooth function f of compact support on $\mathbb{R}$ which satisfy (6.9.9). We can now regard these as operators on $\mathcal{H}^\Delta$. It is usual to write them symbolically in the form

$$\Phi(f) = \int_\mathbb{R} f(x)\phi(x)\,dx$$

$$\dot{\Phi}(f) = \int_\mathbb{R} f(x)\dot{\phi}(x)\,dx,$$

so that ϕ and $\dot{\phi}$ are 'operator-valued distributions' on $\mathbb{R}$. In this notation the relations (6.9.9) become

$$[\phi(x), \phi(y)] = [\dot{\phi}(x), \dot{\phi}(y)] = 0,$$
$$[\dot{\phi}(x), \phi(y)] = -i\delta(x-y). \tag{10.7.2}$$

Furthermore the relation (6.9.8) shows that $\dot{\phi}(x)$ is the time-derivative of $\phi(x)$ for the natural time-evolution on $\mathcal{H}^\Delta$.

---

† Here γ_0 and γ_1 are the 2×2 matrices

$$\gamma_0 = \begin{pmatrix} 0 & 1 \\ 1 & 0 \end{pmatrix}, \quad \gamma_1 = \begin{pmatrix} 0 & 1 \\ -1 & 0 \end{pmatrix}.$$

10.7 QUANTUM FIELD THEORY IN TWO DIMENSIONS

The operators $\dot{\phi}(x)$ and $\phi(x)$ are what one expects to find acting on the state-space of an assembly of bosons. It is easy to see that they preserve 'charge', i.e. the decomposition of $\mathcal{H}^\Delta$ into the 'charge sectors'

$$\mathcal{H}^\Delta_d = \bigoplus_{q-p=d} \hat{\Lambda}^p(H^\Delta_+) \otimes \hat{\Lambda}^q(\bar{H}^\Delta_-).$$

In addition it can be shown [24] that the neutral sector $\mathcal{H}^\Delta_0$ is generated by repeatedly applying the $\phi(x)$ and $\dot{\phi}(x)$ to the vacuum vector $1 \in \hat{\Lambda}^0(H^\Delta_+ \oplus \bar{H}^\Delta_-)$. This is commonly expressed by saying that the neutral sector of the theory of free massive fermions described by (10.7.1) is equivalent to a theory of bosons.

Mathematically one of the interesting things about the representation of the relations (10.7.2) on $\mathcal{H}^\Delta$ is that it is not a 'free-field' theory; in fact it does not arise from *any* Heisenberg representation of the group whose Lie algebra is generated by the operators $\Phi(f)$, $\dot{\Phi}(f)$. It seems to be the only explicitly known representation of the relations which is non-trivial in this sense.

A more striking fact is that physicists believe that the standard time-evolution on $\mathcal{H}^\Delta$ makes the operators $\phi(x)$ and $\dot{\phi}(x)$ evolve according to the highly non-linear sine-Gordon equation

$$\frac{\partial^2 \phi}{\partial t^2} - \frac{\partial^2 \phi}{\partial x^2} = k \sin \phi.$$

This was first observed by Coleman [29], and has been worked out more explicitly by Mandelstam [111]. A mathematically clear formulation of the result seems, however, still not to have been found. Cf. Carey and Ruijsenaars [24].

At the end of Section 6.9 we mentioned how the group $\Lambda \mathbb{T}$ of maps $\mathbb{R} \to \mathbb{T}$ with compact support can be regarded as a subgroup of $U_{\text{res}}(H^\Delta)$. It therefore acts projectively on $\mathcal{H}^\Delta$. Carey and Ruijsenaars [24] have shown that when $m > 0$ this gives a factor representation of $\Lambda \mathbb{T}$ of type III. The same applies to the analogous action of ΛU_n on $\mathcal{H}^{\Delta,(n)}$. These are the only examples we have come upon of representations of ΛG which do not come from representations of LG.

11
THE BOREL–WEIL THEORY

This chapter describes the Borel–Weil approach to the representation theory of loop groups. It provides a systematic construction of all the irreducible representations as spaces of holomorphic sections of line bundles on the homogeneous space $Y = LG/T$, and proves that all representations are completely reducible. The results correspond precisely to those for compact groups, which were summarized in Chapter 2. It should be noticed, however, that the method does not illuminate the action of $\text{Diff}^+(S^1)$ on representations of loop groups, for diffeomorphisms do not act holomorphically on Y, as they do not preserve the subgroup $B^+G_{\mathbb{C}}$ of $LG_{\mathbb{C}}$.

As usual we shall use 'representation' to mean 'smooth representation of positive energy' unless we indicate otherwise. In addition we shall assume that G is *simply connected*.

11.1 The space of holomorphic sections of a homogeneous line bundle

In this section we shall study the spaces of holomorphic sections of line bundles on $Y = LG/T = LG_{\mathbb{C}}/B^+G_{\mathbb{C}}$. We recall that $B^+G_{\mathbb{C}}$ consists of the boundary values of holomorphic maps

$$\gamma: \{z : |z| < 1\} \to G_{\mathbb{C}}$$

such that $\gamma(0)$ belongs to the Borel subgroup of $G_{\mathbb{C}}$ corresponding to the positive roots.

We shall assume given a central extension $\tilde{L}G$ of LG, so that Y can be re-expressed as

$$Y = \tilde{L}G/\tilde{T} = \tilde{L}G_{\mathbb{C}}/\tilde{B}^+G_{\mathbb{C}}.$$

Every character λ of $\tilde{T}$ extends canonically to a holomorphic homomorphism

$$\lambda: \tilde{B}^+G_{\mathbb{C}} \to \mathbb{C}^\times,$$

for $\tilde{B}^+G_{\mathbb{C}}/\tilde{N}^+G_{\mathbb{C}} \cong \tilde{T}_{\mathbb{C}}$. We can therefore define a holomorphic line bundle

$$L_\lambda = (\tilde{L}G_{\mathbb{C}} \times \mathbb{C})/\tilde{B}^+G_{\mathbb{C}}$$

11.1 THE SPACE OF HOLOMORPHIC SECTIONS

on Y by making $\tilde{B}^+G_{\mathbb{C}}$ act on $\mathbb{C}$ via λ. (In other words L_λ is obtained from $\tilde{L}G_{\mathbb{C}} \times \mathbb{C}$ by identifying $(\gamma b, \xi)$ with $(\gamma, b\xi)$ for every $b \in \tilde{B}^+G_{\mathbb{C}}$.) The bundle L_λ is by construction homogeneous under $\tilde{L}G_{\mathbb{C}}$ in the sense that $\tilde{L}G_{\mathbb{C}}$ acts on it compatibly with its action on the base-space Y. The action on Y of the group of rotations of the circle is also clearly covered by an action on L_λ.

We shall denote the space of holomorphic sections of L_λ, with the compact open topology, by Γ_λ. It is obviously a holomorphic representation of $\tilde{L}G_{\mathbb{C}}$† but it may, of course, be zero. In fact we shall prove in Section 11.3 that it is non-zero if and only if the weight λ is antidominant. The main object of this section is to prove

Theorem (11.1.1). *If the representation Γ_λ is non-zero then λ is antidominant and the representation is*
  (i) *of positive energy,*
  (ii) *of finite type,*
  (iii) *essentially unitary, and*
  (iv) *irreducible, with lowest weight λ, where λ is regarded as a character of $\mathbb{T} \times \tilde{T}$ which is trivial on $\mathbb{T}$.*

Furthermore, if μ is any other weight of Γ_λ then $\mu - \lambda$ is a sum of positive roots of $\tilde{L}G$.

Proof of (i) and (ii).

We first observe that the rotation action of $\mathbb{T}$ on L_λ and Y induces an action on the space of sections $\Gamma = \Gamma_\lambda$ which intertwines with the action of $\tilde{L}G_{\mathbb{C}}$.

The space Y has a dense open set U which can be identified with the group $N^-G_{\mathbb{C}}$ (see Section 8.7). The action of $N^-G_{\mathbb{C}}$ defines a trivialization of the line bundle L_λ over U. So the restriction to U of a holomorphic section of L_λ can be regarded as an ordinary holomorphic function on $N^-G_{\mathbb{C}}$, and can be pulled back to a holomorphic function on the Lie algebra $N^-\mathfrak{g}_{\mathbb{C}}$ by the exponential map. Assigning to each section its Taylor series at the base-point gives us an injective map

$$\Gamma \to \prod_{p \geq 0} S^p(N^-\mathfrak{g}_{\mathbb{C}})^*, \qquad (11.1.2)$$

where $S^p(V)^*$ denotes the space of continuous symmetric p-multilinear maps $V \times \ldots \times V \to \mathbb{C}$. The map (11.1.2) is equivariant with respect to rotations of the circle, and the space on the right has positive energy, so Γ has positive energy too. It is not quite true that the right-hand-side of (11.1.2) is of finite type, for $N^-\mathfrak{g}_{\mathbb{C}}$ contains the maximal nilpotent subalgebra $\mathfrak{n}_0^-$ of $\mathfrak{g}_{\mathbb{C}}$ as a subspace of zero energy, and so the zero energy part of $\prod S^p(N^-\mathfrak{g}_{\mathbb{C}})^*$ is the infinite dimensional space $\prod S^p(\mathfrak{n}_0^-)^*$. But to

---
† In particular, *every* element of Γ_λ is a smooth vector in the sense of Section 9.1.

prove that the part $\Gamma(k)$ of energy k in Γ is of finite dimension it is enough to show that the part of it annihilated by $\mathfrak{n}_0^-$ is finite dimensional: for $\Gamma(k)$ is a representation of the compact group G, and each irreducible representation of G is finite dimensional and contains a unique ray annihilated by $\mathfrak{n}_0^-$. Because $N^-\mathfrak{g}_\mathbb{C} = \mathfrak{n}_0^- \oplus L_0^- \mathfrak{g}_\mathbb{C}$, the part of the right-hand-side of (11.1.2) annihilated by $\mathfrak{n}_0^-$ is simply

$$\Pi S^p(L_0^- \mathfrak{g}_\mathbb{C})^*,$$

and this is of finite type.

Before leaving the map (11.1.2) we can read off one further fact. The map is equivariant with respect to $\mathbb{T} \times \tilde{T}$ if one multiplies the natural action on the right-hand-side by the character λ (i.e. by the action of $\mathbb{T} \times \tilde{T}$ on the fibre of L_λ at the base-point). The weights of $\mathbb{T} \times \tilde{T}$ which occur in $(N^- \mathfrak{g}_\mathbb{C})^*$ are the positive roots of LG. It follows that any weight of Γ differs from λ by a sum of positive roots, and hence that λ is antidominant. (If $\langle \lambda, h_\alpha \rangle = m > 0$ for some positive root α then the reflection in W_{aff} corresponding to α would take λ to $\lambda - m\alpha$.)

Proof of (iii).

The natural way to introduce an inner product in Γ would be to consider an L^2 inner product with respect to some invariant measure on Y. But unfortunately no such measure on the infinite dimensional space Y has yet been constructed: we shall return to this question in Section 14.5. We therefore proceed differently.

We begin by observing that Γ must contain at least one non-zero section which is invariant under $N^- G_\mathbb{C}$, or, equivalently, which is annihilated by $N^- \mathfrak{g}_\mathbb{C}$. For the space of sections of minimal energy is annihilated by $L_0^- \mathfrak{g}_\mathbb{C}$, and at the same time is a representation of $G_\mathbb{C}$: any lowest weight vector for $G_\mathbb{C}$ in it is annihilated by $\mathfrak{n}_0^-$ and hence by $N^- \mathfrak{g}_\mathbb{C}$.

On the other hand a section which is invariant under $N^- = N^- G_\mathbb{C}$ is completely determined by its value at the base-point of Y. For two such sections which agree at the base-point must agree everywhere in the N^--orbit of the base-point, which is dense in Y. The space of N^--invariant sections is therefore one-dimensional and is clearly stable under $B^- = B^- G_\mathbb{C}$, which acts on it by the holomorphic homomorphism $\lambda: B^- \to \mathbb{C}^\times$.

Let $\sigma \in \Gamma$ denote the unique N^--invariant section which takes the value 1 at the base-point. Recalling that sections s of L_λ are the same thing as holomorphic maps $s: \tilde{L}G_\mathbb{C} \to \mathbb{C}$ which satisfy

$$s(\gamma b^{-1}) = \lambda(b) s(\gamma)$$

for all $b \in B^+$, we can use σ to define a complex-linear map

$$\beta: \bar{\Gamma}^* \to \Gamma$$

11.1 THE SPACE OF HOLOMORPHIC SECTIONS

by $\beta(\varepsilon).(\gamma) = \varepsilon(\bar{\gamma}.\sigma)$ for $\varepsilon \in \bar{\Gamma}^*$ and $\gamma \in \tilde{L}G_{\mathbb{C}}$. (In checking that $\varepsilon(\bar{\gamma}\bar{b}^{-1}.s) = \lambda(b)\varepsilon(\bar{\gamma}.s)$ for $b \in B^+$ one observes that because $\lambda: \tilde{T}_{\mathbb{C}} \to \mathbb{C}^{\times}$ is the complexification of a homomorphism $\lambda: \tilde{T} \to \mathbb{T}$ it satisfies $\overline{\lambda(t)} = \lambda(\bar{t})^{-1}$.) It is easy to see that β is equivariant with respect to $\tilde{L}G$, though not $\tilde{L}G_{\mathbb{C}}$. It also commutes with rotations of the circle, i.e. it preserves energy levels.

The map β defines a sesquilinear inner product $\langle\ ,\ \rangle$ on $\bar{\Gamma}^*$ by

$$\langle \varepsilon, \varepsilon' \rangle = \varepsilon'(\overline{\beta(\varepsilon)}). \tag{11.1.3}$$

This is Hermitian, i.e.

$$\overline{\langle \varepsilon, \varepsilon' \rangle} = \langle \varepsilon', \varepsilon \rangle. \tag{11.1.4}$$

It is enough to prove (11.1.4) for elements of $\bar{\Gamma}^*$ of the form ε_γ, where, for $\gamma \in \tilde{L}G$, we define $\varepsilon_\gamma(s) = s(\bar{\gamma}^{-1})^-$. For them one has

$$\langle \varepsilon_{\gamma_1}, \varepsilon_{\gamma_2} \rangle = \sigma(\gamma_2 \bar{\gamma}_1^{-1}),$$

and the Hermitian property comes from the identity

$$\sigma(\bar{\gamma}^{-1}) = \overline{\sigma(\gamma)},$$

which holds because $\gamma \mapsto \sigma(\bar{\gamma}^{-1})^-$ is an N^--invariant element of Γ and must therefore coincide with σ. (The elements ε_γ span a dense subspace of $\bar{\Gamma}^*$ because if $\varepsilon_\gamma(s) = 0$ for all γ then $s = 0$.)

The form (11.1.3) is $\tilde{L}G$-invariant, and the energy-eigenspaces $\bar{\Gamma}^*(k)$ for $k \geq 0$ are mutually orthogonal. We shall now prove by using the Casimir operator—following an argument due to Gariand [54]—that the form is positive-definite. It then follows that β maps each finite dimensional space $\bar{\Gamma}^*(k)$ isomorphically to $\Gamma(k)$, and so the space $\check{\Gamma}$ of vectors of finite energy in Γ acquires a positive definite inner product. By continuity the inner product will be positive definite on $\bar{\Gamma}^*$. That will complete the proof of (11.1.1)(iii).

The Casimir operator Δ of Section 9.4 does act on $\bar{\Gamma}^*$, for all vectors in Γ are smooth, and the rotation-action of $\mathbb{T}$ on Γ obviously extends to a $\mathbb{T}^p$-action if $\mathfrak{g}$ is decomposed as a product of p simple or abelian factors.

The element $\varepsilon_1 \in \bar{\Gamma}^*$, given by evaluation of sections at the base point, is a cyclic vector for $\bar{\Gamma}^*$, for $\gamma.\varepsilon_1 = \varepsilon_\gamma$, and the vectors ε_γ span $\bar{\Gamma}^*$. It is also annihilated by $N^-\mathfrak{g}_{\mathbb{C}}$, and invariant under rotations, and has weight $\lambda = (\lambda, h)$ under the action of $\tilde{T}$. We conclude as in (9.4.9) that Δ acts on $\bar{\Gamma}^*$ by multiplication by

$$c_\lambda = \tfrac{1}{2}(\|\lambda - \rho\|^2 - \|\rho\|^2).$$

Notice that $\beta(\varepsilon_1) = \sigma$, so that $\langle \varepsilon_1, \varepsilon_1 \rangle = 1$.

We want to prove that $\langle \varepsilon, \varepsilon \rangle > 0$ for any non-zero element ε of $\bar{\Gamma}^*(k)$. We proceed by induction on k—the result is true when $k = 0$, because $\Gamma(0)$ contains no lowest weight vector for G other than σ, and so is an irreducible representation of G and has an essentially unique G-invariant inner product.

We may as well assume that ε is a weight-vector for the action of $\mathbb{T}^p \times \tilde{T}$, with weight $\boldsymbol{\mu} = (\mathbf{k}, \mu, h)$. (Here $\mathbf{k} = (k_1, \ldots, k_p)$ is the 'multi-energy', and $\sum k_i = k$.) We can assume also that $\boldsymbol{\mu}$ is antidominant, for that can be ensured by transforming ε by an element of W_{aff}. Then we shall have

$$\Delta_\mathfrak{g} \varepsilon = c_\mu \varepsilon = \tfrac{1}{2}(\|\mu - \rho\|^2 - \|\rho\|^2)\varepsilon.$$

We recall from Section 9.4 that

$$\Delta = \Delta_\mathfrak{g} - \sum_{a, n > 0} e_a^n e_a^{-n} + \sum (I + ic_j)\frac{\partial}{\partial \theta_j}.$$

So we have

$$c_\lambda \langle \varepsilon, \varepsilon \rangle = \langle \varepsilon, \Delta \varepsilon \rangle$$

$$= \left(c_\mu - \sum (h + c_j)k_j\right)\langle \varepsilon, \varepsilon \rangle - \sum \langle \varepsilon, e_a^n e_a^{-n} \varepsilon \rangle$$

$$= \left(c_\mu - \sum (h + c_j)k_j\right)\langle \varepsilon, \varepsilon \rangle + \sum \langle e_a^{-n}\varepsilon, e_a^{-n}\varepsilon \rangle$$

$$\geq \left(c_\mu - \sum (h + c_j)k_j\right)\langle \varepsilon, \varepsilon \rangle,$$

because $\langle e_a^{-n}\varepsilon, e_a^{-n}\varepsilon \rangle \geq 0$ by the inductive hypothesis. To conclude that $\langle \varepsilon, \varepsilon \rangle > 0$, and so complete the proof, we need to know that

$$c_\lambda > c_\mu - \sum (h + c_j)k_j.$$

In the notation of (9.4.10) we have

$$2\left\{c_\lambda - c_\mu + \sum (h + c_j)k_j\right\} = \|\lambda - \rho\|^2 - \|\mu - \rho\|^2$$

$$= -\langle \lambda + \mu - 2\rho, \mu - \lambda \rangle.$$

This is strictly positive because $\mu - \lambda$ is a sum of positive roots, and $\langle \lambda + \mu - 2\rho, \alpha_i \rangle < 0$ for each simple root α_i, as λ and μ are antidominant and $\langle \rho, \alpha_i \rangle = 1$.

Proof of (11.1.1)(iv). We now know that $\beta : \bar{\Gamma}^* \to \Gamma$ induces an isomorphism of the parts of finite energy. We also know that $\bar{\Gamma}^*$ is cyclic, generated by the lowest weight vector ε_1, and that Γ contains no lowest weight vector (i.e. no N^--invariant vector) other than scalar multiples of $\sigma = \beta(\varepsilon_1)$. It follows that Γ is irreducible as a representation of $\mathbb{T} \tilde{\times} \tilde{L}G$, for any subrepresentation would have to contain a lowest weight vector for the same reason as Γ itself. By using (9.2.3) we can conclude that Γ is also irreducible as a representation of $\tilde{L}G$.

Remarks (11.1.5).
(i) We have pointed out in the course of the proof that the rotation action of $\mathbb{T}$ on Γ extends to an action of $\mathbb{T}^p$ if $\mathfrak{g}$ is decomposed as $\mathfrak{g}_1 \oplus \ldots \oplus \mathfrak{g}_p$. In view of the proof in the following section that all representations of LG are essentially equivalent to sums of representations of the form Γ we conclude that the $\mathbb{T}$-action on any representation always extends to a $\mathbb{T}^p$-action.

(ii) The subgroup $B^+G_{\mathbb{C}}$ is not preserved by the action of the diffeomorphisms of the circle. It is preserved, however, by the action of the subgroup $PSL_2(\mathbb{R})$, which consists of the diffeomorphisms which extend to holomorphic automorphisms of the unit disc. The $\mathbb{T}$-action on any representation therefore always extends to an action of $PSL_2(\mathbb{R})$.

11.2 The decomposition of representations: complete reducibility

In this section we shall prove that all representations of loop groups are, essentially, sums of representations of the type discussed in the preceding section.

We begin with an arbitrary smooth representation E of positive energy. If a smooth vector $\xi \in E$ is decomposed $\xi = \sum \xi_k$ according to 'energy', so that $\xi_k \in E(k)$, then each component ξ_k is a smooth vector, for it is given by the formula

$$\xi_k = \frac{1}{2\pi} \int_0^{2\pi} e^{ik\theta} R_\theta \xi \, d\theta,$$

and the derivative of $\gamma R_\theta \xi$ with respect to $\gamma \in \tilde{L}G$ is a continuous function of γ and θ. It follows that smooth vectors are dense in each subspace $E(k)$.

The antidual $\bar{E}^*$ is also a representation of positive energy; in fact $\bar{E}^*(k)$ is the antidual of $E(k)$. We may as well suppose that the lowest energy which occurs in E is zero. Then $\bar{E}^*(0)$ is a representation of the compact group G. Let us choose a lowest weight vector ε for G in $\bar{E}^*(0)$; it will have a definite weight λ with respect to the torus $\tilde{T}$ of $\tilde{L}G$. Using ε we can define a map from E to the space of continuous sections of the line bundle $L_\lambda = \tilde{L}G \times_{\tilde{T}} \mathbb{C}$ on $Y = \tilde{L}G/\tilde{T}$ which was studied in the previous section. To $\xi \in E$ we associate $s_\xi : \tilde{L}G \to \mathbb{C}$ given by

$$s_\xi(\gamma) = \varepsilon(\gamma^{-1} \cdot \xi). \qquad (11.2.1)$$

Lemma (11.2.2). *The section s_ξ is actually a holomorphic section of L_λ.*

Proof. First suppose that ξ is a smooth vector. Then the section s_ξ is smooth. We claim that its derivative is complex-linear. From the formula

(11.2.1) and the homogeneity of the complex structure on Y it is enough to consider the derivative at the base-point. This is the map

$$\tilde{L}\mathfrak{g}/\mathfrak{t} \to \mathbb{C}$$

given by $v \mapsto -\varepsilon(v \cdot \xi)$. We must show that when this is extended to a complex-linear map

$$\tilde{L}\mathfrak{g}_{\mathbb{C}}/\mathfrak{t}_{\mathbb{C}} = N^{-}\mathfrak{g}_{\mathbb{C}} \oplus N^{+}\mathfrak{g}_{\mathbb{C}} \to \mathbb{C}$$

it vanishes on $N^{+}\mathfrak{g}_{\mathbb{C}} = \mathfrak{n}_{0}^{+} \oplus L_{0}^{+}\mathfrak{g}_{\mathbb{C}}$. But ε vanishes on $E(k)$ for $k > 0$, so $\varepsilon(v \cdot \xi) = 0$ if $v \in L_{0}^{+}\mathfrak{g}_{\mathbb{C}}$. And if $v \in \mathfrak{n}_{0}^{+}$ then $\varepsilon(v \cdot \xi) = -(\bar{v} \cdot \varepsilon) \cdot (\xi) = 0$ because ε is a lowest weight vector for G and is therefore annihilated by $\bar{v} \in \mathfrak{n}_{0}^{-}$. Thus s_{ξ} is holomorphic if ξ is a smooth vector.

Finally, we notice that $\xi \mapsto s_{\xi}$ is a continuous map from E to the space of continuous sections with the compact-open topology. But the holomorphic sections form a closed subspace of the continuous sections, and so if s_{ξ} is holomorphic when ξ belongs to the dense subset of smooth vectors then it is holomorphic for all ξ.

For each lowest weight vector $\varepsilon \in \bar{E}^{*}$ with weight λ we now have a continuous map $\pi_{\varepsilon} : E \to \Gamma_{\lambda}$. But we know from Section 11.1 that Γ_{λ} is irreducible, so by Schur's lemma we can conclude

Theorem (11.2.3). *Every irreducible representation of $\tilde{L}G$ is essentially equivalent to some Γ_{λ}.*

In order to decompose a representation E which is not irreducible we should now like to produce maps from our standard irreducible representations *into* E. That is also easily done. We know that the smooth vectors in $E(0)$ form a dense G-stable subspace. Let us choose a smooth vector ξ which is a lowest weight vector for G. Then we can define a map $\bar{E}^{*} \to \Gamma_{\lambda}$ by the same formula as (11.2.1), i.e. $\varepsilon \mapsto s_{\varepsilon}$, where $s_{\varepsilon}(\gamma) = \varepsilon(\gamma^{-1}\xi)$. (Now, however, λ is the weight of ξ.) The proof that s_{ε} is holomorphic is slightly easier in this case, as s_{ε} is clearly smooth for all ε. The transpose of $\bar{E}^{*} \to \Gamma_{\lambda}$ is a map

$$\omega_{\xi} : \bar{\Gamma}_{\lambda}^{*} \to E.$$

If ε_{1} is the canonical cyclic vector in $\bar{\Gamma}_{\lambda}^{*}$ (see Section 11.1) then $\omega_{\xi}(\varepsilon_{1}) = \xi$.

Remark (11.2.4). At this point we have proved Proposition (9.2.4), for if ξ is a cyclic vector for E then ω_{ξ} must be an essential equivalence by Schur's lemma; and in constructing ω_{ξ} we have used no properties of E or ξ at all other than the fact that ξ is a smooth vector annihilated by $N^{-}\mathfrak{g}_{\mathbb{C}}$.

If we compose ω_{ξ} with π_{ε} then the resulting map $\bar{\Gamma}_{\lambda}^{*} \to \Gamma_{\lambda}$ must take ε_{1} to an N^{-}-invariant section. Evaluating at the base-point, we find that ε_{1}

11.2 THE DECOMPOSITION OF REPRESENTATIONS

maps to $\varepsilon(\xi)\sigma$, where σ is the canonical section of L_λ. (We are assuming here that ξ and ε have the same weight.) If one begins with a lowest weight vector $\xi \in E$ then one can certainly choose a lowest weight vector $\varepsilon \in \bar{E}^*$, with the same weight, so that $\varepsilon(\xi) = 1$. Then the composite $\pi_\varepsilon \circ \omega_\xi$ is the essential equivalence described in Section 11.1. If E_ξ is the closed subspace of E generated by ξ, and $E_\xi^\perp$ is the kernel of π_ε, then E_ξ is essentially equivalent to Γ_λ, and $E_\xi \oplus E_\xi^\perp$ is dense in E. (In fact because E_ξ is of finite type we must have $E_\xi(k) \oplus E_\xi^\perp(k) = E(k)$ for each k.) By choosing a lowest weight vector in $E_\xi^\perp$ and iterating the argument we obtain

Proposition (11.2.5). *Every representation of $\tilde{L}G$ of finite type is essentially a sum of irreducible representations of the form Γ_λ.*

To prove (9.3.1)(ii)—from which the rest of Theorem (9.3.1) follows at once in view of the known properties of Γ_λ—we have only to remove the restriction to representations of finite type in (11.2.5). We shall now explain how that can be done.

The point is to see that if λ is a lowest weight which occurs in E then one can split off from E the part which is isotypical of type Γ_λ. That suffices, for one can assume the representation E is of a definite level h, and then the possible values of λ form a countable set in which for any energy k there are only finitely many λs with $\Gamma_\lambda(k) \neq 0$.

Suppose that λ is of minimal energy—say of zero energy—in E, and is a lowest weight for G. Let $A \subset E(0)$ be the corresponding weight space. There is a natural $\mathbb{T} \times \tilde{T}$-invariant projection operator $a: E \to A$ which we can think of as an antilinear map $a: E \to \bar{A}$. Let L_A denote the homogeneous holomorphic vector bundle on Y with fibre $\bar{A}$, and let Γ_A denote its space of holomorphic sections. Then our standard formula $\xi \mapsto s_\xi$, where $s_\xi(\gamma) = a(\gamma^{-1}\xi)$ defines a map $\pi_A: E \to \Gamma_A$.

Conversely, for each smooth vector $\xi \in A$ we have a map $\omega_\xi : \bar{\Gamma}_\lambda^* \to E$. This depends antilinearly on ξ, so we can put the maps ω_ξ together to define

$$\omega_A : \bar{A} \otimes \bar{\Gamma}_\lambda^* \to E.$$

The proof will be complete as before if we show that the composite

$$\pi_A \circ \omega_A : \bar{A} \otimes \bar{\Gamma}_\lambda^* \to \Gamma_A$$

is injective with dense image. The injectivity is obvious because it needs to be checked only for finite dimensional subspaces of A. The denseness of the image comes similarly from the routine remark that $\cup \Gamma_B$ is dense in Γ_A when B runs through the finite dimensional subspaces of A.

11.3 The existence of holomorphic sections

In this section we shall prove

Proposition (11.3.1). *The line bundle L_λ on Y possesses non-vanishing holomorphic sections if and only if the weight λ is antidominant.*

We recall that λ is antidominant if $\lambda(h_\alpha) \leq 0$ for each positive coroot h_α of $\tilde{L}G$.

The necessity of the condition is seen very easily.† For each positive root $\alpha = (n, \alpha)$ with $\alpha \neq 0$ there is a homomorphism $i_\alpha : SL_2(\mathbb{C}) \to \tilde{L}G_\mathbb{C}$ (see Section 5.2) whose restriction to the diagonal matrices in $SL_2(\mathbb{C})$ is the coroot $h_\alpha : \mathbb{C}^\times \to \tilde{T}_\mathbb{C}$. (In this discussion we shall regard h_α alternately as a homomorphism $\mathbb{C}^\times \to \tilde{T}_\mathbb{C}$ and as an element of $\tilde{\mathfrak{t}}_\mathbb{C}$. Similarly λ is either a homomorphism $\tilde{T}_\mathbb{C} \to \mathbb{C}^\times$ or an element of $\tilde{\mathfrak{t}}_\mathbb{C}^*$. The composite $\lambda \circ h_\alpha : \mathbb{C}^\times \to \mathbb{C}^\times$ is identified with the integer $\lambda(h_\alpha)$.) The positivity of α implies that $i_\alpha(B_0^+) \subset B^+ G_\mathbb{C}$, where B_0^+ is the upper-triangular matrices in $SL_2(\mathbb{C})$. Thus i_α induces a holomorphic map from the Riemann sphere to Y

$$i_\alpha : S^2 = SL_2(\mathbb{C})/B_0^+ \to Y, \qquad (11.3.2)$$

and the pull-back of L_λ to S^2 is the line bundle associated to the character $\lambda \circ h_\alpha$ of the torus of $SL_2(\mathbb{C})$. This line bundle has non-vanishing holomorphic sections if and only if the integer $\lambda \circ h_\alpha = \lambda(h_\alpha)$ is ≤ 0. On the other hand if L_λ possesses a non-vanishing holomorphic section then by homogeneity it possesses a section which does not vanish at the base-point. The base-point belongs to $i_\alpha(S^2)$, so $i_\alpha^* L_\lambda$ has non-vanishing sections, and $\lambda(h_\alpha) \leq 0$.

Conversely we shall prove that if λ is antidominant the bundle L_λ has a section σ which is invariant under $N^- = N^- G_\mathbb{C}$ and takes the value 1 at the base-point. We recall from Section 8.7 that the orbit of the base-point under N^- is a dense open subset U in its connected component in Y, and that Y is covered by the translates wU of U by the elements w of the affine Weyl group W_{aff}. (We are supposing here that a representative $w \in \tilde{L}G$ has been chosen for each element $w \in W_{\text{aff}}$.) The action of N^- defines a holomorphic trivialization of $L_\lambda | U$, in terms of which the desired section σ will be represented by the constant function 1. Similarly wU is the orbit of a point $y_w \in Y$ under $wN^- w^{-1}$, and the action of the latter group defines a trivialization of $L_\lambda | wU$. In terms of these trivializations a section σ of L_λ is the same thing as a compatible collection of holomorphic functions $\phi_w : wN^- w^{-1} \to \mathbb{C}$.

---

† In fact it has already been proved in (11.1.1), but we shall give an alternative argument here which is also instructive.

11.3 THE EXISTENCE OF HOLOMORPHIC SECTIONS

Now (see Section 8.7) we have
$$wN^-w^{-1} = N_w^- \cdot A_w,$$
where $N_w^- = N^- \cap wN^-w^{-1}$, and $A_w = N^+ \cap wN^-w^{-1}$. If the section σ is N^--invariant then ϕ_w will be invariant under the left action of N_w^-, so that ϕ_w is really a function on the finite dimensional group A_w, whose dimension $\ell(w)$ is the codimension of the stratum $\Sigma_w = N_w^- y_w = N^- y_w$ of Y which corresponds to w, i.e. is the number of negative roots made positive by w. We shall construct the functions ϕ_w by induction on $\ell(w)$ beginning with $\phi_1 = 1$.

The only w with $\ell(w) = 0$ is the identity. We know from Section 8.7 that the intersection of wU with the union of the $w'U$ with $\ell(w') < \ell(w)$ is just $wU - \Sigma_w$, so for the inductive step we can assume that ϕ_w is already defined and N_w^--invariant on $N_w^- \cdot (A_w - \{1\})$.

There are now two cases. If $\ell(w) = \dim(A_w) > 1$ then by Hartogs's theorem any holomorphic function on $A_w - \{1\}$ extends automatically to A_w. On the other hand if $\ell(w) = 1$ then w is the reflection in a simple affine root α, and we can perform an explicit calculation. The subgroup A_w is the one-parameter subgroup generated by the root vector e_α, and the point $\exp(xe_\alpha) \cdot y_w$ of wU belongs to U if $x \neq 0$. Indeed from the relation
$$\begin{pmatrix} 1 & x \\ 0 & 1 \end{pmatrix}\begin{pmatrix} 0 & 1 \\ -1 & 0 \end{pmatrix} = \begin{pmatrix} 1 & 0 \\ x^{-1} & 1 \end{pmatrix}\begin{pmatrix} -x & 0 \\ 0 & -x^{-1} \end{pmatrix}\begin{pmatrix} 1 & -x^{-1} \\ 0 & 1 \end{pmatrix}$$
in $SL_2(\mathbb{C})$ we find on applying the homomorphism i_α that
$$\exp(xe_\alpha) \cdot w = \exp(x^{-1}e_{-\alpha}) \cdot h_\alpha(-x) \cdot \exp(-x^{-1}e_\alpha).$$
This means that the point
$$(\exp(xe_\alpha).w \, , \, \xi) \in \tilde{L}G_\mathbb{C} \times_{B^+} \mathbb{C} = L_\lambda,$$
is the same as
$$(\exp(x^{-1}e_{-\alpha}), \lambda(h_\alpha(-x))\xi),$$
from which we deduce that
$$\phi_w(\exp(xe_\alpha)) = \lambda(h_\alpha(-x))^{-1} = (-x)^{-\lambda(h_\alpha)}.$$
Thus ϕ_w extends holomorphically across the point $x = 0$ if and only if $\lambda(h_\alpha) \leq 0$, i.e. if and only if the weight λ is antidominant. That completes the proof of the existence of the holomorphic section σ.

Reformulation in terms of bi-invariant functions on $\tilde{L}G_\mathbb{C}$

An N^--invariant section σ of L_λ is the same thing as a holomorphic function $\sigma : \tilde{L}G_\mathbb{C} \to \mathbb{C}$ which satisfies
$$\sigma(\gamma b^{-1}) = \lambda(b)\sigma(\gamma) \quad \text{for} \quad b \in B^+,$$

and
$$\sigma(n\gamma) = \sigma(\gamma) \qquad \text{for} \quad n \in N^-.$$

If G is simply connected, so that $N^-.\tilde{T}_\mathbb{C}.N^+$ is a dense subset of $\tilde{L}G_\mathbb{C}$, then we can equivalently say that σ is a holomorphic function $\tilde{L}G_\mathbb{C} \to \mathbb{C}$ which is invariant under the left action of N^- and the right action of N^+, and whose restriction to $\tilde{T}_\mathbb{C}$ is λ. Notice that $\sigma(\gamma)$ is the 'vacuum expectation value' of γ in the unitary representation of $\tilde{L}G$ corresponding to λ: i.e.

$$\sigma(\gamma) = \langle \Omega, \gamma.\Omega \rangle,$$

where Ω is the essentially unique N^--invariant vector in the representation. It is worth restating what we have proved in terms of the algebra F of such $N^- \times N^+$-invariant functions.

To be quite precise, we define F as a *graded* algebra. The centre $\mathbb{T}$ of $\tilde{L}G$ acts on $\tilde{L}G_\mathbb{C}$ commuting with both left and right multiplications, and so acts on the $N^- \times N^+$-invariant functions. Let F_h denote the space of $N^- \times N^+$-invariant holomorphic functions on $\tilde{L}G_\mathbb{C}$ on which $u \in \mathbb{T}$ acts as u^h; and let F denote $\bigoplus_{h \geq 0} F_h$. Then F is a dense subalgebra of the algebra of *all* $N^- \times N^+$-invariant holomorphic functions. We can now assert

Proposition (11.3.3). *If G is simply connected then*
  (i) *each dominant weight $\lambda: \tilde{T} \to \mathbb{T}$ extends uniquely to a holomorphic function $\sigma_\lambda \in F$, and*
  (ii) *the σ_λ form a vector space basis for the algebra F.*

This result is an exact analogue of the finite dimensional situation. Consider, for example, the group $SL_n(\mathbb{C})$. If $N^\pm$ denote as usual the groups of strictly upper and lower triangular matrices then there are $n-1$ basic $N^- \times N^+$-invariant holomorphic functions $\sigma_1, \ldots, \sigma_{n-1}$ on $SL_n(\mathbb{C})$: if $A \in SL_n(\mathbb{C})$ we have $\sigma_k(A) = \det(A_k)$, where A_k is the leading $k \times k$ submatrix of A—i.e. σ_k is a matrix element of the basic irreducible representation $\Lambda^k(\mathbb{C}^n)$. The polynomial algebra $\mathbb{C}[\sigma_1, \ldots, \sigma_n]$ is a dense subalgebra of the algebra of $N^- \times N^+$-invariant holomorphic functions on $SL_n(\mathbb{C})$, and obviously has a vector space basis indexed by the positive weights.

The case of $\tilde{L}SU_n$

In the case of the basic representation of $\tilde{L}SU_n$ we can write down the holomorphic function σ rather more explicitly, for $\tilde{L}SU_n$ is (cf. Section 6.7) a subgroup of the identity component of $\widetilde{GL}_{\text{res}}$, and σ extends to the larger group.

We recall that an element of $\widetilde{GL}_{\text{res},0}$ is an equivalence class of pairs

(A, q) with

$$A = \begin{pmatrix} a & b \\ c & d \end{pmatrix} \in GL_{\text{res}}$$

and $q \in GL(H_+)$ such that $aq^{-1} - 1$ is of trace class. We can define a holomorphic function $\sigma : \widetilde{GL}_{\text{res},0} \to \mathbb{C}$ by

$$\sigma(A, q) = \det(aq^{-1})$$

—it vanishes precisely when a is not invertible. It is immediate that this function is invariant under the left action of GL^- and the right action of GL^+, where GL^- is the subgroup of $\widetilde{GL}_{\text{res}}$ consisting of elements of the form

$$\left(\begin{pmatrix} a & 0 \\ c & d \end{pmatrix}, a \right),$$

and GL^+ is defined analogously. That implies that the restriction of σ to $\tilde{L}SU_n$ is the function we want.

11.4 The smoothness condition

We have been restricting our attention to representations of loop groups which are *smooth* as well as being of positive energy. The hypothesis of smoothness is probably redundant: it is certainly so when $G = SU_n$. We shall now sketch a rather roundabout proof of this fact.

We begin with a simply connected group G. Then the homogeneous space $Y = LG/T$ has a dense subspace Y_{pol} which is the orbit of the base-point y_1 under the polynomial loop group $L_{\text{pol}}G$ of Section 3.5. We have seen in Chapter 8 that Y_{pol} is the union of a family of finite dimensional cells C_w indexed by the elements of the affine Weyl group W_{aff}. The closure $\bar{C}_w$ is a compact complex algebraic variety, usually with singularities.

Having chosen a homogeneous line bundle L_λ on Y let Γ_{alg} denote the space of sections of $L_\lambda | Y_{\text{pol}}$ which are *algebraic* on $\bar{C}_w$ for all w. We give Γ_{alg} the topology of uniform convergence on $\bar{C}_w$ for each w. There is a continuous map $\Gamma \to \Gamma_{\text{alg}}$, where Γ is the space of holomorphic sections of L_λ, for on a compact algebraic variety every holomorphic function is algebraic. The polynomial loop group $\tilde{L}_{\text{pol}}G_\mathbb{C}$ acts on Γ_{alg}.

Proposition (11.4.1). *The map* $\Gamma \to \Gamma_{\text{alg}}$ *is injective, and its image is dense.*

Proof. The injectivity holds because Y_{pol} is dense in Y, so it is enough to show that Γ_{alg} is an irreducible representation of $\tilde{L}_{\text{pol}}G$. This is proved in exactly the same way as the irreducibility of Γ; it depends on three

11 THE BOREL-WEIL THEORY

observations:
  (i)   the space Γ_{alg} is of positive energy,
  (ii) the polynomial part of $N^-G_{\mathbb{C}}$ acts on Y_{pol} with a dense orbit, and
  (iii) the polynomial Lie algebra $L_{\text{pol}}\mathfrak{g}_{\mathbb{C}}$ acts on Γ_{alg}, and hence also the Casimir operator.

To prove that there are no other representations of $\check{L}G$ than the ones we have been studying, i.e. to prove that all positive energy representations are smooth, it is enough, following the method of Section 11.2, to show that for any representation E one can construct maps

$$\pi_\varepsilon : E \to \Gamma_{\text{alg}}$$

and

$$\omega_\xi : \bar{E}^* \to \Gamma_{\text{alg}}$$

analogous to the maps into Γ constructed in the smooth case.†

If the element w of W_{aff} is written as a product $w = \rho_{\alpha_1}\rho_{\alpha_2}\ldots\rho_{\alpha_k}$ of reflections in root-planes corresponding to simple roots α_i of LG then the closed cell $\bar{C}_w$ is the image of the map

$$SU_2 \times \ldots \times SU_2 \to Y \qquad (11.4.2)$$

given by

$$(g_1, \ldots, g_k) \mapsto i_{\alpha_1}(g_1) \ldots i_{\alpha_k}(g_k) \cdot y_1.$$

The map (11.4.2) evidently factorizes through

$$Z_w = SU_2 \times_{\mathbb{T}} SU_2 \times_{\mathbb{T}} \ldots \times_{\mathbb{T}} SU_2/\mathbb{T},$$

which is an iterated fibration of 2-spheres over S^2. (Note that Z_w depends on the factorization of w which was chosen.) Furthermore Z_w is naturally a complex manifold, because it can also be described as

$$Z_w = P_{\alpha_1} \times_{B^+} P_{\alpha_2} \times_{B^\times} \ldots \times_{B^+} P_{\alpha_k}/B^+,$$

where

$$P_{\alpha_i} = i_{\alpha_i}(SL_2(\mathbb{C})) \cdot B^+ \subset LG_{\mathbb{C}}.$$

The surjection $q: Z_w \to \bar{C}_w$ is a birational equivalence of algebraic varieties, and there is also a surjective holomorphic map

$$p: SL_2(\mathbb{C}) \times \ldots \times SL_2(\mathbb{C}) \to Z_w.$$

---

† That will tell us that, as a representation of $\check{L}_{\text{pol}}G$, E is essentially equivalent to a product of spaces of the form Γ_{alg}. But then the dense subspace $\check{E}$ of E is isomorphic to a product of spaces $\check{\Gamma}_{\text{alg}} = \check{\Gamma}$. It follows that E has a unique Hilbert space completion which is a Hilbert space sum of the standard representations which we know, at first as a representation of $\check{L}_{\text{pol}}G$, but then, by continuity, as a representation of $\check{L}G$.

11.4 THE SMOOTHNESS CONDITION

To define $\pi_\varepsilon : E \to \Gamma_{\text{alg}}$ we have to show that the formula (11.2.1) for s_ξ defines an element of Γ_{alg}. By continuity and linearity we may as well assume that ξ has a definite weight under the action of the torus $\tilde{T}$ of $\tilde{L}G$. But then ξ transforms according to a finite dimensional representation of each of the groups $i_{\alpha_i}(SU_2)$, and so does $\gamma \cdot \xi$ for any $\gamma \in \tilde{L}G$. As any finite dimensional representation of SU_2 extends holomorphically to $SL_2(\mathbb{C})$ it follows that the restriction of s_ξ to $\bar{C}_w$ is holomorphic when pulled back by the composite

$$qp : SL_2(\mathbb{C}) \times \ldots \times SL_2(\mathbb{C}) \to Z_w \to \bar{C}_w.$$

It is therefore holomorphic, or equivalently algebraic, on Z_w, for p is a holomorphic fibration. In order to conclude that $s_\xi \mid \bar{C}_w$ is algebraic we need to know that the variety $\bar{C}_w$ is *normal* in the sense of algebraic geometry. If that is true then the proof is complete, for exactly the same reasoning produces a map $\omega_\xi : \bar{E}^* \to \Gamma_{\text{alg}}$.

In general it is unknown whether the closed Bruhat cells $\bar{C}_w$ are normal. But when $G = SU_n$ the normality has been proved by Lusztig (cf. [106], which one must combine with the results of [95]), and so we can assert

Proposition (11.4.3). *Every representation of $\tilde{L}SU_n$ of positive energy is essentially equivalent to a smooth representation.*

12
THE SPIN REPRESENTATION

The orthogonal group O_{2n} has a very interesting irreducible projective unitary representation of dimension 2^n, called the *spin representation*. In the first two sections of this chapter we shall describe it as explicitly as possible, and then we shall go on to discuss its infinite dimensional analogue, which among other things provides a construction of the basic irreducible representation of the loop group LO_{2n}.

12.1 The Clifford algebra

If we identify the space $\mathbb{R}^{2n}$ on which O_{2n} acts with $\mathbb{C}^n$ then we can think of the unitary group U_n as a subgroup of O_{2n}. The group U_n has its natural action on $\mathbb{C}^n$, and it acts irreducibly on each exterior power $\Lambda^k(\mathbb{C}^n)$ for $0 \leq k \leq n$. These actions do not, of course, extend from U_n to O_{2n}; but if one takes them together then the action of U_n on the 2^n-dimensional space $\Lambda(\mathbb{C}^n) = \bigoplus_{k=0}^{n} \Lambda^k(\mathbb{C}^n)$ does extend to a *projective* representation of O_{2n}. This is the spin representation.

The simplest and most traditional way to describe the action of O_{2n} on the exterior algebra $\Lambda(\mathbb{C}^n)$ is by means of the Clifford algebra. If V is any real vector space with an inner product B then the Clifford algebra $C(V)$ of V is an algebra over the real numbers which contains V as a sub-vector-space and is generated by the elements of V subject to the relations

$$v_1 v_2 + v_2 v_1 = 2B(v_1, v_2) \tag{12.1.1}$$

for $v_1, v_2 \in V$. Thus if $\{e_1, \ldots, e_m\}$ is an orthonormal basis of V then the e_i anticommute in $C(V)$, and $e_i^2 = 1$. The 2^m elements $e_{i_1} e_{i_2} \ldots e_{i_k}$, for $i_1 < i_2 < \ldots < i_k$, form a vector space basis of $C(V)$.

The Clifford algebra (at least if V is finite dimensional) contains the Lie algebra $\mathfrak{o}(V)$ of the orthogonal group $O(V)$ as a sub-Lie-algebra, i.e as a sub-vector-space closed under the bracket operation $(a, b) \mapsto [a, b] = ab - ba$. Indeed, with respect to an orthonormal basis as above, the Lie algebra $\mathfrak{o}(V)$ consists of the $m \times m$ real skew matrices. The matrices $E_{ij} - E_{ji}$ (for $i < j$), where E_{ij} has 1 in the $(i, j)^{\text{th}}$ place and zero elsewhere, form a basis for the Lie algebra; and it is easy to check that the elements $\frac{1}{2} e_i e_j$ of $C(V)$ have exactly the same bracket relations as the $E_{ij} - E_{ji}$. (For a more elegant and invariant description of this isomorphism one would first identify the Lie algebra of $O(V)$ with $\Lambda^2(V)$, and then map $\Lambda^2(V)$ into $C(V)$ by $v_1 \wedge v_2 \mapsto \frac{1}{2}(v_1 v_2 - v_2 v_1)$.) The embedding of $\mathfrak{o}(V)$ in $C(V)$

12.1 THE CLIFFORD ALGEBRA

is characterized by the relation

$$[a, v] = A(v) \tag{12.1.2}$$

in $C(V)$, where a is the element of $C(V)$ corresponding to $A \in \mathfrak{o}(V)$.

From the preceding remarks it follows that the Lie algebra of $O(V)$ acts on any vector space M on which the algebra $C(V)$ acts; and if M is finite dimensional we can then exponentiate and get a (possibly multiple-valued) representation of the identity component $SO(V)$ of $O(V)$. In fact any representation of $SO(V)$ so obtained is two-valued, for while $\exp 2\pi(E_{ij} - E_{ji}) = 1$, we have $(\frac{1}{2}e_i e_j)^2 = -\frac{1}{4}$ in $C(V)$, and hence $\exp(2\pi \cdot \frac{1}{2} e_i e_j) = -1$.

The elements of the Lie algebra of $O(V)$ can be exponentiated inside the finite-dimensional algebra $C(V)$. The exponentials generate a subgroup $\mathrm{Spin}(V)$ of the group of invertible elements of $C(V)$. If $g \in \mathrm{Spin}(V)$ and $v \in V$ then

$$gvg^{-1} = T_g(v) \tag{12.1.3}$$

in $C(V)$, where T_g is the element of $SO(V)$ corresponding to g. (This follows at once from (12.1.2).)

It is well known that the orthogonal group $O(V)$ is generated by the reflections in the hyperplanes of V. But if $u \in V$ is non-zero then $uvu^{-1} = -\rho_u(v)$ in $C(V)$, where ρ_u is the reflection in the hyperplane perpendicular to u. Thus if $\dim V$ is even, and we define $\mathrm{Pin}(V)$ as the subgroup generated by the unit vectors of V in the group of invertible elements of $C(V)$, then there will be a surjective homomorphism $\mathrm{Pin}(V) \to O(V)$ characterized by the property (12.1.3). It is not hard to show that $\mathrm{Pin}(V)$ is a double covering of $O(V)$, and that its identity component is $\mathrm{Spin}(V)$. We shall not pursue this approach, however, because it is not very useful in infinite dimensions.

We must now describe how $C(\mathbb{R}^{2n})$ acts on $\Lambda(\mathbb{C}^n)$. Let $J: V \to V$ be the map such that $J^2 = -1$ which defines the complex structure of $V = \mathbb{R}^{2n}$. We can extend J to a complex-linear map $J: V_\mathbb{C} \to V_\mathbb{C}$. The complexified space $V_\mathbb{C}$ breaks up

$$V_\mathbb{C} = W \oplus \bar{W} \tag{12.1.4}$$

into the $+i$ and $-i$ eigenspaces of J. The complex vector space defined by (V, J) is best identified with W: i.e. it is more convenient to think of it as a complex subspace of $V_\mathbb{C}$ than to identify it as a set with V. We shall assume that J preserves the inner product B of V, and shall extend the inner product in the obvious way to a complex bilinear symmetric form on $V_\mathbb{C}$—not to a hermitian form. It is clear how to define the complex Clifford algebra $C(V_\mathbb{C})$ of the complex vector space $V_\mathbb{C}$: it will have $V_\mathbb{C}$ as a complex subspace which generates it subject to the relations (12.1.1) for v_1 and v_2 in $V_\mathbb{C}$. This algebra is simply the complexification of the original algebra $C(V)$. We shall prove

12 THE SPIN REPRESENTATION

Proposition (12.1.5). *The exterior algebra $\Lambda(W)$ is a module for $C(V_{\mathbb{C}})$. In fact $C(V_{\mathbb{C}})$ can be identified with the $2^n \times 2^n$ matrix algebra of all complex linear transformations of $\Lambda(W)$.*

Proof. We first observe that W is an isotropic subspace of $V_{\mathbb{C}}$, i.e. that $B(w_1, w_2) = 0$ for all $w_1, w_2 \in W$. (For $B(w_1, w_2) = B(Jw_1, Jw_2) = B(iw_1, iw_2) = -B(w_1, w_2)$.) The conjugate subspace $\bar{W}$ is also isotropic. Thus $C(V_{\mathbb{C}})$ contains the exterior algebras $\Lambda(W)$ and $\Lambda(\bar{W})$ as subalgebras. Furthermore $C(V_{\mathbb{C}})$ is generated by the elements of W and $\bar{W}$ subject to the anticommutation relation

$$\bar{w}_1 w_2 + w_2 \bar{w}_1 = 2B(\bar{w}_1, w_2) \tag{12.1.6}$$

for $\bar{w}_1 \in \bar{W}$ and $w_2 \in W$.

Now the algebra $\Lambda(W)$ acts on itself by left multiplication. On the other hand for any α in the dual space W^* of W there is an antiderivation

$$D_\alpha : \Lambda(W) \to \Lambda(W)$$

which lowers degree by one and is characterized by the two properties

(i) $D_\alpha(\xi \wedge \eta) = (D_\alpha \xi) \wedge \eta + (-1)^{\deg(\xi)} \xi \wedge D_\alpha \eta,$ (12.1.7)

and

(ii) $D_\alpha(w) = \alpha(w)$ for $w \in W$.

It is easy to check that D_α and D_β anticommute for any $\alpha, \beta \in W^*$. The bilinear form on $V_{\mathbb{C}}$ identifies $\bar{W}$ with W^*. We define an action of $\Lambda(\bar{W})$ on $\Lambda(W)$ by making $\bar{w}$ act as D_α, where $\alpha = 2B(\bar{w}, \cdot)$. Then the relation (12.1.7) becomes (12.1.6), and so the actions of $\Lambda(W)$ and $\Lambda(\bar{W})$ combine to define an action of $C(V_{\mathbb{C}})$. (The operations of W and $\bar{W}$ on $\Lambda(W)$ are familiar as 'creation and annihilation operators' in quantum field theory.)

To see that $C(V_{\mathbb{C}})$ is the algebra of all endomorphisms of $\Lambda(W)$ we notice that $C(V_{\mathbb{C}})$ and $\text{End}(\Lambda(W))$ have the same dimension as vector spaces, and that the $2^n \times 2^n$ elements

$$w_{i_1} \ldots w_{i_p} \bar{w}_1 \bar{w}_2 \ldots \bar{w}_n w_{j_1} w_{j_2} \ldots w_{j_q}$$

of $C(V_{\mathbb{C}})$, where $\{w_1, \ldots, w_n\}$ is a basis of W such that $2B(\bar{w}_i, w_j) = \delta_{ij}$, and $\{i_1, \ldots, i_p\}$ and $\{j_1, \ldots, j_q\}$ run through the subsets of $\{1, \ldots, n\}$, correspond to the natural basis of $\text{End}(W)$.

It follows almost at once from (12.1.5) that the representation of $\text{Pin}(V)$ on $\Lambda(W)$ is irreducible (cf. Section 12.5); we shall give a completely different proof of the irreducibility in Section 12.3. If we separate out the even and odd exterior powers, writing

$$\Lambda(W) = \Lambda^{\text{even}}(W) \oplus \Lambda^{\text{odd}}(W),$$

12.2 CONSTRUCTION OF THE SPIN REPRESENTATION

then the connected group Spin(V) preserves the decomposition, and $\Lambda^{\text{even}}(W)$ and $\Lambda^{\text{odd}}(W)$ are inequivalent irreducible representations of Spin(V): they are inequivalent even as projective representations of the subgroup $U(W)$ of $O(V)$.

Remark. The double covering of $U(W)$ which is the restriction of the covering Spin(V) of $U(W)$ can be described explicitly as

$$U(W)^{\sim} = \{(A, u) \in U(W) \times \mathbb{T} : u^2 = \det A\}.$$

It acts on $\Lambda(W)$ preserving each $\Lambda^k(W)$, and (A, u) acts on $\Lambda^k(W)$ by $u^{-1}\Lambda^k(A)$. Projectively, of course, this is just the standard action of $U(W)$ on $\Lambda(W)$.

To conclude this section let us notice that $\Lambda(W)$ has a natural hermitian inner product which makes it into a (finite dimensional) complex Hilbert space. For $w_1, w_2 \in W$ we define

$$\langle w_1, w_2 \rangle = 2B(\bar{w}_1, w_2), \tag{12.1.8}$$

and in general

$$\langle w_1 \wedge w_2 \wedge \ldots \wedge w_k, w_1' \wedge \ldots \wedge w_k' \rangle = \det(\langle w_i, w_j' \rangle).$$

The operations of $w \in W$ and $\bar{w} \in \bar{W}$ on $\Lambda(W)$ are then *adjoint* with respect to $\langle \ , \ \rangle$:

$$\langle \eta, w.\xi \rangle = \langle \bar{w}.\eta, \xi \rangle$$

for $\xi, \eta \in \Lambda(W)$. It follows that the elements of V act by self-adjoint operators, and the elements of the Lie algebra of $O(V)$ are skew. The spin representation is therefore unitary.

12.2 Second construction of the spin representation

Our next object is to describe a construction of the spin representation which is global rather than infinitesimal, and which will generalize to the infinite dimensional case. In the preceding section, by introducing a complex structure on $V = \mathbb{R}^{2n}$, we decomposed $V_{\mathbb{C}}$ as $W \oplus \bar{W}$, where W was an isotropic subspace of $V_{\mathbb{C}}$, and we realized the spin representation on the exterior algebra $\Lambda(W)$. There is therefore a canonical ray $\text{Pf}_{\bar{W}}$ in the representation space consisting of the multiples of the unit in the exterior algebra, i.e. $\text{Pf}_{\bar{W}} = \Lambda^0(W)$. (The reason for the notation $\text{Pf}_{\bar{W}}$ will be explained in Section 12.3.) If we think of the representation space $S \cong \Lambda(W)$ as a module for the Clifford algebra $C(V_{\mathbb{C}})$ we can describe $\text{Pf}_{\bar{W}}$ as the unique 'vacuum state' corresponding to $\bar{W}$, i.e.

$$\text{Pf}_{\bar{W}} = \{\xi \in S : \bar{w}\xi = 0 \text{ for all } \bar{w} \in \bar{W}\}. \tag{12.2.1}$$

We could have constructed S equally well starting from any complex structure on V: it contains a distinguished ray for each complex structure. The construction we shall now describe takes the existence of the rays $\text{Pf}_{\bar{W}}$ as its starting point.

We begin by assembling the main facts about the space $\mathscr{J}(V)$ of complex structures on V. A point of $\mathscr{J}(V)$ is by definition an orthogonal transformation $J: V \to V$ such that $J^2 = -1$. Any two choices of J are conjugate in the orthogonal group $O(V) \cong O_{2n}$, and so $\mathscr{J}(V)$ can be identified with the homogeneous space O_{2n}/U_n. On the other hand to give J is precisely equivalent to giving a complex n-dimensional isotropic subspace W of $V_{\mathbb{C}}$ (cf. (12.1.4)). This means that $\mathscr{J}(V)$ can be regarded as a complex algebraic subvariety of the Grassmannian $\text{Gr}_n(V_{\mathbb{C}})$ of all n-dimensional complex subspaces of $V_{\mathbb{C}}$, and is a homogeneous space of the complexified group $O_{2n}(\mathbb{C})$. We shall usually think of the points of $\mathscr{J}(V)$ as isotropic subspaces W rather than as operators J. The space $\mathscr{J}(V)$ has complex dimension $\frac{1}{2}n(n-1)$, and consists of two connected components $\mathscr{J}_{\pm}(V)$: a complex structure on V defines an orientation of V, and the components correspond to the two possible orientations.

If $W \in \mathscr{J}(V)$ then the graph W_S of a linear map $S: W \to \bar{W}$ belongs to $\mathscr{J}(V)$—i.e. is isotropic—if and only if S is skew:

$$B(w, Sw') = -B(Sw, w')$$

for all $w, w' \in W$. These graphs form a dense open set U_W of the connected component of $\mathscr{J}(V)$ containing W, and the U_W form an atlas for $\mathscr{J}(V)$: an element $Y \in \mathscr{J}(V)$ belongs to U_W if and only if $Y \cap \bar{W} = 0$. In general—with respect to a given $W \in \mathscr{J}(V)$—we can express Y as $Y_0 \oplus \bar{W}_1$, where $\bar{W}_1 = Y \cap \bar{W}$ and Y_0 is the graph of a skew map $W_0 \to \bar{W}_0$, where $W_0 = W \cap \bar{W}_1^{\perp}$. (Thus $W = W_0 \oplus W_1$.) The space Y can clearly be connected continuously to $W_0 \oplus \bar{W}_1$ in $\mathscr{J}(W)$, and $W_0 \oplus \bar{W}_1$ belongs to the same connected component of $\mathscr{J}(V)$ as W if and only if $\dim(W_1)$ is even. This gives us

Proposition (12.2.2). *Two spaces W and Y belong to the same connected component of $\mathscr{J}(V)$ if and only if $\dim(Y \cap \bar{W})$ is even.*

Suppose that an element A of the complex orthogonal group $O(V_{\mathbb{C}})$ is written as a 2×2 matrix

$$A = \begin{pmatrix} a & b \\ c & d \end{pmatrix}$$

with respect to the decomposition $V_{\mathbb{C}} = \bar{W} \oplus W$. (We shall write the factor $\bar{W}$ first in this section because the vacuum vector in $\Lambda(W)$ corresponds to $\bar{W}$ rather than W.) Proposition (12.2.2) implies that A belongs to the identity component of $O(V_{\mathbb{C}})$ if and only if the nullity of a is even.

12.2 CONSTRUCTION OF THE SPIN REPRESENTATION

If $\bar{W}_S \in U_{\bar{W}}$ is the graph of $S: \bar{W} \to W$ then $A(\bar{W}_S)$ belongs to $U_{\bar{W}}$ if and only if $a + bS$ is invertible, in which case $A(\bar{W}_S) = \bar{W}_T$, where

$$T = (c + dS)(a + bS)^{-1},$$

which is necessarily skew. To see this, observe that $\bar{W}_S$ is the image of

$$\begin{pmatrix} 1 \\ S \end{pmatrix}: \bar{W} \to \bar{W} \oplus W$$

and so $A(\bar{W}_S)$ is the image of

$$\begin{pmatrix} a + bS \\ c + dS \end{pmatrix}$$

and hence of

$$\begin{pmatrix} 1 \\ (c + dS)(a + bS)^{-1} \end{pmatrix}.$$

We shall see—and in any case it follows simply from the characterization (12.2.1)—that when the spin representation is realized as $\Lambda(W)$ the ray corresponding to the graph $\bar{W}_S$ of a skew operator $S: \bar{W} \to W$ is generated by the vector $e^{\frac{1}{2}S}$. Here we have identified $\bar{W}$ with the dual of W, so that S can be regarded as an element of $\Lambda^2(W)$: if $\{w_1, \ldots, w_n\}$ is an orthonormal basis of W for the inner product $\langle \, , \, \rangle$ of (12.1.8), and $S: \bar{W} \to W$ is represented by the matrix (s_{ij}) with respect to the bases $\{\bar{w}_i\}$ and $\{w_i\}$, then S is identified with $\sum s_{ij} w_i \wedge w_j$ in $\Lambda^2(W)$. We need the following important calculation.

Proposition (12.2.3). *If $S, T \in \Lambda^2(W)$ then*

$$\langle e^{\frac{1}{2}S}, e^{\frac{1}{2}T} \rangle = \det(1 - \bar{S}T)^{\frac{1}{2}}.$$

Here one must choose the branch of the square root which becomes $+1$ when S or T is zero. We can be more explicit than this, however, by using the notion of *Pfaffian*, which we shall now briefly recall.

The determinant of a skew $n \times n$ matrix $S = (s_{ij})$ is the square of a polynomial with integer coefficients in the entries of S. This polynomial is called the Pfaffian of S, written $\text{Pf}(S)$. It is zero if n is odd, and if $n = 2r$ then

$$\text{Pf}(S) = \frac{1}{r! \, 2^r} \sum \text{sign}(i) s_{i_1 i_2} s_{i_3 i_4} \cdots s_{i_{n-1} i_n},$$

where the sum is over all permutations $i = (i_1, \ldots, i_n)$ of $(1, \ldots, n)$. Thus if S is identified with $\sum s_{ij} w_i \wedge w_j \in \Lambda^2(W)$ then

$$\frac{1}{2^r \cdot r!} \underbrace{S \wedge \ldots \wedge S}_{r} = \text{Pf}(S) \cdot w_1 \wedge w_2 \wedge \ldots \wedge w_n. \qquad (12.2.4)$$

12 THE SPIN REPRESENTATION

It is less well known that if S and T are skew $n \times n$ matrices then $\det(1 - ST)$ is also the square of a polynomial with integer coefficients in the entries of S and T, which we shall again denote by $\text{Pf}(1 - ST)$. Furthermore

$$\text{Pf}(1 - ST) = \sum_{\sigma} \text{Pf}(S_\sigma)\text{Pf}(T_\sigma), \qquad (12.2.5)$$

where σ runs through the subsets of $\{1, 2, \ldots, n\}$, and S_σ denotes the skew submatrix made from the rows and columns of S indexed by σ. We shall prove these results about Pfaffians in the Appendix at the end of this section.

Proof of Proposition (12.2.3). It follows at once from the definition (12.2.4) that

$$e^{\frac{1}{2}S} = \sum_{\sigma} \text{Pf}(S_\sigma) w_\sigma, \qquad (12.2.6)$$

where σ runs through the subsets of $\{1, \ldots, n\}$, and $w_\sigma = w_{i_1} \wedge \ldots \wedge w_{i_k}$ if $\sigma = \{i_1, \ldots, i_k\}$. Thus Proposition (12.2.3) is a consequence of (12.2.5).

We can now reconstruct the even part of the spin representation from the rays corresponding to the points $\bar{W}_S$ of the space $\mathcal{J}(V)$ which are the graphs of skew operators $S: \bar{W} \to W$. Let F denote the abstract free complex vector space whose basis is a set of symbols $\{\Omega_S\}$ in one-to-one correspondence with the operators S. We define an inner product in F by prescribing it on basis elements by

$$\langle \Omega_S, \Omega_T \rangle = \text{Pf}(1 - \bar{S}T). \qquad (12.2.7)$$

There is a linear map $F \to \Lambda(W)$ which takes Ω_S to $e^{\frac{1}{2}S}$. It preserves inner products—from this it follows in particular that the inner product in F is positive-semidefinite. The Hilbert space $\hat{F}$ obtained by completing F with respect to the seminorm defined by its inner product can automatically be identified with a subspace of $\Lambda(W)$. In fact it is precisely $\Lambda^{\text{even}}(W)$, for the elements $e^{\frac{1}{2}S}$ lie in $\Lambda^{\text{even}}(W)$ and span it: if k is even then

$$e^{w_{i_1} \wedge w_{i_2} + \ldots + w_{i_{k-1}} \wedge w_{i_k}} = w_{i_1} \wedge \ldots \wedge w_{i_k} + \text{lower terms}.$$

If F is made smaller by allowing only the basis elements Ω_S for which S lies in some open set U of the space of skew matrices, then the completion $\hat{F}$ does not change, for the vectors $e^{\frac{1}{2}S}$ for $S \in U$ still span $\Lambda^{\text{even}}(W)$. That is obvious if U contains $S = 0$; but it is true when U is a neighbourhood of any point S_0, because $\Lambda^{\text{even}}(W)$ is a commutative algebra in which $e^{\frac{1}{2}(S_0 + S_1)} = e^{\frac{1}{2}S_0} e^{\frac{1}{2}S_1}$, and $e^{\frac{1}{2}S_0}$ is an invertible element.

To construct the spin representation, however, we need not only the vector space $\hat{F}$ but also the projective action of the orthogonal group on

12.2 CONSTRUCTION OF THE SPIN REPRESENTATION

it. To give this we begin with an explicit description of the elements of the complex spin group $\text{Spin}(V_C)$, the unique double covering of $SO(V_C)$.

Let us choose a decomposition $V_C = \bar{W} \oplus W$, and write elements of $O(V_C)$ as 2×2 matrices

$$A = \begin{pmatrix} a & b \\ c & b \end{pmatrix} \qquad (12.2.8)$$

as usual. If $A \in SO(V_C)$ and a is invertible then the two elements of $\text{Spin}(V_C)$ which lie above A correspond precisely to the two square-roots of $\det(a)$. But because $\det(a)$ may vanish a complete description of $\text{Spin}(V_C)$ requires a little more subtlety.

Let us consider, for given $A \in O(V_C)$, the function $S \mapsto \det(a + bS)$ on the space of skew operators $S: \bar{W} \to W$. If we introduce an orthonormal basis for W and regard S as an $n \times n$ matrix then $\det(a + bS)$ is a polynomial in the entries of S. We know that it vanishes precisely when $A(\bar{W}_S)$ does not belong to the open set $U_{\bar{W}}$ of $\mathcal{J}(V)$, so it is identically zero when A has determinant -1, and not identically zero for $A \in SO(V_C)$. If $A_1 A_2 = A_3$ in $O(V_C)$, and

$$A_i = \begin{pmatrix} a_i & b_i \\ c_i & d_i \end{pmatrix},$$

then

$$a_3 + b_3 S = (a_1 + b_1 S')(a_2 + b_2 S), \qquad (12.2.9)$$

where $S' = (c_2 + d_2 S)(a_2 + b_2 S)^{-1}$, providing $\det(a_2 + b_2 S) \neq 0$. The crucial fact for the construction of $\text{Spin}(V_C)$ is

Proposition (12.2.10). *For $A \in SO(V_C)$ as in (12.2.8), the function $S \mapsto \det(a + bS)$, regarded as an element of the ring $\mathbb{C}[s_{ij}]$ of polynomials in the entries of S, is a perfect square.*

We shall give the proof of this in the appendix to this section. It provides us with a new concrete definition of the spin group.

Definition (12.2.11). *An element of $\text{Spin}(V_C)$ is a pair (A, f), where A belongs to $O(V_C)$, and f is a polynomial function on the space of skew maps $S: \bar{W} \to W$ such that*

$$f(S)^2 = \det(a + bS).$$

The multiplication is defined, in the notation of (12.2.9), by $(A_1, f_1)(A_2, f_2) = (A_3, f_3)$, where

$$f_3(S) = f_1(S') f_2(S).$$

(Notice that $f_1(S')$ is a rational function of the entries of S.)

12 THE SPIN REPRESENTATION

Remarks.
(i) If a is invertible then f is determined by choosing a square-root of $\det(a)$.
(ii) It is worth noticing that this description of $\mathrm{Spin}(V_\mathbb{C})$ is purely algebraic, and makes sense for the orthogonal groups over any field not of characteristic 2.

After all these preliminaries we can at last describe the spin representation. Suppose that (A, f) belongs to the compact group $\mathrm{Spin}(V) = \{(A, f) \in \mathrm{Spin}(V_\mathbb{C}) : A \in O(V)\}$. We define the action of (A, f) on the basis element Ω_S of the free vector space F by

$$(A, f) \cdot \Omega_S = f(S)\Omega_{S'},$$

where $S' = (c + dS)(a + bS)^{-1}$. (This makes sense only if $a + bS$ is invertible.) Provided that we check that

$$\langle f(S)\Omega_{S'}, f(T)\Omega_{T'}\rangle = \langle \Omega_S, \Omega_T \rangle, \qquad (12.2.12)$$

it follows immediately that (A, f) induces a unitary transformation of the completion $\hat{F}$, which we know to be isomorphic to $\Lambda^{\mathrm{even}}(W)$. It does not matter that $(A, f) \cdot \Omega_S$ is not defined for all S, for we have seen that the same space $\hat{F}$ is obtained when S is restricted to lie in any open set of the skew operators.

To prove (12.2.12) we must show that

$$\overline{f(S)}f(T)\mathrm{Pf}(1 - \bar{S}'T') = \mathrm{Pf}(1 - \bar{S}T). \qquad (12.2.13)$$

Apart from a factor ± 1—which cannot depend on S or T, as both sides are rational functions in the entries of S and T—this follows from

$$\det(1 - \bar{S}'T') = \det(1 - \bar{S}T)\overline{\det(a + bS)}^{-1}\det(a + bT)^{-1},$$

which holds because

$$\det(1 - \bar{S}'T') = \det\begin{pmatrix}1\\S'\end{pmatrix}^*\begin{pmatrix}1\\T'\end{pmatrix}$$

$$= \det\left\{A\begin{pmatrix}1\\S\end{pmatrix}(a+bS)^{-1}\right\}^*\left\{A\begin{pmatrix}1\\T\end{pmatrix}(a+bT)^{-1}\right\}$$

$$= \det\left\{(a+bS)^{*-1}\begin{pmatrix}1\\S\end{pmatrix}^*\begin{pmatrix}1\\T\end{pmatrix}(a+bT)^{-1}\right\}$$

$$= \overline{\det(a+bS)}^{-1}\det(1 - \bar{S}T)\det(a+bT)^{-1}.$$

(Here * denotes the complex-conjugate transpose of a matrix, and we are using the fact that $A^*A = 1$ for $A \in O(V)$.)

Returning to the question of sign in (12.2.13), the formula is valid when a is invertible because both sides reduce to $+1$ when $S = T = 0$. But the elements A of $SO(V)$ for which (12.2.13) holds form a subgroup, and

12.2 CONSTRUCTION OF THE SPIN REPRESENTATION

the subgroup must be the whole group because the elements with a invertible form a neighbourhood of the identity element.

We have now completed the construction of the spin representation of $SO(V)$ on the even part of the exterior algebra of a maximal isotropic subspace W. It remains to make the whole of $O(V)$ act on $\Lambda(W)$ so that elements of determinant -1 interchange the even and odd parts. This presents no problem, but one should notice that it can be done in two different ways, corresponding to the fact that there are two different double coverings of $O(V)$ which restrict to $\text{Spin}(V)$ over $SO(V)$, and one may wish to take either of them as '$\text{Pin}(V)$'. The distinction between them is whether the elements of $\text{Pin}(V)$ which cover reflections in a hyperplane of V have order 2 or 4, but the two ways of making the odd elements of $O(V)$ act on $\Lambda(W)$ differ only by multiplication by $\pm i$. The simplest way to extend the representation to $O(V)$ is to embed $O(V)$ in $SO(V \oplus \mathbb{R}^2)$ by

$$A \mapsto \begin{pmatrix} A & & \\ & \det(A) & \\ & & 1 \end{pmatrix}$$

and to observe that $SO(V \oplus \mathbb{R}^2)$ acts on $\Lambda^{\text{even}}(W \oplus \mathbb{C})$, which is canonically isomorphic to $\Lambda(W)$.

Remark. The construction of the spin representation which we have just described was not purely algebraic, as it involved completing F to obtain $\hat{F}$. A purely algebraic variant, which has the advantage of displaying the action of the whole complex group $\text{Spin}(V_\mathbb{C})$, runs as follows.

Construct alongside F the free vector space $\tilde{F}$ on a set of symbols $\tilde{\Omega}_S$ corresponding to the space of skew maps $S: W \to \bar{W}$. Define a complex bilinear form $\beta: \tilde{F} \times F \to \mathbb{C}$ by

$$\beta(\tilde{\Omega}_S, \Omega_{S'}) = \text{Pf}(1 - SS').$$

This form is invariant under $\text{Spin}(V_\mathbb{C})$. Define $\hat{F} = F/F_0$, where $F_0 = \{\xi \in F : \beta(\eta, \xi) = 0 \text{ for all } \eta \in \tilde{F}\}$.

We conclude this section by returning to the map $\mathcal{I}(V) \to P(\Lambda(W))$ which assigns to an isotropic subspace Y of $V_\mathbb{C}$ the corresponding ray Pf_Y in $\Lambda(W)$. Almost by definition this was given on the open set $U_{\bar{W}}$ of $\mathcal{I}(V)$ by $\bar{W}_S \mapsto e^{\frac{1}{2}S}$. It can then be extended uniquely to all of $\mathcal{I}(V)$ so as to be equivariant with respect to $O(V)$. For a general subspace $Y \in \mathcal{I}(V)$ let $Y \cap W = W_1$, and let W_2 be the orthogonal complement of W_1 in W (for the inner product $\langle \, , \, \rangle$). It is easy to see that Y is necessarily of the form $W_1 \oplus (\bar{W}_2)_S$, where $(\bar{W}_2)_S$ is the graph of a skew map $S: \bar{W}_2 \to W_2$. Let $\{w_1, \ldots, w_k\}$ be a basis for W_1. We have

Proposition (12.2.14). *The ray Pf_Y in $\Lambda(W)$ associated to the isotropic*

subspace $Y = W_1 \oplus (\bar{W}_2)_S$ contains the vector

$$w_1 \wedge \ldots \wedge w_k \wedge e^{\frac{1}{2}S}.$$

Proof. One may as well suppose that the vectors $\{w_1, \ldots, w_k\}$ are orthonormal. Let $u_i = w_i + \bar{w}_i$. Then $\{u_1, \ldots, u_k\}$ are orthonormal in V. The subspace Y is obtained from $\tilde{Y} = \tilde{W}_1 \oplus (\bar{W}_2)_S$ by the successive reflections $\rho_{u_1}, \ldots, \rho_{u_k}$. Now $\tilde{Y}$ corresponds to the ray $\mathbb{C} \cdot e^{\frac{1}{2}S}$, and the reflection ρ_{u_i} corresponds to Clifford multiplication by u_i; the result (12.2.14) follows at once.

It follows from (12.2.14) that the holomorphic map $Y \mapsto \mathrm{Pf}_Y$ is an embedding of $\mathcal{J}(V)$ in the projective space $P(\Lambda(W))$ of rays in $\Lambda(W)$: to see that it is one-to-one it is enough, in view of its equivariance, to observe that if $Y \neq \bar{W}$ then $\mathrm{Pf}_Y \neq \mathrm{Pf}_{\bar{W}}$.

As a particular case of (12.2.14) we have the following. The Grassmannian $\mathrm{Gr}(W)$ of all complex subspaces of W can be regarded as a submanifold of $\mathcal{J}(V)$ by assigning to a subspace $W_1 \subset W$ the isotropic subspace $W_1 \oplus (W_1^\perp)^-$ of $W \oplus \bar{W} = V_\mathbb{C}$. Now there is the obvious Plücker embedding of $\mathrm{Gr}(W)$ in $P(\Lambda(W))$ which assigns to a subspace W_1 with basis $\{w_1, \ldots, w_k\}$ the ray containing $w_1 \wedge \ldots \wedge w_k$.

Proposition (12.2.15). *The embedding of $\mathcal{J}(V)$ in $P(\Lambda(W))$ restricts to the Plücker embedding of $\mathrm{Gr}(W)$.*

Appendix: Pfaffians

To show that $\mathrm{Pf}(S)^2 = \det(S)$ for a skew $n \times n$ matrix $S = (s_{ij})$ we shall begin from the fact that there is an invertible matrix P such that PSP^t is of the form $J \oplus J \oplus \ldots \oplus J \oplus 0$, where

$$J = \begin{pmatrix} 0 & 1 \\ -1 & 0 \end{pmatrix}.$$

From this we deduce that if $P^{-1} = Q = (q_{ij})$ and $\hat{S} = \sum s_{ij} w_i \wedge w_j$—it will be convenient in this appendix to distinguish between S and $\hat{S}$—then

$$\tfrac{1}{2}\hat{S} = q_1 \wedge q_2 + q_3 \wedge q_4 + \ldots + q_{r-1} \wedge q_r,$$

where r is the rank of S, and $q_i = \sum q_{ki} w_k$. Thus if $n = 2m$ we have $\hat{S}^m = 0$ if $r < n$, and

$$\frac{1}{2^m m!} \hat{S}^m = q_1 \wedge q_2 \wedge \ldots \wedge q_n = \det(Q) \cdot w_1 \wedge \ldots \wedge w_n$$

if $r = n$. Hence $\mathrm{Pf}(S) = \det(Q)$. But $\det(PSP^t) = 1$ if $r = n$, so that $\det(S) = \det(Q)^2 = \mathrm{Pf}(S)^2$.

We turn next to the definition of $\mathrm{Pf}(1 - ST)$, where S and T are skew $n \times n$ matrices. It is enough to consider the case when n is even, for if n is odd we can regard S and T as $(n+1) \times (n+1)$ matrices by adding a row and column of zeros. Let us think of the entries of S and T as indeterminates, so that $\det(1 - ST)$ is an element of the polynomial ring $\mathbb{Z}[s_{ij}, t_{ij}]$. To see that $\det(1 - ST)$ is a perfect square it is enough to prove it is a square in the quotient field of the polynomial ring, for the ring is a unique factorization domain. But in the quotient field we have

$$\det(1 - ST) = \det(S)\det(S^{-1} - T),$$

and both S^{-1} and $S^{-1} - T$ are skew matrices. The square-root of $\det(1 - ST)$ which becomes 1 when $S = T = 0$ is denoted by $\text{Pf}(1 - ST)$. We have

$$\text{Pf}(1 - ST) = \text{Pf}(S)\text{Pf}(S^{-1} - T). \tag{12.2.16}$$

To prove that

$$\text{Pf}(1 - ST) = \sum_\sigma \text{Pf}(S_\sigma)\text{Pf}(T_\sigma) \tag{12.2.17}$$

we first show that when R and T are skew we have

$$\text{Pf}(R + T) = \sum_\sigma \varepsilon_\sigma \, \text{Pf}(R_{\sigma'})\text{Pf}(T_\sigma), \tag{12.2.18}$$

where σ' is the complement of σ in $\{1, \ldots, n\}$, and ε_σ is the sign of the shuffle permutation (σ', σ). The formula (12.2.18) holds because

$$\frac{1}{2^n \cdot n!} (\hat{R} + \hat{T}) \wedge \ldots \wedge (\hat{R} + \hat{T})$$

$$= \sum_{k=0}^{n} \left\{ \frac{1}{2^{n-k}(n-k)!} \hat{R} \wedge \ldots \wedge \hat{R} \right\} \wedge \left\{ \frac{1}{2^k \cdot k!} \hat{T} \wedge \ldots \wedge \hat{T} \right\}$$

$$= \sum_\sigma \text{Pf}(R_{\sigma'})w_{\sigma'} \wedge \text{Pf}(T_\sigma)w_\sigma.$$

To obtain (12.2.17) we apply (12.2.18) with $R = -S^{-1}$, and then use (12.2.16). We need to know that

$$\text{Pf}((S^{-1})_{\sigma'}) = (-1)^{n-k}\varepsilon_\sigma \, \text{Pf}(S_\sigma)\text{Pf}(S)^{-1} \tag{12.2.19}$$

when σ has $2k$ elements. The corresponding result for determinants is well known ([147] Section V.3): it is simply the expression in terms of coordinates of the naturality of the isomorphism

$$\Lambda^n(W) \otimes \Lambda^{n-k}(W^*) \to \Lambda^k(W).$$

The sign in (12.2.19) is checked first when $\sigma = \{1, 2, \ldots, 2k\}$ by considering the example $S = J \oplus J \oplus \ldots \oplus J$; the general case is reduced to this by replacing S by PSP^t, where P is a permutation matrix—we notice that $\text{Pf}(PSP^t) = \text{Pf}(S) \cdot \det(P)$.

Our last task in this appendix is to prove Proposition (12.2.10), that the polynomial $\det(a + bS)$ is a square in $\mathbb{C}[s_{ij}]$ when

$$A = \begin{pmatrix} a & b \\ c & d \end{pmatrix}$$

is a complex special orthogonal matrix. The orthogonality of a implies that ab^t is skew. As before, we can suppose that n is even. Then if a is invertible $a^{-1}b$ is skew, and $\det(a + bS) = \det(a)\det(1 + a^{-1}bS)$, which is a square by our previous discussion. But from (12.2.9) we see that the elements of $SO(V_\mathbb{C})$ for which $\det(a + bS)$ is a square form a subgroup of $SO(V_\mathbb{C})$; by the usual argument this must be all of $SO(V_\mathbb{C})$.

12.3 The spin representation as the sections of a holomorphic line bundle

The holomorphic embedding $Y \mapsto \text{Pf}_Y$ of $\mathcal{J}(V)$ in the projective space $P(\Lambda(W))$ defines a holomorphic line bundle Pf on $\mathcal{J}(V)$: it is the subbundle of the trivial bundle $\mathcal{J}(V) \times \Lambda(W)$ whose fibre at Y is Pf_Y. As $\mathcal{J}(V)$ is a submanifold of the Grassmannian $\text{Gr}(V_\mathbb{C})$ the most obvious projective embedding of $\mathcal{J}(V)$ is the Plücker embedding in $P(\Lambda(V_\mathbb{C}))$, which corresponds to the determinant line bundle Det on $\text{Gr}(V_\mathbb{C})$. The relation between Pf and Det, and the explanation of the notation Pf, is given by

Proposition (12.3.1). *The line bundle* Pf *is a square-root of* $\text{Det} \mid \mathcal{J}(V)$.

This means that there is a fibre-preserving holomorphic map $\mathbf{sq}: \text{Pf} \to \text{Det}$ such that $\mathbf{sq}(\lambda\xi) = \lambda^2 \mathbf{sq}(\xi)$ for $\lambda \in \mathbb{C}$ and $\xi \in \text{Pf}$.

Proof. Let us choose an orthonormal basis $\{w_1, \ldots, w_n\}$ of W. If $Y \in \mathcal{J}(V)$ is the graph of $S: \bar{W} \to W$ then Pf_Y is the ray containing

$$e^{\frac{1}{2}S} = \sum_\sigma \text{Pf}(S_\sigma) w_\sigma,$$

in the notation of (12.2.6). Now $\text{Pf}(S_\sigma)$ is the square-root of $\det(S_\sigma)$, which is the determinant of the $n \times n$ matrix formed from the rows of the $2n \times n$ matrix

$$\begin{pmatrix} 1 \\ S \end{pmatrix} \qquad (12.3.2)$$

indexed by $\tilde{\sigma} = \bar{\sigma}' \cup \sigma$, where σ' is the complement of σ in $\{1, 2, \ldots, n\}$. (Here we are thinking of the rows of (12.3.2) as being indexed by $\{\bar{1}, \bar{2}, \ldots, \bar{n}; 1, 2, \ldots, n\}$, and $\bar{\sigma}' = \{\bar{k} : k \in \sigma'\}$.) In other words $\det(S_\sigma)$ is the $\tilde{\sigma}^{\text{th}}$ Plücker coordinate of Y. Thus if we define $\mathbf{sq}: P(\Lambda(W)) \to$

12.3 SPIN REPRESENTATION AS SECTIONS OF Pf

$P(\Lambda(V_\mathbb{C}))$ by squaring coordinates, i.e. by

$$\mathbf{sq}\left(\sum_\sigma x_\sigma w_\sigma\right) = \sum_\sigma x_\sigma^2 w_\sigma \wedge \bar{w}_{\sigma'},$$

then the diagram

$$\begin{array}{ccc} \mathcal{J}(V) & \to & P(\Lambda(W)) \\ \downarrow & & \downarrow \mathbf{sq} \\ \mathrm{Gr}(V_\mathbb{C}) & \to & P(\Lambda(V_\mathbb{C})), \end{array}$$

where the bottom map is the Plücker embedding, is commutative. That proves (12.3.1).

When the bundle Pf is restricted to the subspace $\mathrm{Gr}(W)$ of $\mathcal{J}(V)$, however, we know from (12.3.1) that it becomes the determinant bundle of $\mathrm{Gr}(W)$.

We saw in Section 2.9 that a holomorphic embedding $\mathcal{J}(V) \to P(\Lambda(W))$ defines a linear map from the dual space $\Lambda(W)^*$ to the space $\Gamma(\mathrm{Pf}^*)$ of holomorphic sections of the line bundle dual to Pf. From the nature of the construction of Pf the group $\mathrm{Pin}(V)$ acts on it, and the map $\Lambda(W)^* \to \Gamma(\mathrm{Pf}^*)$ is equivariant.

Proposition (12.3.3). *The map $\Lambda(W)^* \to \Gamma(\mathrm{Pf}^*)$ is an isomorphism.*

Proof. Because $\mathrm{Pf}^* | \mathrm{Gr}(W)$ is Det^*, whose space of holomorphic sections is $\Lambda(W^*)$ by (2.9.2), the operation of restricting sections to $\mathrm{Gr}(W)$ gives a map $\Gamma(\mathrm{Pf}^*) \to \Lambda(W^*) = \Lambda(W)^*$; and the composition $\Lambda(W)^* \to \Gamma \to \Lambda(W)^*$ is the identity. On the other hand we know a priori (cf. (2.9.1)) that $\Gamma(\mathrm{Pf}^*)$ is an irreducible representation of $\mathrm{Pin}(V)$; so $\Lambda(W)^* \to \Gamma$ must be surjective.

Corollary (12.3.4).
 (i) *The spin representation is irreducible.*
 (ii) *It extends to a holomorphic representation of $\mathrm{Pin}(V_\mathbb{C})$.*

The holomorphic line bundle Pf on $\mathcal{J}(V)$ is uniquely determined by the fact that $\mathrm{Pf}^{\otimes 2} \cong \mathrm{Det}$, for the space $\mathcal{J}(V) \cong O_{2n}/U_n$ is simply connected. So if we show directly that Det on $\mathcal{J}(V)$ possesses a square-root then we can *define* the spin representation as the dual of $\Gamma(\mathrm{Pf}^*)$. Restriction of sections to $\mathrm{Gr}(W)$ will then give a map to $\Lambda(W)^*$. The disadvantage of this approach, however, is that it is not obvious that the restriction is an isomorphism. The existence of the square-root Pf can be seen in at least three different ways.

 (i) One can show that the first Chern class of Det is divisible by two in $H^2(\mathcal{J}(V); \mathbb{Z})$. That is easy because $S^2 = \mathcal{J}_+(\mathbb{R}^4) \to \mathcal{J}_+(V)$ induces an isomorphism of H^2.
 (ii) One can show that $\pi_1(SO(V))$ has two elements. Then $\mathrm{Spin}(V)$ is

defined as the universal covering of $SO(V)$, and $\mathrm{Pf}\,|\,\mathcal{J}_+(V)$ is defined as $\mathrm{Spin}(V) \times_{\tilde{U}(W)} \mathbb{C}$, where

$$\tilde{U}(W) = \{(A, u) \in U(W) \times \mathbb{C}^\times : u^2 = \det A\}$$

is the induced double covering of $U(W)$, which acts on $\mathbb{C}$ by $(A, u) \mapsto u$.

(iii) One can observe that the transition functions for Det have canonical square-roots. This approach reduces essentially to Proposition (12.2.10).

12.4 The infinite dimensional spin representation

The construction of the spin representation in Section 12.2 was framed so as to apply without essential change to the restricted orthogonal group of an infinite dimensional Hilbert space. We need to do little more than list the appropriate definitions: we shall not give proofs of facts about the orthogonal group which are exactly parallel to ones discussed for the unitary group in Chapter 6.

In this section H will be a real Hilbert space, with complexification $H_\mathbb{C}$. The hermitian inner product in $H_\mathbb{C}$ will be denoted by $\langle\ ,\ \rangle$. The complex orthogonal group $O(H_\mathbb{C})$ is the group of invertible $\mathbb{C}$-linear maps $H_\mathbb{C} \to H_\mathbb{C}$ which preserve the bilinear form B defined by

$$2B(\xi, \eta) = \langle \bar{\xi}, \eta \rangle;$$

it contains the usual orthogonal group $O(H)$. We shall suppose that a preferred complex structure J has been chosen on H, so that $H_\mathbb{C} = \bar{W} \oplus W$ for some isotropic subspace W. The *restricted* orthogonal group $O_{\mathrm{res}}(H_\mathbb{C})$ is defined as the subgroup of $O(H_\mathbb{C})$ consisting of elements A such that $[A, J]$ is Hilbert–Schmidt, i.e. those of the form

$$A = \begin{pmatrix} a & b \\ c & d \end{pmatrix},$$

with respect to the decomposition $\bar{W} \oplus W$, in which the operators b and c are Hilbert–Schmidt. In other words $O_{\mathrm{res}}(H_\mathbb{C}) = O(H_\mathbb{C}) \cap GL_{\mathrm{res}}(H_\mathbb{C})$. Similarly we define $O_{\mathrm{res}}(H) = O(H) \cap O_{\mathrm{res}}(H_\mathbb{C})$. We give $O_{\mathrm{res}}(H)$ and $O_{\mathrm{res}}(H_\mathbb{C})$ the topology induced from $GL_{\mathrm{res}}(H_\mathbb{C})$.

The spin representation is a projective unitary representation of $O_{\mathrm{res}}(H)$ on the Hilbert space completion of the exterior algebra $\Lambda(W)$. In Chapter 10 we constructed an irreducible representation of the restricted unitary group $U_{\mathrm{res}}(H)$ of a *complex* Hilbert space H equipped with a polarization $H = H_+ \oplus H_-$. If $J: H \to H$ is $+i$ on H_+ and $-i$ on H_-, and we forget the original complex structure on H and define $O_{\mathrm{res}}(H_\mathbb{R})$ by using the complex structure J—here $H_\mathbb{R}$ is H regarded as a real vector space—then $U_{\mathrm{res}}(H)$ is a subgroup of $O_{\mathrm{res}}(H_\mathbb{R})$, and the representation of

12.4 INFINITE DIMENSIONAL SPIN REPRESENTATION 245

Chapter 10 is the restriction of the spin representation of $O_{\text{res}}(H_\mathbb{R})$. Notice that the complexification $(H_\mathbb{R})_\mathbb{C}$ is canonically isomorphic to $H \oplus \bar{H}$ by the map

$$\lambda \otimes \xi \mapsto \lambda\xi \oplus \bar\lambda\xi \qquad (12.4.1)$$

for $\lambda \in \mathbb{C}$, $\xi \in H_\mathbb{R}$, and under this isomorphism W corresponds to $H_+ \oplus \bar{H}_-$.

Returning to the real Hilbert space H, we define the space $\mathcal{J}(H)$ of complex structures by allowing all complex structures J' such that $[J', J]$ is Hilbert–Schmidt; or, equivalently, we define

$$\mathcal{J}(H) = \{Y \in \text{Gr}(H_\mathbb{C}) : Y \text{ is isotropic for } B, \text{ and } \bar{Y} \oplus Y = H_\mathbb{C}\}.$$

Here the Grassmannian $\text{Gr}(H_\mathbb{C})$ is defined as in Chapter 7, using the polarization $H_\mathbb{C} = \bar{W} \oplus W$. It is immediate that any $Y \in \mathcal{J}(H)$ has virtual dimension zero (relative to $\bar{W}$). Furthermore just as $\text{Gr}(H_\mathbb{C})$ has the same homotopy type as its dense subspace $\bigcup_{n \geq 0} \text{Gr}(\mathbb{C}^n)$, so $\mathcal{J}(H)$ has the same homotopy type as $\bigcup_{n \geq 0} \mathcal{J}(\mathbb{R}^{2n}) = \bigcup_{n \geq 0} O_{2n}/U_n = O_\infty/U_\infty$. In particular it has two connected components $\mathcal{J}_\pm(H)$. The Hilbert space $\mathcal{S}k(\bar{W})$ of skew Hilbert–Schmidt operators $S : \bar{W} \to W$ can be identified with a dense open subset of $\mathcal{J}_+(H)$ by associating to S its graph.

The group $O_{\text{res}}(H)$ acts transitively on $\mathcal{J}(H)$, and the isotropy group of $\bar{W}$ is $U(\bar{W})$. Because this is contractible we have

Proposition (12.4.2). *The group $O_{\text{res}}(H)$ has the same homotopy type as $\mathcal{J}(H)$ or O_∞/U_∞. It has two connected components, and each is simply connected.*

The spin representation is a projective representation of $O_{\text{res}}(H)$, but an important difference from the finite dimensional situation is that it cannot be normalized so as to be only two-valued. We shall construct an extension $\text{Spin}^\mathbb{C}(H)$ of the connected component $SO_{\text{res}}(H)$ by $\mathbb{C}^\times$ on which it is well-defined. In fact the definition makes sense for the complex group $SO_{\text{res}}(H_\mathbb{C})$.

Definition (12.4.3) (cf. (12.2.11)). *An element of $\text{Spin}^\mathbb{C}(H_\mathbb{C})$ is a pair (A, f), where $A \in SO_{\text{res}}(H_\mathbb{C})$ and $f : \mathcal{S}k(\bar{W}) \to \mathbb{C}$ is a holomorphic function such that f is proportional to $S \mapsto \text{Pf}(a + bS)$.*

This definition needs to be explained, as $a + bS$ will not normally be an operator with a determinant, and still less with a Pfaffian. One can choose $S_0 \in \mathcal{S}k(\bar{W})$ so that $a + bS_0$ is invertible. The conditions that A is orthogonal imply that the Hilbert–Schmidt operator $(a + bS_0)^{-1}b$ is skew. So

$$(a + bS_0)^{-1}(a + bS) = 1 + (a + bS_0)^{-1}b \cdot (S - S_0) \qquad (12.4.4)$$

is an operator with a determinant, being of the form $1 - PQ$, with P, Q

skew and Hilbert–Schmidt. But such an operator even has a Pfaffian, given by

$$\text{Pf}(1 - PQ) = \sum_\sigma \text{Pf}(P_\sigma)\text{Pf}(Q_\sigma), \tag{12.4.5}$$

where σ runs through the finite subsets of an orthonormal basis for $\bar{W}$, and P_σ and Q_σ are the corresponding finite skew submatrices of P and Q. The formula (12.4.5) is obtained from (12.2.5) by writing P and Q as the limits of sequences of operators of finite rank: one proves first that

$$\sum_\sigma |\text{Pf}(P_\sigma)|^2 = \text{Pf}(1 + P^*P), \tag{12.4.6}$$

and similarly for Q.

Thus in Definition (12.4.3) we are to interpret $\text{Pf}(a + bS)$ as meaning the Pfaffian of (12.4.4) for any suitable S_0: up to a constant multiple the choice of S_0 is immaterial.

From this point onwards the construction of the spin representation is exactly as in the finite dimensional case, and we need to say no more about it. The interpretation of the representation as the space of holomorphic sections of the line bundle Pf on $\mathcal{J}(H)$ can also be taken over unchanged from Section 12.3; apart, of course, from the need to distinguish between the space Γ of sections, its antidual $\bar{\Gamma}^*$, and the Hilbert space which lies between them, which we have discussed fully in Chapter 10. The representation is irreducible because, as we saw in Chapter 10, it is irreducible even as a representation of the subgroup $U_{\text{res}}(H)$.

12.5 The basic representation of LO_{2n}

The loop group LO_{2n} acts orthogonally on the real Hilbert space H of L^2 functions on the circle with values in $\mathbb{R}^{2n}$. If $\{e_1, \ldots, e_{2n}\}$ is the usual basis of $\mathbb{R}^{2n}$, and $\varepsilon_k = \frac{1}{2}(e_{2k-1} + ie_{2k}) \in \mathbb{C}^n$, then the elements $\varepsilon_k z^m$ and $\bar{\varepsilon}_k z^m$, for $k = 1, \ldots, n$ and $m \in \mathbb{Z}$, are a basis of $H_\mathbb{C}$. We define a complex structure on H by writing $H_\mathbb{C} = \bar{W} \oplus W$, where W is spanned by the $\varepsilon_k z^m$ for $m \geq 0$ and $\bar{\varepsilon}_k z^m$ for $m > 0$. The argument of (6.3.1) proves

Proposition (12.5.1). *The groups LO_{2n} and $LO_{2n}(\mathbb{C})$ are subgroups of $O_{\text{res}}(H)$ and $O_{\text{res}}(H_\mathbb{C})$ respectively.*

The spin representation of $O_{\text{res}}(H)$ can therefore be restricted to give a projective representation of LO_{2n} on $\Lambda(W)$. We shall call this the *basic representation* of LO_{2n}. It restricts to the basic representation of LU_n, and is therefore irreducible. The subgroup LSO_{2n} also acts irreducibly on $\Lambda(W)$. It has two connected components, corresponding to $\pi_1(SO_{2n})$.

12.5 THE BASIC REPRESENTATION OF LO_{2n}

The identity component acts separately on $\Lambda^{\text{even}}(W)$ and $\Lambda^{\text{odd}}(W)$ by two inequivalent irreducible representations.

We mentioned in Sections 6.4 and 8.8 the Bott periodicity theorem, which asserts that the inclusion of ΩU_n in $U_{\text{res}}(H)$ induces an isomorphism of homotopy groups up to dimension $2n - 2$. Exactly the same arguments can be used to prove the following.

Proposition (12.5.2). *The inclusion $\Omega O_{2n} \to O_{\text{res}}(H)$ induces an isomorphism of homotopy groups up to dimension $2n - 3$.*

We shall not give the proof of this here. But let us notice that in view of (12.4.2) it asserts that ΩO_∞ is homotopy equivalent to O_∞/U_∞. That is an important part of the Bott periodicity theorem for the orthogonal group; in fact together with the unitary periodicity theorem it implies the orthogonal periodicity theorem. Cf. [2].

12.6 An analogy: the 'extra-special 2-group'

In Chapter 10 we constructed an isomorphism between an exterior algebra and a sum of symmetric algebras by realizing the unique irreducible projective representation of the abelian group $L\mathbb{T}$ (with its basic cocycle) in two ways, first as a spin representation, and secondly as (roughly) the space of holomorphic functions on the group itself. We shall now describe a finite-dimensional analogue of this isomorphism.

In the group O_{2n} there is a maximal abelian subgroup A of order 2^{2n} consisting of the diagonal matrices with entries ± 1. The double covering $\text{Pin}_{2n} \to O_{2n}$ restricts to a central extension $\tilde{A}$ of A by the group with two elements. This extension is sometimes called an 'extra-special 2-group' ([65] Section 5.3, [123]).

Proposition (12.6.1). *The restriction of the spin representation $\Lambda(\mathbb{C}^n)$ to $\tilde{A}$ is irreducible.*

Proof. The group $\tilde{A}$ can be identified with the multiplicative subgroup of the Clifford algebra $C(\mathbb{R}^{2n})$ which is generated by the elements $\pm e_i$ for $i = 1, \ldots, 2n$. (The e_i are the standard basis vectors of $\mathbb{R}^{2n}$.) Thus the elements of $\tilde{A}$ generate $C(\mathbb{R}^{2n})$ additively, and by (12.1.5) they generate additively the full matrix ring of endomorphisms of $\Lambda(\mathbb{C}^n)$. That implies that the representation of $\tilde{A}$ is irreducible.

Remark. The preceding proposition gives us, of course, a new proof of the irreducibility of the spin representation.

Corollary (12.6.2). *The representation on $\Lambda(\mathbb{C}^n)$ is the only irreducible representation of $\tilde{A}$ except for the 2^{2n} one-dimensional representations of A.*

12 THE SPIN REPRESENTATION

Proof. The sum of the squares of the dimensions of the irreducible representations of a finite group is equal to the order of the group.

Now let us think of the spin representation as the space of holomorphic sections of the line bundle Pf* on $\mathcal{J}(\mathbb{R}^{2n})$. Let X be the orbit of the base-point of $\mathcal{J}(\mathbb{R}^{2n})$ under A. It consists of 2^n points, and can be identified with A/B, where $B = A \cap U_n$. The restriction of Pf* to X is evidently trivial, so we can identify its space of sections $\Gamma(\text{Pf}^* | X)$ with the space of complex-valued functions on $X = A/B$. But the bundle Pf* $| X$ is not trivial as a homogeneous bundle under $\tilde{A}$, and $\Gamma(\text{Pf}^* | X)$ is more precisely described as the representation of $\tilde{A}$ induced from any one-dimensional representation λ of $\tilde{B}$ such that $\lambda | C$ is non-trivial. (Here C is the kernel of $\tilde{A} \to A$, and $\tilde{B}$ is the subgroup of $\tilde{A}$ such that $\tilde{B}/C = B$.)

To understand $\Gamma(\text{Pf}^* | X)$ better we should be more precise about the structure of $\tilde{A}$. Let us write the group A additively: it is a vector space over the field $\mathbb{F}_2$ with two elements. The extension

$$C \to \tilde{A} \to A$$

is completely described by giving the map $q: A \to C \cong \mathbb{F}_2$ such that $q(a) = \tilde{a}^2$, where $\tilde{a}$ is a lift of a in $\tilde{A}$. In fact q is a quadratic form, and the associated bilinear form $b: A \times A \to \mathbb{F}_2$ defined by

$$b(a, a') = q(a + a') - q(a) - q(a'),$$

which is characterized by the relation

$$\tilde{a} \cdot \tilde{a}' = (-1)^{b(a,a')} \tilde{a}' \cdot \tilde{a},$$

is easily seen to be non-degenerate. (This is equivalent to the assertion that the centre of $\tilde{A}$ is C, which is clear if one thinks of $\tilde{A}$ as a subset of the Clifford algebra.) Now $B = A \cap U_n$ is a maximal isotropic subspace of A for the form b: it is isotropic because it is contained in a torus of SO_{2n}, necessarily covered by a torus in Spin$_{2n}$. Thus we can write $A = B \oplus B'$, where B' is another isotropic subspace of A. As a projective representation of A the space $\Gamma(\text{Pf}^* | X)$ can be regarded as the space of complex-valued functions on B', with the elements of B' acting by translations, and the element $a \in B$ acting by multiplication by the function $(-1)^{b(a, \cdot)}$. This is precisely analogous to the situation in Section 9.5. For if $\tilde{V}$ is the Heisenberg group associated to a real vector space V with a skew bilinear form S then the standard representation of $\tilde{V}$ can be realized either on the space of holomorphic functions on A, where A is V with a chosen complex structure, or else, equivalently, on the space of L^2 functions on V_1, where $V = V_1 \oplus V_2$ is a decomposition into dual isotropic subspaces, and V_1 acts on $L^2(V_1)$ by translation, and $v \in V_2$ acts by multiplication by $e^{iS(v, \cdot)}$.

13
'BLIPS' OR 'VERTEX OPERATORS'

In Chapter 10 we saw that the irreducible projective representation $\mathcal{H}$ of the group $L\mathbb{T}$ could be realized as a completed exterior algebra $\hat{\Lambda}(W)$. In this chapter we shall obtain that realization in a completely different and more direct way. The idea is that when the group $L\mathbb{T}$ acts projectively on a suitable Hilbert space $\mathcal{H}$ one can define a singular 'blip' or 'vertex operator' ψ_θ for each point θ of the circle. This operator is the representative of a 'limiting' element of $L\mathbb{T}$ which winds rapidly once around $\mathbb{T}$ in an infinitesimal neighbourhood of θ, and is the identity elsewhere. The operators ψ_θ anticommute among themselves. By 'smearing' the products $\psi_{\theta_1}\psi_{\theta_2}\ldots\psi_{\theta_k}$ in the sense of quantum field theory and applying them to a vacuum vector in $\mathcal{H}$ we obtain a map $\hat{\Lambda}(W) \to \mathcal{H}$. (More precisely, we obtain an action of an infinite dimensional Clifford algebra on $\mathcal{H}$.)

The idea of obtaining fermionic field operators such as the ψ_θ from a projective representation of a commutative group $L\mathbb{T}$, i.e. from bosonic fields, by considering the representatives of elements which are localized but topologically non-trivial, seems to have been first suggested by Skyrme [139] (cf. Finkelstein and Rubinstein [44]), who called the representatives 'kinks'. But the precise operators we shall consider were introduced in 'dual resonance' theory [80] for completely different and ungeometrical reasons.

In the representation theory of loop groups a number of slightly different uses have been made of blips. Apart from the role outlined above they can be used to construct the representations of $\tilde{L}G$ of level 1 for any simply laced group G. That is described in Section 13.3. It is important for at least two reasons. In the first place the construction is very explicit, and sheds light on the structure of the representations. It enables one to write down at once a very manageable formula for their characters which is quite different from the one given by Kac's character formula. (We shall give both formulae in Section 14.3.) Secondly, the blip construction is natural with respect to diffeomorphisms of the circle, and enables one to prove in general that any positive energy representation of a loop group admits an intertwining action of $\text{Diff}^+(S^1)$. This argument is given in Section 13.4.

In Section 13.5 we have explained the relationship between the two different kinds of blips which were used in Sections 13.1 and 13.3.

13 'BLIPS' OR 'VERTEX OPERATORS'

13.1 The fermionic operators on $\mathcal{H}$

The basic central extension of the group $L\mathbb{T}$ is defined by the cocycle c given by

$$c(e^{if}, e^{ig}) = e^{iS(f,g)}, \qquad (13.1.1)$$

where (see Section 4.7)†

$$S(f, g) = \frac{1}{4\pi} \int_0^{2\pi} f(\theta)g'(\theta)\, d\theta + \tfrac{1}{2}(\hat{f}\Delta_g + \Delta_f \hat{g}) - \tfrac{1}{2}\Delta_f g(0). \qquad (13.1.2)$$

Here f and g are smooth functions $\mathbb{R} \to \mathbb{R}$ such that

$$f(\theta + 2\pi) = f(\theta) + 2\pi \Delta_f$$

and

$$g(\theta + 2\pi) = g(\theta) + 2\pi \Delta_g,$$

where Δ_f and Δ_g are the winding numbers of e^{if} and e^{ig}, and $\hat{f}$ and $\hat{g}$ denote the mean values of f and g on $[0, 2\pi]$, i.e.

$$\hat{f} = \frac{1}{2\pi} \int_0^{2\pi} f(\theta)\, d\theta$$

and

$$\hat{g} = \frac{1}{2\pi} \int_0^{2\pi} g(\theta)\, d\theta.$$

In this section we shall suppose given a projective unitary representation of $L\mathbb{T}$ with the cocycle (13.1.1) on a Hilbert space $\mathcal{H}$. We shall suppose the representation has *positive energy* in the sense of Section 9.2. The operator representing $\phi \in L\mathbb{T}$ will be denoted by $U(\phi)$.

The most important property of the cocycle (13.1.1), for our present purposes, is the following. We shall say that two elements ϕ and ψ of $L\mathbb{T}$ have *disjoint supports* if for each $\theta \in S^1$ either $\phi(\theta) = 1$ or $\psi(\theta) = 1$.

Proposition (13.1.3). *If ϕ and ψ have disjoint supports then*

$$U(\phi)U(\psi) = (-1)^{\Delta_\phi \Delta_\psi} U(\psi)U(\phi),$$

where Δ_ϕ and Δ_ψ are the winding numbers of ϕ and ψ.

This follows directly from the formula (13.1.2). Notice in particular that $U(\phi)$ and $U(\psi)$ anticommute if ϕ and ψ have winding numbers ± 1.

The cocycle (13.1.1) is not invariant under rotations of S^1. This reflects

† The part of (13.1.2) involving the winding numbers looks complicated and unnatural. One has considerable freedom in describing the extension by a cocycle, and a simpler-looking possible choice would have been $\tfrac{1}{2}f(0)\Delta_g$. The choice (13.1.2) makes S as nearly as possible rotation-invariant: if f and g are rotated through α then $S(f, g)$ changes by $\alpha \Delta_f \Delta_g$.

13.1 THE FERMIONIC OPERATORS ON $\mathcal{H}$

the fact that the representative $U(\phi)$ of ϕ cannot be chosen equivariantly with respect to rotations of S^1: with the choice corresponding to (13.1.2) we have

$$R_\alpha U(e^{if}) R_\alpha^{-1} = e^{-\frac{1}{2} i \alpha \Delta_f (\Delta_f - 1)} U(R_\alpha e^{if})$$

where R_α represents rotation through the angle α. In particular we have

$$R_\alpha U(\phi) R_\alpha^{-1} = U(R_\alpha \phi) \qquad (13.1.4)$$

when ϕ has winding number 0 or 1.

We shall now consider a sequence $\{\phi_k\}$ of elements of $L\mathbb{T}$ which tend to a 'blip' at the point $\alpha \in S^1$. In other words, each ϕ_k has winding number $+1$, and $\phi_k(\theta) \to 1$ as $k \to \infty$ when $\theta \neq \alpha$. In addition we shall suppose that $\phi_k = e^{if_k}$, where the functions $\{f_k\}$ are *uniformly bounded* on $[0, 2\pi]$. We shall see that the sequence of unitary operators $\{U(\phi_k)\}$ tends to zero in the sense that $\langle \eta, U(\phi_k) \xi \rangle \to 0$ for each $\xi, \eta \in \mathcal{H}$.

The blip operator ψ_α is a renormalized limit of the $U(\phi_k)$: we define

$$\psi_\alpha = \lim_{k \to \infty} \varepsilon_k^{-\frac{1}{2}} U(\phi_k), \qquad (13.1.5)$$

where ε_k is the 'width' of the approximation ϕ_k, as defined below, and the limit is formed pointwise in the space of operators $\mathcal{\breve{H}} \to \mathcal{\breve{H}}$, where $\mathcal{\breve{H}} = \prod_k \mathcal{H}(k)$. (We recall that $\mathcal{H}(k)$ denotes the vectors of energy k in $\mathcal{H}$, and that $\mathcal{\breve{H}}$ denotes the vectors of finite energy, i.e. the algebraic direct sum of the spaces $\mathcal{H}(k)$.) This definition needs a certain amount of commentary.

Our aim is to define an 'operator-valued distribution', so we want the smeared operator

$$\psi(p) = \frac{1}{2\pi} \int_0^{2\pi} p(\alpha) \psi_\alpha \, d\alpha$$

to make sense rather than ψ_α itself. (Here p is a smooth real-valued function on S^1.) But even $\psi(p)$ is potentially an unbounded operator, and hence not everywhere defined. We shall construct it so that its domain always includes the space $\mathcal{\breve{H}}$ of vectors of finite energy. In fact to begin with we shall content ourselves with defining $\psi(p)$ when p is a trigonometrical polynomial. To do this it is enough to define operators $\psi_\alpha : \mathcal{\breve{H}} \to \mathcal{\breve{H}}$ such that

$$R_\alpha \psi_\beta R_\alpha^{-1} = \psi_{\alpha + \beta}; \qquad (13.1.6)$$

for then the smeared operator $\psi(p)$ will map $\mathcal{\breve{H}}$ to $\mathcal{\breve{H}}$. (It follows from (13.1.6) that if $p(\theta) = e^{in\theta}$ then $\psi(p)$ maps $\mathcal{H}(k)$ to $\mathcal{H}(k+n)$.) We shall see later (after Proposition (13.1.13)) that $\psi(p): \mathcal{H} \to \mathcal{H}$ can be defined for any smooth function p, and is bounded.

The next point is to explain what is meant by the 'width' of the approximations ϕ_k. Let us write $\phi = e^{if}$, and decompose f in the usual way

$$f(\theta) = \theta + a + f^+(\theta) + f^-(\theta) = \theta + \sum a_n e^{in\theta} \qquad (13.1.7)$$

where $a = a_0$, and f^+ and f^- are the positive and negative parts of the Fourier series of the periodic function $f(\theta) - \theta$. We shall define the width $\varepsilon(\phi)$ of ϕ in an ad hoc and unilluminating way by

$$\varepsilon(\phi) = e^{-2iS(f^+, f^-)} = e^{-\sum_{n>0} n|a_n|^2}. \qquad (13.1.8)$$

In the language of quantum field theory [78] this means that the operator ψ_α has been defined by renormalizing the $U(\phi_k)$ by the process of 'normal ordering', for

$$\varepsilon(\phi)^{-\frac{1}{2}} U(\phi) = U(e^{if^+}) U(e^{if^-}) U(e^{i(\theta + a)}). \qquad (13.1.9)$$

The operator $U(e^{if^+}) U(e^{if^-})$ is called the *normal ordering* of $U(e^{i(f^+ + f^-)})$, and is commonly denoted by $:U(e^{i(f^+ + f^-)}):$.

We shall see presently the reason for emphasizing the role of $\varepsilon(\phi)$ as the 'width' of ϕ. Meanwhile we shall give an example and a proposition to make the terminology seem reasonable.

Example. The usual approach to blips uses a specific approximating family $\{\phi_\lambda\}$ given by

$$\phi_\lambda(\theta) = \frac{e^{i(\theta - \alpha)} - \lambda}{\lambda e^{i(\theta - \alpha)} - 1},$$

where $\lambda \to 1-$. Geometrically $\phi_\lambda(\theta)$ is the angle subtended at the point $P = e^{i\theta}$ of the circle by the points $A = \lambda e^{i\alpha}$ and $B = \lambda^{-1} e^{i\alpha}$. (See Fig. 4.)

Intuitively this is certainly a blip of length about $2(1 - \lambda)$. On the other hand we have

$$f_\lambda^+(\theta) = \sum_{n>0} \frac{\lambda^n}{in} e^{in(\theta - \alpha)},$$

and

$$f_\lambda^-(\theta) = \sum_{n<0} \frac{\lambda^{-n}}{in} e^{in(\theta - \alpha)},$$

so that

$$\varepsilon(\phi_\lambda) = e^{-\sum_{n>0} n|a_n|^2} = 1 - \lambda^2 \sim 2(1 - \lambda).$$

A more cogent reason for referring to $\varepsilon(\phi)$ as the width of ϕ is given by the following result. Suppose that $\{\phi_k\}$ is a sequence in $L\mathbb{T}$ which tends to the blip at α in the sense we have described. If h is a diffeomorphism of the circle we define $h_* \phi_k \in L\mathbb{T}$ as $\phi_k \circ h^{-1}$. Then the sequence $\{h_* \phi_k\}$ clearly tends to the blip at $h(\alpha)$.

13.1 THE FERMIONIC OPERATORS ON $\mathcal{H}$

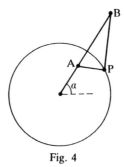

Fig. 4

Proposition (13.1.10). *We have*

$$\lim_{k\to\infty} \varepsilon(h_*\phi_k)/\varepsilon(\phi_k) = h'(\alpha).$$

We shall postpone the proof of this result, which we do not need for the present, to the end of this section.

Our task now is to show that the limit (13.1.5) exists. For this we must look at the operator

$$U(e^{if^+})U(e^{if^-})$$

with care. The first point to notice is that because e^{if^+} and e^{if^-} belong to $L\mathbb{C}^\times$ rather than $L\mathbb{T}$ the operators $U(e^{if^\pm})$ are not defined in $\mathcal{H}$ but only in some larger space, say, $\mathcal{H}_{\text{hol}}$ (see Section 9.5). On that space the equation (13.1.9) makes sense and is valid. We also know that

$$U(e^{if}) = \exp(i\mathfrak{a}(f))$$

when f is real, where $\mathfrak{a}(f)$ depends linearly on f and can be regarded as either a genuine operator in $\mathcal{H}_{\text{hol}}$ or an unbounded operator in $\mathcal{H}$. If f is a (complex) trigonometric polynomial then $\mathfrak{a}(f)$ is defined as an operator $\mathcal{H} \to \mathcal{H}$, and $A_n = \mathfrak{a}(e^{in\theta})$ raises energy by n. Thus we have

$$U(e^{if^+}) = \exp(i\mathfrak{a}(f^\pm))$$
$$= \exp i \sum_{m>0} a_{\pm m} A_{\pm m}$$
$$= 1 + B_{\pm 1}(f^\pm) + B_{\pm 2}(f^\pm) + \ldots, \qquad (13.1.11)$$

where $B_n(f^\pm): \mathcal{H} \to \mathcal{H}$ raises energy by n and depends polynomially on the Fourier coefficients a_m of $f^\pm$ for $|m| \leq |n|$. This means that the operator $U(e^{if^-})$ maps $\mathcal{H}$ to $\mathcal{H}$, while $U(e^{if^+})$ maps $\mathcal{H}$ to $\mathcal{H}$.

If now $\phi = e^{if}$ runs through a sequence tending to the blip at α then f converges in the L^1 sense to the step function s_α, where

$$s_\alpha(\theta) = 0 \quad \text{when} \quad 0 \leq \theta < \alpha$$
$$= 2\pi \quad \text{when} \quad \alpha < \theta \leq 2\pi. \qquad (13.1.12)$$

The Fourier expansion of s_α is

$$s_\alpha(\theta) = \theta + (\pi - \alpha) + \sum_{n\neq 0} \frac{1}{in} e^{in(\theta - \alpha)}.$$

The coefficient a_n of f therefore tends to $(1/in)e^{-in\alpha}$, and $B_n(f^\pm)$ in (13.1.11) tends to a well-defined operator $B_{n,\alpha}: \mathcal{H} \to \mathcal{H}$. This means that of the factors on the right-hand-side of (13.1.9) we can say that $U(e^{if^-})$ converges pointwise as an operator $\mathcal{H} \to \mathcal{H}$ and $U(e^{if^+})$ converges pointwise as an operator $\mathcal{H} \to \mathcal{H}$. The remaining factor $U(e^{i(\theta+a)})$ presents no problem. It is unitary, and preserves $\mathcal{H}$. As $f \to s_\alpha$ it tends to $U(e^{i(\theta-\alpha+\pi)})$.

That completes the proof of the existence of $\psi_\alpha: \mathcal{H} \to \mathcal{H}$. Its equivariance property (13.1.6) follows at once from (13.1.4).

Alongside the blip ψ_α there is its formal adjoint ψ_α^*. This is the limit of $\varepsilon_k^{-\frac{1}{2}} U(\phi_k)^{-1}$, and so it corresponds to a loop which has winding number -1 in an infinitesimal neighbourhood of α. The crucial property of these operators is their anticommutation relations.

Proposition (13.1.13). *We have*

$$\psi_\alpha \psi_\beta + \psi_\beta \psi_\alpha = 0,$$
$$\psi_\alpha^* \psi_\beta^* + \psi_\beta^* \psi_\alpha^* = 0, \quad \text{and}$$
$$\psi_\alpha \psi_\beta^* + \psi_\beta^* \psi_\alpha = 2\pi \delta(\alpha - \beta).$$

These relations are, of course, to be understood in the sense of distributions. The expressions make sense after they have been smeared with a trigonometrical polynomial in α and β. Thus the third relation means that

$$\psi(p)\psi(q)^* + \psi(q)^*\psi(p) = \frac{1}{2\pi} \int_0^{2\pi} p(\theta)\overline{q(\theta)}\, d\theta \qquad (13.1.14)$$

for any trigonometric polynomials p and q.

Putting $p = q$ in (13.1.14) we find that the operator $\psi(p)$ is bounded, and in fact that $\|\psi(p)\| \leq \|p\|$, where $\|p\|$ denotes the L^2 norm. It follows from this that $\psi(p)$ is really an operator $\mathcal{H} \to \mathcal{H}$, and that it can be defined for any L^2 function p.

Proof of (13.1.13). We shall give the proof of the third relation. The other two are similar but easier. We have

$$\psi_\alpha = \lim_{f \to s_\alpha} U(e^{if^+}) U(e^{if^-}) U(e^{i(\theta-\alpha+\pi)})$$

and

$$\psi_\beta^* = \lim_{g \to s_\beta} U(e^{i(\theta-\beta+\pi)})^{-1} U(e^{-ig^+}) U(e^{-ig^-}).$$

13.1 THE FERMIONIC OPERATORS ON $\mathcal{H}$

One can use any convenient approximating functions f and g, but it is sensible to assume that f is a function of $\theta - \alpha$ and g a function of $\theta - \beta$, for then one can smear in α and β before letting $f \to s_\alpha$ and $g \to s_\beta$. With that assumption

$$\psi_\alpha \psi_\beta^* + \psi_\beta^* \psi_\alpha = \lim\{e^{i(\alpha-\beta)}U(e^{if^+})U(e^{if^-})U(e^{-ig^+})U(e^{-ig^-})$$
$$+ U(e^{-ig^+})U(e^{-ig^-})U(e^{if^+})U(e^{if^-})\}U(e^{-i(\alpha-\beta)})$$
$$= F(f, g)U(e^{i(f^+ - g^+)})U(e^{i(f^- - g^-)})U(e^{-i(\alpha-\beta)}), \quad (13.1.15)$$

where

$$F(f, g) = e^{-2iS(f^-, g^+) + i(\alpha-\beta)} + e^{-2iS(g^-, f^+)}.$$

These equations are intended to make sense only after smearing with trigonometric polynomials in α and β. On the right hand side of (13.1.15), however, the product

$$U(e^{i(f^+ - g^+)})U(e^{i(f^- - g^-)})U(e^{-i(\alpha-\beta)})$$

tends to a limit

$$U(e^{i(s_\alpha^+ - s_\beta^+)})U(e^{i(s_\alpha^- - s_\beta^-)})U(e^{-i(\alpha-\beta)})$$

which, as an operator $\mathcal{H} \to \mathcal{H}$, depends smoothly on α and β, and takes the value 1 when $\alpha = \beta$. To prove (13.1.13), therefore, it suffices to show that the expression $F(f, g)$ in (13.1.15), which is a smooth function of α and β, tends to $2\pi\delta(\alpha - \beta)$.

Suppose that the Fourier expansions of f and g are

$$f(\theta) = \theta - \alpha + \pi + \sum \frac{f_n}{in} e^{in(\theta-\alpha)}$$

and

$$g(\theta) = \theta - \beta + \pi + \sum \frac{g_n}{in} e^{in(\theta-\beta)}.$$

Then

$$2iS(f^-, g^+) = \sum_{n>0} \frac{\bar{f}_n g_n}{n} e^{in(\alpha-\beta)},$$

so $e^{-2iS(f^-, g^+)}$ can be expanded in positive powers of $e^{i(\alpha-\beta)}$ with coefficients which are polynomials in the $\bar{f}_n g_n$. As $f \to s_\alpha$ and $g \to s_\beta$ the coefficients f_n and g_n tend to 1, so the coefficients of $e^{-2iS(f^-, g^+)}$ tend to those of

$$\exp\left\{-\sum_{n>0} \frac{1}{n} e^{in(\alpha-\beta)}\right\} = (1 - e^{i(\alpha-\beta)})^{-1}$$
$$= \sum_{n \geq 0} e^{in(\alpha-\beta)}.$$

Treating the other half of the expression $F(f, g)$ in the same way we find that the complete expression converges coefficient by coefficient to

$$\sum_{n \in \mathbb{Z}} e^{in(\alpha - \beta)} = 2\pi\delta(\alpha - \beta).$$

This is the sense of convergence that we require to prove (13.1.13), as we are always smearing with trigonometric polynomials.

Apart from their anticommutation relations and their compatibility with rotations of the circle (13.1.6) we shall need to know how the blip operators ψ_α interact with the representation of $L\mathbb{T}$ from which we began.

Proposition (13.1.16). *For any $\phi \in L\mathbb{T}$ we have*

$$U(\phi)\psi_\alpha U(\phi)^{-1} = (-1)^{\Delta_\phi}\phi(\alpha)\psi_\alpha$$

and

$$U(\phi)\psi_\alpha^* U(\phi)^{-1} = (-1)^{\Delta_\phi}\phi(\alpha)^{-1}\psi_\alpha^*.$$

Remark. By differentiating with respect to ϕ these relations give

$$[A_\beta, \psi_\alpha] = 2\pi\delta(\beta - \alpha)\psi_\alpha$$

and

$$[A_\beta, \psi_\alpha^*] = -2\pi\delta(\beta - \alpha)\psi_\alpha^*,$$

where A_β denotes the operator-valued distribution defined by

$$a(f) = \frac{1}{2\pi} \int_0^{2\pi} f(\beta) A_\beta \, d\beta.$$

Proof of (13.1.16). Consider the relation

$$U(e^{ig})U(e^{if})U(e^{ig})^{-1} = e^{iS(g,f) - iS(f,g)}U(e^{if}). \qquad (13.1.17)$$

If g is smooth and f tends to the step-function s_α then

$$S(g, f) - S(f, g) \to g(\alpha) - \pi\Delta_g.$$

Comparing (13.1.17) with the defintion of ψ_α gives us (13.1.16).

We have now done all the work needed to exhibit the action of the Clifford algebra $C(H_\mathbb{R})$ on $\mathcal{H}$. Here H is the usual Hilbert space of L^2 complex-valued functions on S^1 and $H_\mathbb{R}$ is the underlying real vector space. For $f \in H$ we define

$$C(f) = \psi(f) + \psi(f)^*,$$

which depends on f in a real-linear way. Then the relations (13.1.13)

13.1 THE FERMIONIC OPERATORS ON $\mathcal{H}$

become

$$C(f)C(g) + C(g)C(f) = \frac{1}{2\pi}\int (\bar{f}g + f\bar{g})$$
$$= 2\operatorname{Re}\langle f, g\rangle.$$

The complexification of $H_{\mathbb{R}}$ is identified in the usual way (cf. (12.4.1)) with $H \oplus \bar{H}$. The complex Clifford algebra of $H \oplus \bar{H}$ acts on $\mathcal{H}$ by operators

$$C(f, \bar{g}) = \psi(f) + \psi(g)^*.$$

It contains the exterior algebras on H and $\bar{H}$ as subalgebras.

Because of the equivariance relation (13.1.6) the operator $\psi(f)$ raises energy by n if f has energy n. Let us now write $H = H_+ \oplus H_-$, where H_+ is spanned, as usual, by $e^{in\theta}$ for $n \geq 0$. Let $\mathcal{V}$ be the subspace of $\mathcal{H}$ consisting of all vectors which are annihilated by all the energy-lowering operators $\psi(f)$ for $f \in H_-$ and $\psi(f)^*$ for $f \in e^{i\theta}H_+$. This subspace is certainly non-zero, for it contains every vector of minimal energy. It is preserved by the complex Clifford algebra generated by $\psi(1)$ and $\psi(1)^*$, which is isomorphic to the algebra of 2×2 matrices. In fact we can write

$$\mathcal{V} = \mathcal{V}_0 \oplus \mathcal{V}_1,$$

where $\mathcal{V}_\varepsilon$ is the ε-eigenspace of the operator $\psi(1)\psi(1)^*$, which is idempotent. Vectors in $\mathcal{V}_0$ are annihilated by $\psi(f)$ for $f \in H_-$, and by $\psi(f)^*$ for $f \in H_+$.

Let us choose a vector $\Omega \in \mathcal{V}_0$ such that $\|\Omega\| = 1$, and define a map

$$i_\Omega : \Lambda(H_+ \oplus \bar{H}_-) \to \mathcal{H}$$

by

$$f_1 \wedge \ldots \wedge f_k \wedge \bar{g}_1 \wedge \ldots \wedge \bar{g}_m \mapsto \psi(f_1)\ldots\psi(f_k)\psi(g_1)^*\ldots\psi(g_m)^*\Omega.$$

A simple calculation using the commutation relations shows that this map preserves inner products. (For the definition of the inner product in $\Lambda(H_+ \oplus \bar{H}_-)$, see Section 10.2.) For example

$$\langle \psi(f_1)\Omega, \psi(f_2)\Omega\rangle = \langle \Omega, \psi(f_1)^*\psi(f_2)\Omega\rangle$$
$$= \langle f_1, f_2\rangle\langle \Omega, \Omega\rangle - \langle \Omega, \psi(f_2)\psi(f_1)^*\Omega\rangle$$
$$= \langle f_1, f_2\rangle.$$

It follows that i_Ω extends to give an isometric embedding of the Hilbert space completion $\hat{\Lambda}(H_+ \oplus \bar{H}_-)$ in $\mathcal{H}$. But the argument proves more. If we put together the maps i_Ω for $\Omega \in \mathcal{V}_0$ we obtain an isometry

$$\Lambda(H_+ \oplus \bar{H}_-) \otimes \mathcal{V}_0 \to \mathcal{H}. \tag{13.1.18}$$

Proposition (13.1.19). *The map of* (13.1.18) *extends to give an isometric* *isomorphism*†

$$\hat{\Lambda}(H_+ \oplus \bar{H}_-) \otimes \mathcal{V}_0 \to \mathcal{H}. \tag{13.1.20}$$

Proof. The only point is to show that the map is surjective. Let its image be $\mathcal{I}$. If $\mathcal{I} \neq \mathcal{H}$ then $\mathcal{I}^\perp$ is non-zero, and so must contain a vector of minimal energy, for $\mathcal{I}^\perp$ is stable under rotations of S^1. Such a vector belongs to $\mathcal{V}$, and hence to $\mathcal{I}$, giving a contradiction. (Notice that $\mathcal{V}_1 = \psi(1)\mathcal{V}_0$.)

Remark. We know from Chapter 10, of course, that $\mathcal{V}_0 = \mathbb{C}$ if $\mathcal{H}$ is irreducible.

To conclude this section we shall return once again to the action of $\text{Diff}^+(S^1)$. The isomorphism found in Chapter 10 between the Heisenberg construction and the spin construction of the basic irreducible representation of $\tilde{L}\mathbb{T}$ must by Schur's lemma be compatible with the action of the diffeomorphisms of the circle. The actual construction in Chapter 10, however, was not at all natural from that point of view: it made use of a specific parametrization of the circle. An interesting feature of the blip construction is that it shows very clearly that the Hilbert space H whose Clifford algebra acts on $\mathcal{H}$ is naturally the space of *half-densities* on S^1. For the operator ψ_α was constructed by renormalizing the loop operator $U(\phi)$ by dividing it by the square-root of its width, and so because $U(\phi)$ behaves naturally under diffeomorphisms the blip ψ_α transforms as a half-density.

To make the preceding statement quite precise requires, unfortunately, some more discussion. For ψ_α is an operator $\mathcal{H} \to \tilde{\mathcal{H}}$, and $\text{Diff}^+(S^1)$ does not act on either $\mathcal{H}$ or $\tilde{\mathcal{H}}$. We have seen that $\psi(f): \mathcal{H} \to \mathcal{H}$ can be defined for any L^2 function f, and we should like to assert that for any $h \in \text{Diff}^+(S^1)$ we have

$$\psi(h^*f \cdot (h')^{\frac{1}{2}}) = T_h^{-1} \cdot \psi(f) \cdot T_h, \tag{13.1.21}$$

where $h^*f = f \circ h$, and T_h is the operator representing the action of h on $\mathcal{H}$. (We know from (9.5.14) that $\text{Diff}^+(S^1)$ acts projectively on any positive energy representation of $\tilde{L}\mathbb{T}$: thus T_h is defined up to a scalar multiplier.) The formula (13.1.21) would follow from (13.1.10) if we knew that

$$\lim_{k \to \infty} \frac{1}{2\pi\varepsilon_k^{\frac{1}{2}}} \int_0^{2\pi} f(\alpha) U(\phi_{k,\alpha}) \xi \, d\alpha = \psi(f) \cdot \xi \tag{13.1.22}$$

for all smooth functions f on S^1 and all ξ belonging to some dense $\text{Diff}^+(S^1)$-invariant subspace of $\mathcal{H}$, where $\{\phi_k\}$ is a sequence tending to the blip at 0, and $\phi_{k,\alpha}(\theta) = \phi_k(\theta - \alpha)$. Our definition of $\psi(f)$ ensures

† Here $\otimes$ denotes the Hilbert space tensor product.

13.1 THE ACTION OF DIFF⁺(S¹) ON REPRESENTATIONS

that (13.1.22) is true when f is a trigonometric polynomial and ξ is of finite energy. We shall now explain how this can be improved.

Let $\mathscr{F}$ denote the space of distributions on S^1 of the form $f = \sum a_k e^{ik\theta}$ with a_k rapidly decreasing as $k \to +\infty$. In other words $f = f^+ + f^-$, where the positive energy part f^+ is a smooth function. We topologize $\mathscr{F}$ by using the C^∞ topology on f^+ and the distribution topology on f^-.

Lemma (13.1.23). *The action of* $\mathrm{Diff}^+(S^1)$ *preserves* $\mathscr{F}$.

Proof. Let P denote the projection operator $f \mapsto f^+$. We saw in the proof of (6.8.2) that if h is a diffeomorphism then
$$h^* \circ P \circ (h^*)^{-1} - P$$
is an operator with a smooth kernel, which accordingly maps distributions to smooth functions. In fact the action map $\mathrm{Diff}^+(S^1) \times \mathscr{F} \to \mathscr{F}$ is smooth.

Definition (13.1.24). *Let* $\mathscr{H}_{\text{smooth}}$ *denote the subspace of vectors* $\xi \in \mathscr{H}$ *such that* $f \mapsto U(e^{if}) \cdot \xi$ *extends to a holomorphic map* $\mathscr{F} \to \mathscr{H}$.

(Notice that $U(e^{if}) \cdot \xi$ is defined initially only when f is a real smooth function on S^1.)

Lemma (13.1.23) implies that $\mathscr{H}_{\text{smooth}}$ is preserved by $\mathrm{Diff}^+(S^1)$.

Remark. The space $\mathscr{H}_{\text{smooth}}$ so defined is contained in the subspace of vectors $\xi \in \mathscr{H}$ which are smooth for the action of $\tilde{L}\mathbb{T}$. Probably the two subspaces coincide, and also coincide with the subspace of vectors which are smooth for the action of $\mathbb{T}$ on $\mathscr{H}$, i.e. the vectors ξ such that $\{\xi(k)\}$ is rapidly decreasing. (Here $\xi(k)$ is the component of ξ of energy k.) But we do not need to decide these questions. In any case we have

Lemma (13.1.25). $\check{\mathscr{H}} \subset \mathscr{H}_{\text{smooth}}$.

Proof. This follows from the result (9.5.15) that vectors of finite energy are smooth for $\tilde{L}\mathbb{T}$, together with the observation that if $\xi \in \check{\mathscr{H}}$ then $U(\phi^-) \cdot \xi$ depends on only finitely many Fourier coefficients of ϕ^-.

The question of the behaviour of blips under diffeomorphisms is now settled by

Proposition (13.1.26). *The relation (13.1.22) holds for all* $\xi \in \mathscr{H}_{\text{smooth}}$ *and all smooth* f.

Proof. The essential point is to show that the limit on the left-hand-side of (13.1.22) exists. Now a sequence $\{\xi_k\}$ in a Hilbert space converges if $\langle \xi_k, \xi_m \rangle$ converges as $k, m \to \infty$. It is therefore enough to show that

$$(\varepsilon_k \varepsilon_m)^{-\frac{1}{2}} \langle U(\phi_{k,\alpha}) \cdot \xi, U(\phi_{m,\beta}) \cdot \xi \rangle \tag{13.1.27}$$

converges as a distribution in (α, β) as $k, m \to \infty$. For this we use an argument like that of (13.1.13). If we write $\phi_{k,\alpha} = e^{if}$ and $\phi_{m,\beta} = e^{ig}$ then (13.1.27) can be rewritten as

$$e^{-2iS(f^-, g^+)} \langle U(e^{i(f^- - g^-)}).\xi,\, U(e^{i(\alpha-\beta)})U(e^{-i(f^- - g^-)}).\xi \rangle.$$

Here $e^{-2iS(f^-, g^+)}$ converges as a distribution to $\sum_{n \geq 0} e^{in(\alpha - \beta)}$, while the inner product converges to a smooth function of (α, β), for $f^- - g^-$ tends to a limit in $\mathscr{F}$ which depends smoothly on (α, β).

To prove (13.1.26) we need to know that the limit not only exists but is continuous in ξ. But by the argument of (13.1.13) we find that the limit is dominated by $\|f\|^2 \cdot \|\xi\|^2$.

We return finally to the omitted proof of (13.1.10).

Proof of (13.1.10). If $\phi = e^{if}$ and $f = \theta + a + f^+ + f^-$ then $\varepsilon(\phi)$ is defined as $e^{-2iS(f^+, f^-)}$. If J is the Hilbert transform of (6.3.2), so that $Jf^\pm = \pm f^\pm$, then $S(f^+, f^-) = -\tfrac{1}{2}S(\tilde{f}, J\tilde{f})$, where $\tilde{f} = f^+ + f^-$. A little manipulation gives

$$-2iS(f^+, f^-) = \frac{1}{4\pi^2} \int_0^{2\pi} f'(\theta)f'(\phi)\log|2\sin\tfrac{1}{2}(\theta - \phi)|\, d\theta\, d\phi$$

$$= F(f), \quad \text{say}. \tag{13.1.28}$$

What we want to prove is that if h is a diffeomorphism of S^1—we shall think of h as a map $\mathbb{R} \to \mathbb{R}$ such that $h(\theta + 2\pi) = h(\theta) + 2\pi$—then

$$F(f \circ h^{-1}) - F(f) \to \log h'(\alpha)$$

when f tends to the blip at α. From (13.1.28) we find that

$$F(f \circ h^{-1}) - F(f) = \frac{1}{4\pi^2} \int_0^{2\pi} f'(\theta)f'(\phi)k(\theta, \phi)\, d\theta\, d\phi,$$

where $k(\theta, \phi)$ is a smooth function on $S^1 \times S^1$; and if f tends to the blip at α then $f' \to 2\pi \delta_\alpha$ in the sense of distributions. So

$$F(f \circ h^{-1}) - F(f) \to k(\alpha, \alpha)$$
$$= \log h'(\alpha).$$

13.2 The action of $U_{\text{res}}(H)$ and $O_{\text{res}}(H_\mathbb{R})$ on $\mathscr{H}$

Because the group $L\mathbb{T}$ has a *unique* irreducible representation with the cocycle (13.1.1) it follows indirectly from the results of Chapter 10 that the action of $L\mathbb{T}$ on $\mathscr{H}$ can be extended to an action of $U_{\text{res}}(H)$. It would certainly be of interest to be able to say *explicitly* how the elements of $U_{\text{res}}(H)$ act. The blip operators constructed in the preceding section provide a partial answer to this question by giving us representatives for at least a class of elements of the Lie algebra of $U_{\text{res}}(H)$.

13.2 THE ACTION OF $U_{\text{res}}(H)$ AND $O_{\text{res}}(H_\mathbb{R})$ ON $\mathcal{H}$

The group $GL_{\text{res}}(H)$ contains the subgroup of all invertible operators in H of the form $1 + K$, where K is an operator with a smooth kernel, i.e.

$$(K\xi)(\theta) = \frac{1}{2\pi} \int_0^{2\pi} k(\theta, \theta')\xi(\theta')\,d\theta' \tag{13.2.1}$$

for $\xi \in H$, where k is a smooth function on $S^1 \times S^1$. The Lie algebra $\mathscr{S}m(H)$ of this subgroup consists of all operators of the form (13.2.1). Let us associate to such an operator K the operator A_K in $\mathcal{H}$ defined by

$$A_k = \frac{1}{4\pi^2} \int \int k(\theta, \theta')\psi_\theta \psi_{\theta'}^* \, d\theta\, d\theta'. \tag{13.2.2}$$

We notice that A_K is a well-defined operator in $\mathcal{H}$ providing that the θ' integration is performed first. From the commutation relations (13.1.13) we deduce

$$[A_{K_1}, A_{K_2}] = A_{[K_1, K_2]},$$

so we have an action of the Lie algebra $\mathscr{S}m(H)$ on $\mathcal{H}$. It is not hard to see that this can be exponentiated on the unitary subalgebra to obtain a representation of a subgroup of $U_{\text{res}}(H)$.

In the formula (13.2.2) we could just as well have replaced $\psi_\theta \psi_{\theta'}^*$ by $-\psi_{\theta'}^* \psi_\theta$, or more generally by

$$(1 + \lambda)\psi_\theta \psi_{\theta'}^* + \lambda \psi_{\theta'}^* \psi_\theta$$

for any λ. If we had done so then A_K would have been changed to $A_K + \lambda \operatorname{trace}(K)$, and the associated group representation would have been multiplied by the character $\det^\lambda$.

We know (cf. Section 12.4) that the action of $U_{\text{res}}(H)$ on $\mathcal{H}$ even extends to an action of the orthogonal group $O_{\text{res}}(H_\mathbb{R})$ of the real vector space underlying H. The elements Q of the Lie algebra of the latter group which have smooth kernels can be written in the form

$$(Q\xi)(\theta) = \frac{1}{2\pi} \int_0^{2\pi} \{a(\theta, \theta')\xi(\theta') + b(\theta, \theta')\overline{\xi(\theta')}\}\,d\theta',$$

where we require

$$a(\theta', \theta) = -\overline{a(\theta, \theta')} \quad \text{and}$$
$$b(\theta', \theta) = -b(\theta, \theta').$$

If we associate to Q the operator A_Q on $\mathcal{H}$ defined by

$$A_Q = \frac{1}{4\pi^2} \int \int \{\tfrac{1}{2}a(\theta, \theta')(\psi_\theta \psi_{\theta'}^* - \psi_{\theta'}^* \psi_\theta)$$
$$+ b(\theta, \theta')(\psi_\theta \psi_{\theta'} + \psi_{\theta'}^* \psi_\theta^*)\}\,d\theta\, d\theta'$$

then we have a representation of the smooth part of the Lie algebra of $O_{\text{res}}(H_{\mathbb{R}})$.

The constructions we have just described are, as they stand, of limited interest because they apply only to small subgroups of $U_{\text{res}}(H)$ and $O_{\text{res}}(H_{\mathbb{R}})$. On the other hand general operators in H can be written in the form (13.2.1) providing one allows sufficiently singular kernels k, and in many cases one can make sense of the expression (13.2.2) even when k is singular. Consider, for example, the action of the Lie algebra of $L\mathbb{T}$ itself, given by the usual operators

$$\mathfrak{a}(f) = \frac{1}{2\pi} \int_0^{2\pi} f(\theta) A_\theta \, d\theta.$$

For $\mathfrak{a}(f)$ the appropriate kernel function is $2\pi f(\theta)\delta(\theta - \theta')$, and indeed in the Pickwickian but suggestive language of quantum field theory we have

Proposition (13.2.3). $A_\theta = {:}\psi_\theta \psi_\theta^*{:}$

Here the 'normal ordered' expression on the right is to be interpreted as the limit of

$$\int_0^{2\pi} f(\theta')\{\psi_\theta \psi_{\theta'}^* - S(\theta - \theta')\} \, d\theta'$$

when f tends to the delta-function δ_θ at θ, and $S(\theta - \theta')$ is the 'vacuum expectation value' of $\psi_\theta \psi_{\theta'}^*$, which is given by

$$S(\theta - \theta') = \sum_{n>0} e^{in(\theta - \theta')},$$

i.e. $S(\theta - \theta')$ is the positive energy part of the anticommutator

$$\psi_\theta \psi_{\theta'}^* + \psi_{\theta'}^* \psi_\theta = 2\pi\delta(\theta - \theta').$$

(In this case normal ordering means that we *subtract* the vacuum expectation of $\psi_\theta \psi_{\theta'}^*$, whereas in defining ψ_θ itself we *divided* by the vacuum expectation.) The formula of (13.2.3) looks more surprising if we reflect that in the same notation the definition of ψ_θ is

$$\psi_\theta = {:}e^{is_\theta}{:} = {:}\exp\left(i\int_0^\theta A_{\theta'} \, d\theta'\right){:}. \tag{13.2.4}$$

To a physicist the formula (13.2.3) is very natural and suggestive. For A_θ is the local infinitesimal generator of the 'gauge group' $L\mathbb{T}$—as is expressed by the commutation relation $[A_\theta, \psi_{\theta'}] = 2\pi\delta(\theta - \theta')\psi_{\theta'}$—while $\psi_\theta \psi_\theta^*$ is the operator corresponding to the 'charge density' of the fermionic field. Thus (13.2.3) asserts that the charge density generates the gauge transformations.

13.3 REPRESENTATIONS OF $\tilde{L}G$ WHEN G IS SIMPLY LACED

Proof of (13.2.3). We have

$$\psi_\alpha \psi_\beta^* = e^{i(\alpha-\beta)} e^{i a(s_\alpha^+)} e^{i a(s_\alpha^-)} e^{-i a(s_\beta^+)} e^{-i a(s_\beta^-)} e^{-i a(\alpha-\beta)}$$

$$= S(\alpha - \beta) e^{i a(s_\alpha^+ - s_\beta^+)} e^{i a(s_\alpha^- - s_\beta^-)} e^{-i a(\alpha-\beta)}.$$

So

$$\psi_\alpha \psi_\beta^* - S(\alpha - \beta) = S(\alpha - \beta)\{e^{i a(s_\alpha^+ - s_\beta^+)} e^{i a(s_\alpha^- - s_\beta^-)} e^{-i a(\alpha-\beta)} - 1\}.$$

Each of the preceding three equations is intended to make sense and be valid only after multiplying by a smooth function $f(\beta)$ and integrating with respect to β. When that is done all the expressions represent operators $\mathcal{H} \to \mathcal{H}$. But the right-hand-side of the last equation depends smoothly on α and β, and because

$$S(\alpha - \beta) \sim i/(\alpha - \beta)$$

as $\alpha \to \beta$, its value when $\alpha \to \beta$ is

$$-\frac{d}{d\alpha} a(s_\alpha^+ + s_\alpha^- - \alpha) = a(2\pi\delta_\alpha) = A_\alpha.$$

Multiplying by f and integrating, and then letting f tend to a delta function at α, we obtain (13.2.3).

One could write down formulae analogous to (13.2.3) for the action of LU_n and LO_{2n}, but there seems little point in doing so. In general such 'explicit' formulae in terms of renormalized combinations of field operators involve so much ad hoc interpretation that they are not really explicit at all, and do not help to understand operators whose existence actually follows from quite simple general considerations.

13.3 The level 1 representations of $\tilde{L}G$ when G is simply laced

In this section we shall show how blips can be used to give an explicit construction of the most important irreducible representations of the loop group of a *simply laced* group G. Let T be a maximal torus of G. We begin with a positive energy faithful unitary representation $\gamma \mapsto U(\gamma)$ of the subgroup $\tilde{L}T$ of $\tilde{L}G$ on a Hilbert space $\mathcal{H}$, where $\tilde{L}G$ is the basic central extension of LG described in Chapter 4. We shall use blips to define an action of the complexified Lie algebra $\tilde{L}\mathfrak{g}_\mathbb{C}$ of $\tilde{L}G$ on $\mathcal{H}$, extending the given action of the Lie algebra of $\tilde{L}T$. Finally, the representation of $\tilde{L}\mathfrak{g}$ is exponentiated to give a representation of the group $\tilde{L}G$.

If $\mathcal{H}$ is an irreducible representation of $\tilde{L}T$ then we clearly obtain an irreducible representation of $\tilde{L}G$. The faithful irreducible representations of $\tilde{L}T$ were classified in Section 9.5, and we know already that they correspond one-to-one to the level 1 representations of $\tilde{L}G$.

13 'BLIPS' OR 'VERTEX OPERATORS'

To give the action of $\tilde{L}\mathfrak{g}_{\mathbb{C}}$ on $\mathcal{H}$ we must associate to each $\xi \in L\mathfrak{g}_{\mathbb{C}}$ an (unbounded) operator $J(\xi)$ in $\mathcal{H}$ in such a way that $J(\bar{\xi}) = J(\xi)^*$ and

$$[J(\xi), J(\eta)] = -iJ([\xi, \eta]) - i\omega(\xi, \eta), \qquad (13.3.1)$$

where ω is the cocycle (4.2.2) defining the extension $\tilde{L}\mathfrak{g}_{\mathbb{C}}$. If $\{e_i\}$ is a basis for $\mathfrak{g}_{\mathbb{C}}$ we can write $\xi = \sum \xi_i e_i$, where the ξ_i are smooth complex-valued functions on the circle, and

$$J(\xi) = \sum J^i(\xi_i),$$

where $J^i(f) = J(fe_i)$. Then J^i is an operator-valued distribution, which we write symbolically in the form

$$J^i(f) = \frac{1}{2\pi} \int_0^{2\pi} f(\theta) J_\theta^i \, d\theta.$$

The relation (13.3.1) translates into

$$[J_\theta^k, J_\phi^m] = -2\pi i \sum_r c_{kmr} \delta(\theta - \phi) J_\theta^r - 2\pi i g_{km} \delta'(\theta - \phi), \qquad (13.3.2)$$

where the c_{kmr} are the structural constants of $\mathfrak{g}$ (i.e. $[e_k, e_m] = \sum c_{kmr} e_r$), and $g_{km} = \langle e_k, e_m \rangle$.

We want, in other words, to associate a 'field operator' J_θ^k to each point θ of the circle and each basis vector e_k of $\mathfrak{g}_{\mathbb{C}}$. The given projective representation of $L\mathfrak{t}$ on $\mathcal{H}$ provides operators A_θ^h for each $h \in \mathfrak{t}$, and so we need only construct the J_θ^k for the basis elements of $\mathfrak{g}_{\mathbb{C}}/\mathfrak{t}_{\mathbb{C}}$. The natural choice for such a basis consists of *root vectors* e_α indexed by the roots α of G (see Section 2.4). For each root α there is a corresponding coroot which is a homomorphism $\eta_\alpha: \mathbb{T} \to T$. Using this we can define J_θ^α as the blip representing an element of LT which is the identity except in an infinitesimal neighbourhood of θ, where it traverses the loop η_α in T with infinite speed.

The crucial property of a simply laced group G is that there is an invariant inner product $\langle \, , \, \rangle$ on $\mathfrak{g}$ such that
 (i) $\langle \lambda, \mu \rangle \in \mathbb{Z}$ when λ and μ belong to the lattice $\check{T} = \text{Hom}(\mathbb{T}; T)$,
 (ii) $\langle \lambda, \lambda \rangle \in 2\mathbb{Z}$ for $\lambda \in \check{T}$,
 (iii) $\langle \lambda, \lambda \rangle = 2$ if and only if λ is a coroot.†

The cocycle on LT corresponding to the basic central extension $\tilde{L}G$ is given, as we saw in Section 4.8, by

$$c(\exp \xi, \exp \eta) = (-1)^{b(\Delta_\xi, \Delta_\eta)} e^{iS(\xi, \eta)}, \qquad (13.3.3)$$

† Here the coroot is regarded as an element of $\mathfrak{t}$. As such we denote it by h_α. Thus $\eta_\alpha(e^{i\theta}) = \exp(\theta h_\alpha)$.

13.3 REPRESENTATIONS OF $\tilde{L}G$ WHEN G IS SIMPLY LACED

where $b: \check{T} \times \check{T} \to \mathbb{Z}/2$ is a certain bilinear form, and

$$S(\xi, \eta) = \frac{1}{4\pi} \int_0^{2\pi} \langle \xi(\theta), \eta'(\theta) \rangle \, d\theta + \tfrac{1}{2}\langle \hat{\xi}, \Delta_\eta \rangle + \tfrac{1}{2}\langle \Delta_\xi, \hat{\eta} - \eta(0) \rangle. \tag{13.3.4}$$

Here $\hat{\xi}$ and $\hat{\eta}$ denote the average values of ξ and η on $[0, 2\pi]$. We should emphasize, however, that the rather complicated calculations of Section 4.8 by which the cocycle (13.3.3) was obtained are *not* essential for the present construction. If one simply *guesses* the formula (13.3.3) then one finds that it leads to a projective representation of LG with the desired Lie algebra cocycle; and it follows retrospectively that the guess was correct. The cocycle was indeed originally discovered in this way in [49] and [131], and, surprisingly enough, the elegant description (2.5.1) of the Lie algebra of a simply laced group was discovered at the same time. Thus the role of Section 4.8 is only as an additional check.

The operation of rotation through the angle θ intertwines (cf. (4.8.5)) with the operator $U(\gamma)$ representing $\gamma \in \tilde{L}T$ according to

$$R_\theta U_\gamma R_\theta^{-1} = e^{-\frac{1}{2}i\theta \langle \Delta_\gamma, \Delta_\gamma \rangle} U(R_\theta \gamma). \tag{13.3.5}$$

For each coroot $\eta_\alpha : \mathbb{T} \to T$ the extension of the subgroup $(L\mathbb{T})_\alpha = \eta_\alpha(L\mathbb{T})$ of LT induced by (13.3.3) is given by

$$\tilde{c}(e^{if}, e^{ig}) = (-1)^{\Delta_f \Delta_g} e^{i\tilde{S}(f, g)}, \quad \text{where}$$

$$\tilde{S}(f, g) = \frac{1}{2\pi} \int_0^{2\pi} f(\theta) g'(\theta) \, d\theta + \hat{f}\Delta_g + \Delta_f(\hat{g} - g(0)). \tag{13.3.6}$$

This is essentially twice the cocycle considered earlier in this chapter. The new extension has the property that the representatives of loops of winding numbers ± 1 with disjoint supports *commute* instead of anticommuting.

The blip J_θ^α of type α at the point $\theta \in S^1$ is defined as the limit of $e^{-i\theta} \varepsilon_k^{-1} U(\eta_\alpha \circ \phi_k)$, where $\{\phi_k\}$ is a sequence of approximating elements of $L\mathbb{T}$ just as in Section 13.1. We have ε_k^{-1} instead of $\varepsilon_k^{-\frac{1}{2}}$ because the cocycle has been doubled. This means that as far as diffeomorphisms of the circle are concerned J_θ^α behaves like a density rather than a half-density: thus

$$J^\alpha(f) = \frac{1}{2\pi} \int_0^{2\pi} f(\theta) J_\theta^\alpha \, d\theta \tag{13.3.7}$$

is naturally defined when f is a *function* on S^1, just as we should hope. The factor $e^{-i\theta}$ is included in the definition of J_θ^α to compensate for the lack of rotational invariance of the cocycle (cf. (13.3.5)), i.e. to ensure that

$$R_\theta J_\theta^\alpha R_\theta^{-1} = J_{\theta+\theta'}^\alpha.$$

The existence of the blip is proved exactly as in Section 13.1, and needs no further discussion. We must, however, verify the relations of the form (13.3.2) needed to ensure that we have a representation of $\tilde{L}\mathfrak{g}_\mathbb{C}$. The relations are as follows (see Section 2.5).

Proposition (13.3.8). *We have*
(i) $[A_\theta^h, J_\phi^\alpha] = 2\pi\delta(\theta - \phi)\alpha(h)J_\phi^\alpha$ *(for each root α, and $h \in \mathfrak{t}$)*;
(ii) $[J_\theta^\alpha, J_\phi^{-\alpha}] = 2\pi\delta(\theta - \phi)A_\phi^\alpha - 2\pi i\delta'(\theta - \phi)$;
(iii) $[J_\theta^\alpha, J_\phi^\beta] = 0$ *if $\alpha + \beta$ is neither zero nor a root*;
(iv) $[J_\theta^\alpha, J_\phi^\beta] = (-1)^{b(\alpha,\beta)}2\pi i\delta(\theta - \phi)J_\phi^{\alpha+\beta}$ *if $\alpha + \beta$ is a root.*

Note. On the right-hand-side of (ii) we have identified the root α with an element of $\mathfrak{t}$, by using the quadratic form $\langle\ ,\ \rangle$.

Proof. The proof of (i) is exactly the same as that of (13.1.16) and needs no further discussion. We shall treat (ii), (iii), and (iv) simultaneously. The argument is parallel to that of (13.1.13) so we shall be a little briefer. We shall approximate the blips at θ and ϕ by e^{if} and e^{ig}, where

$$f(\theta') = q_\theta(\theta') + \sum \frac{\lambda^{|n|}}{in} e^{in(\theta'-\theta)} \quad \text{and}$$

$$g(\theta') = q_\phi(\theta') + \sum \frac{\lambda^{|n|}}{in} e^{in(\theta'-\phi)}$$

where $q_\theta(\theta') = \theta' - \theta + \pi$, and $0 < \lambda < 1$. We can write $\eta_\alpha \circ e^{if} = \exp(f\alpha)$ and $\eta_\beta \circ e^{ig} = \exp(g\beta)$, where once again we are regarding α and β as elements of $\mathfrak{t}$. We calculate

$$e^{-i(\theta+\phi)}\varepsilon(e^{if})^{-1}\varepsilon(e^{ig})^{-1}\{U(\eta_\alpha \circ e^{if})U(\eta_\beta \circ e^{ig}) - U(\eta_\beta \circ e^{ig})U(\eta_\alpha \circ e^{if})\}$$
$$= F_\lambda(\theta, \phi)U_\lambda(\theta, \phi), \quad (13.3.9)$$

where

$$U_\lambda(\theta, \phi) = U(\exp(f^+\alpha + g^+\beta))U(\exp(f^-\alpha + g^-\beta))U(\exp(q_\theta\alpha + q_\phi\beta))$$

and

$$F_\lambda(\theta, \phi) = e^{-i(\theta+\phi)}\{(-1)^{b(\alpha,\beta)}e^{i\langle\alpha,\beta\rangle(\tilde{S}(f^-,g^+)+\frac{1}{2}\tilde{S}(q_\theta,q_\phi))}$$
$$- (-1)^{b(\beta,\alpha)}e^{i\langle\beta,\alpha\rangle(\tilde{S}(g^-,f^+)+\frac{1}{2}\tilde{S}(q_\phi,q_\theta))}\}.$$

We find

$$F_\lambda(\theta, \phi) = e^{-i(\theta+\phi)}\{(-1)^{b(\alpha,\beta)}[-ie^{-i\theta}(1 - \lambda^2 e^{i(\theta-\phi)})]^{\langle\alpha,\beta\rangle}$$
$$- (-1)^{b(\beta,\alpha)}[-ie^{-i\phi}(1 - \lambda^2 e^{-i(\theta-\phi)})]^{\langle\beta,\alpha\rangle}\}.$$

Now we distinguish the three cases. If $\beta = -\alpha$ then $\langle\alpha, \beta\rangle = -2$, and $F_\lambda(\theta, \phi)$ becomes

$$\lambda^{-2}\sum_{n\in\mathbb{Z}} n\lambda^{2|n|}e^{in(\theta-\phi)}.$$

13.3 REPRESENTATIONS OF $\tilde{L}G$ WHEN G IS SIMPLY LACED

This tends to the distribution $-2\pi i\delta'(\theta-\phi)$ as $\lambda \to 1$. The operator $U_\lambda(\theta, \phi)$ in (13.3.9) tends to a limit $U(\theta, \phi)$ which is a smooth function of θ and ϕ. If we use the relation

$$\delta'(\theta-\phi)U(\theta,\phi) = \delta'(\theta-\phi)U(\theta,\theta) - \delta(\theta-\phi)\frac{\partial U}{\partial \theta}\bigg|_{\theta=\phi}$$

and observe that $U_\lambda(\theta, \theta) = 1$ and

$$\frac{\partial U_\lambda}{\partial \theta}\bigg|_{\theta=\phi} = iA\left(-\alpha \sum \lambda^{|n|}e^{in(\theta'-\theta)}\right)$$

$$\to -iA_0^\alpha \quad \text{as} \quad \lambda \to 1$$

then we obtain (13.3.8) (ii).

If $\alpha+\beta$ is neither zero nor a root then $\langle \alpha, \beta \rangle = 0$, and so $F_\lambda(\theta, \phi) = 0$. That proves (13.3.8) (iii).

Finally, if $\alpha+\beta$ is a root then $\langle \alpha, \beta \rangle = -1$ and $b(\beta, \alpha) = b(\alpha, \beta)+1$. In this case

$$F_\lambda(\theta, \phi) = (-1)^{b(\alpha,\beta)}ie^{-i\theta}\left\{\sum_{n\leq 0} \lambda^{2|n|}e^{in(\theta-\phi)} + \sum_{n>0} \lambda^{2n-2}e^{in(\theta-\phi)}\right\},$$

which tends to $(-1)^{b(\alpha,\beta)}2\pi ie^{-i\theta}\delta(\theta-\phi)$ as $\lambda \to 1$, proving (13.3.8) (iv).

We have now constructed a representation of the Lie algebra $\tilde{L}_{\text{pol}}\mathfrak{g}_\mathbb{C}$ on $\mathcal{H}$. (It is a representation of the polynomial loops because the construction of the blips permitted smearing only by trigonometric polynomials.) We must extend it to all of $\tilde{L}\mathfrak{g}_\mathbb{C}$, and prove that it can be exponentiated to give a representation of the group $\tilde{L}G$. Both steps can be accomplished by the same simple observation.

The representation of $\tilde{L}_{\text{pol}}\mathfrak{g}$ on $\mathcal{H}$ is clearly unitary. If we restrict it to the constant loops $\mathfrak{g} \subset \tilde{L}_{\text{pol}}\mathfrak{g}$ then it can certainly be exponentiated to a representation of G: that is true for any unitary representation of the Lie algebra of a compact group. (Indeed we need this fact really only for finite dimensional representations, for if we assume that $\mathcal{H}$ is an irreducible representation of $\tilde{L}T$ then each of the spaces $\mathcal{H}(k)$ is finite dimensional, and they are preserved by the action of $\mathfrak{g}$.) We now make the observation that in $\tilde{L}_{\text{pol}}\mathfrak{g}$ the element $J^\alpha(f)+J^\alpha(f)^*$ can be conjugated by the adjoint action of G into $A^\alpha(f+\bar{f})$. The latter can be defined as an unbounded operator for any smooth function f, since we began with a representation of the smooth loop group $\tilde{L}T$. It follows that $J^\alpha(f)+J^\alpha(f)^*$, and hence $J^\alpha(f)$, can be defined for any smooth function f. This gives us a representation of $\tilde{L}\mathfrak{g}_\mathbb{C}$.

But we also know that $A^\alpha(f+\bar{f})$ can be exponentiated. It follows that $J^\alpha(f)+J^\alpha(f)^*$ can be exponentiated. The one-parameter subgroups so obtained generate all of $\tilde{L}G$. We need to know that the exponentiated operators satisfy all the relations which hold in $\tilde{L}G$. That is true because by construction they have the 'correct' intertwining relations with the operators representing $\tilde{L}\mathfrak{g}_{\mathbb{C}}$. In other words, if

$$\exp(\xi_1)\exp(\xi_2)\ldots\exp(\xi_r) = 1$$

in $\tilde{L}G$ then

$$\prod = \exp iJ(\xi_1)\exp iJ(\xi_2)\ldots\exp iJ(\xi_r)$$

commutes with the operators of $\tilde{L}\mathfrak{g}_{\mathbb{C}}$. It therefore acts as a scalar on any irreducible component $\mathcal{H}_0$ of $\mathcal{H}$. So we should have a projective representation of $\tilde{L}G$ on $\mathcal{H}_0$. The projective multiplier, however, is trivial on the Lie algebra, and hence trivial globally. So $\prod = 1$ on $\mathcal{H}_0$; and as that is true for an arbitrary irreducible component $\mathcal{H}_0$ we have $\prod = 1$ on $\mathcal{H}$.

We ought also to mention the behaviour of the blips J_θ^α and the smeared operators $J^\alpha(f)$ with respect to $\mathrm{Diff}^+(S^1)$. Repeating the discussion at the end of Section 13.1 we can define

$$J^\alpha(f): \mathcal{H}_{\text{smooth}} \to \mathcal{H}$$

for any $f \in \mathcal{F}$ by a limit formula like (13.1.22). With that definition the naturality under $\mathrm{Diff}^+(S^1)$ is clear. But however $J^\alpha(f)$ is defined it possesses the intertwining property

$$U(\gamma)J^\alpha(f)U(\gamma)^{-1} = J^\alpha(\eta_\alpha(\gamma)\cdot f)$$

with the operators representing $\tilde{L}T$ (cf. (13.1.16) and (13.3.8) (i)), and this property characterizes it completely.

It is worth emphasizing that the 'bosonic' blips J_θ^α lead to *unbounded* operators even when they are smeared with smooth functions. The smeared 'fermionic' operators of Section 13.1 were bounded. We shall return to this point in Section 13.5.

13.4 The action of $\mathrm{Diff}^+(S^1)$ on positive energy representations of loop groups

The group $\mathrm{Diff}^+(S^1)$ acts projectively by the metaplectic representation on the Hilbert space $\mathcal{H}$ of a positive energy representation of $\tilde{L}T$ such as we considered in the previous section. (Cf. Section 9.5.) We have already pointed out that the blip construction is natural with respect to diffeomorphisms, i.e. if f is a smooth function on S^1 and ϕ is a diffeomorphism which acts on $\mathcal{H}$ by an operator T_ϕ, then the operator $J^\alpha(f)$ of (13.3.7) satisfies

$$T_\phi^{-1} J^\alpha(f) T_\phi = J^\alpha(f \circ \phi). \tag{13.4.1}$$

It follows from (13.4.1) that the action of $\mathrm{Diff}^+(S^1)$ on $\mathcal{H}$ intertwines with

13.4 THE ACTION OF DIFF$^+(S^1)$ ON REPRESENTATIONS

the action of $\tilde{L}G$ which we have constructed. In other words we have proved that when G is simply laced the level 1 representations of $\tilde{L}G$ admit an intertwining action of Diff$^+(S^1)$. It turns out that this is enough to enable one to prove the following quite general theorem.

Theorem (13.4.2). *If E is a positive energy representation of a loop group $\tilde{L}G$ then E admits a projective intertwining action of* Diff$^+(S^1)$.

It is enough (by (9.3.1)) to prove this when E is irreducible. The proof depends on the following lemma.

Lemma (13.4.3). *Suppose that E and F are positive energy representations of $\tilde{L}G$, and that E is irreducible. Then E admits a projective intertwining action of* Diff$^+(S^1)$ *if $E \oplus F$ does.*

Proof. Let $\tilde{D}$ be the central extension of Diff$^+(S^1)$ which acts on $E \oplus F$, and let

$$T_\phi : E \oplus F \to E \oplus F$$

be the action of $\phi \in \tilde{D}$. Notice that T_ϕ is an $\tilde{L}G$-equivariant map

$$E \oplus F \to \phi^* E \oplus \phi^* F,$$

where $\phi^* E$ denotes E with its $\tilde{L}G$-action twisted by the automorphism ϕ of $\tilde{L}G$. Define $S_\phi : E \to E$ by

$$S_\phi = p \circ T_\phi \circ i,$$

where i and p are the inclusion and projection of the summand E in $E \oplus F$. Because S_ϕ is an $\tilde{L}G$-equivariant map $E \to \phi^* E$ we know from Schur's lemma that it is either zero or an isomorphism, for E and $\phi^* E$ are irreducible. If ϕ is sufficiently close to the identity then S_ϕ must be an isomorphism, for the action of $\tilde{D}$ on $E \oplus F$ is assumed to be continuous. For such ϕ we can multiply S_ϕ by a scalar to obtain a unitary operator R_ϕ in E.

If now ϕ_1 and ϕ_2 in $\tilde{D}$ are such that R_{ϕ_1}, R_{ϕ_2}, and $R_{\phi_2\phi_1}$ are defined then $R_{\phi_2\phi_1}$ and $R_{\phi_2}R_{\phi_1}$ can differ only by multiplication by a scalar of modulus 1, for both are intertwining operators between E and $\phi_1^*\phi_2^* E$. Because $\tilde{D}$ is connected, and hence generated by any neighbourhood of its identity element, we can conclude that $\phi \mapsto R_\phi$ extends to a projective representation of $\tilde{D}$ (and hence of Diff$^+(S^1)$) on E.

Proof of (13.4.2). Let us first suppose that G is a simply laced group. Then we saw in (9.3.9) that any irreducible representation of $\tilde{L}G$ is a summand in a representation $m_n^* F$, where F is a representation of level 1 and m_n is the endomorphism of $\tilde{L}G$ induced by composing loops $S^1 \to G$

with the map $\pi_n : S^1 \to S^1$ given by $z \mapsto z^n$. The blip construction proves that $\text{Diff}^+(S^1)$ acts projectively on F. It follows that it acts projectively on $m_n^* F$, for if $\phi \in \text{Diff}^+(S^1)$ and $\gamma \in LG$ then

$$m_n(\phi_*(\gamma)) = \psi_*(m_n(\gamma)),$$

where $\psi \in \text{Diff}^+(S^1)$ is such that $\pi_n \circ \psi = \phi \circ \pi_n$. (For given ϕ there are n choices of ψ differing by rotations of angles $2\pi k/n$.) Lemma (13.4.3) therefore completes the proof of (13.4.2) when G is simply laced.

The general case of (13.4.2) follows from the simply laced case in virtue of the following lemma.

Lemma (13.4.4). *For any simply connected group G and any positive energy irreducible representation E of $\tilde{L}G$ there is an embedding $i : \tilde{L}G \to \tilde{L}G'$, where G' is a simply laced group, and an irreducible representation E' of $\tilde{L}G'$, such that E is a summand in $i^* E'$.*

Proof. Because irreducible representations correspond to antidominant weights (cf. (9.3.5)) it is enough to show that any antidominant weight of $\tilde{L}G$ is the restriction of an antidominant weight of $\tilde{L}G'$. It is also enough to consider the case where G is simple, i.e. we have to consider the cases $G = \text{Spin}_{2n+1}, \text{Sp}_n, G_2, F_4$.

In the standard description of the maximal torus T of Spin_{2n+1} (cf. [20], Chapter 6 Planche II) the lattice of weights $\hat{T}$ is identified with the subset of $\mathbb{R}^n$ consisting of the points $\lambda = (\lambda_1, \ldots, \lambda_n)$ such that either all λ_i are integers or all λ_i are half-integers; and λ is antidominant if

$$\lambda_1 \leq \lambda_2 \leq \ldots \leq \lambda_n \leq 0.$$

The antidominant weights of $\tilde{L}\text{Spin}_{2n+1}$ of level h correspond to those λ such that $\lambda_1 + \lambda_2 \geq -h$. Let us consider the inclusion of Spin_{2n+1} in the simply laced group Spin_{2n+2}. The weight lattice $\hat{T}'$ of Spin_{2n+2} is contained in $\mathbb{R}^{n+1}$, the restriction $\hat{T}' \to \hat{T}$ being given by forgetting the last coordinate. An element $(\lambda_1, \ldots, \lambda_{n+1})$ of $\hat{T}'$ is antidominant if

$$\lambda_1 \leq \lambda_2 \leq \ldots \leq \lambda_{n+1},$$

and

$$\lambda_n + \lambda_{n+1} \leq 0.$$

The antidominant weights of $\tilde{L}\text{Spin}_{2n+1}$ of level h are again picked out by the condition that $\lambda_1 + \lambda_2 \geq -h$. Thus an antidominant weight of $\tilde{L}\text{Spin}_{2n+1}$ can be extended to one of $\tilde{L}\text{Spin}_{2n+1}$ simply by taking $\lambda_{n+1} = 0$.

The case of Sp_n is very similar: one embeds it in SU_{2n}. The exceptional group G_2 is embedded in Spin_8. The only delicate case is F_4. This can be embedded in E_6, but to obtain all the antidominant weights of $\tilde{L}F_4$ it seems to be necessary to twist the embedding $\tilde{L}F_4 \to \tilde{L}E_6$ by the outer automorphism of $\tilde{L}E_6$. It does not seem worthwhile, however, to give all the details here. One hopes that a more satisfactory proof of (13.4.2) can be found.

13.5 General remarks: the case of a general maximal abelian subgroup of LG.

In Section 12.2 we constructed a representation of $\tilde{U}_{\text{res}}(H)$, and hence one of $\tilde{L}U_n$, from a representation of the group $L\mathbb{T}$. In Section 13.3 we constructed a representation of $\tilde{L}G$ for any simply laced group G from a representation of $\tilde{L}T$. Now the group U_n is simply laced, and both $L\mathbb{T}$ and LT are maximal abelian subgroups of LU_n, so it is natural to ask whether the two constructions are particular cases of a more general procedure. In fact the generalization is fairly obvious.

Let us recall from Section 3.6 that a maximal abelian subgroup A of LG is obtained by choosing for each $\theta \in S^1$ a maximal torus T_θ of G, depending smoothly on θ, and defining

$$A = \{\gamma \in LG : \gamma(\theta) \in T_\theta \text{ for all } \theta\}.$$

The subgroup $L\mathbb{T}$ of LU_n is obtained by taking $T_\theta = \lambda(\theta)T\lambda(\theta)^{-1}$, where λ is a path in U_n from 1 to the permutation matrix corresponding to the cyclic permutation $(12\ldots n)$.

The blip construction for extending a projective representation of LT to one of LG is local with respect to the circle. But for loops with support in a proper subset of S^1 any maximal abelian subgroup A looks exactly the same as LT.† Thus the construction serves equally well to extend representations from A to LG. Let us reconsider the case $A = L\mathbb{T} \subset LU_n$ in this light.

Let e_{pq} be the $(p, q)^{\text{th}}$ root vector in $\mathfrak{gl}_n(\mathbb{C})$, i.e. the $n \times n$ matrix with $+1$ in the $(p, q)^{\text{th}}$ place, -1 in the $(q, p)^{\text{th}}$ place, and zeros elsewhere. The corresponding coroot is the homomorphism $\mathbb{T} \to T = \mathbb{T}^n$ which has degree $+1$ on the pth factor and -1 on the qth factor. Let $\{\phi_k\}$ be a sequence in A which tends to the blip at θ of type (p, q) with respect to the torus T_θ. Using the formulae (6.5.1) we can regard the ϕ_k as elements of $L\mathbb{T}$, and then ϕ_k tends to the product of the blips ψ_{θ_p} at $\theta_p = (\theta + 2\pi(p-1))/n$ and $\psi^*_{\theta_q}$ at $\theta_q = (\theta + 2\pi(q-1))/n$. An argument like that of (13.2.3) shows that $B^{pq}_\theta = \psi_{\theta_p}\psi^*_{\theta_q}$ does indeed make sense as a distribution in θ, and that the resulting operators define an action of $\tilde{L}\mathfrak{gl}_n(\mathbb{C})$. (Because ψ_{θ_p} and $\psi^*_{\theta_q}$ anticommute there seems at first to be an ambiguity between $\psi_{\theta_p}\psi^*_{\theta_q}$ and $\psi^*_{\theta_q}\psi_{\theta_p}$. But in replacing the double blip by $\psi_{\theta_p}\psi^*_{\theta_q}$ we omitted a projective multiplier equal to the sign of $q - p$.)

† As we saw in Section 9.2 it makes sense to speak of 'positive energy' representations of A. To make the present remarks quite cogent we should need to check that this notion reduces to the untwisted one on the subgroup of loops with support in an interval. Cf. (9.2.6).

13 'BLIPS' OR 'VERTEX OPERATORS'

In fact if $(r, s) \neq (q, p)$ the desired relation

$$[B_\theta^{pq}, B_\phi^{rs}] = \delta(\theta - \phi)(\delta_{qr}B_\phi^{ps} - \delta_{ps}B_\phi^{rq})$$

follows from (13.1.13), but the case $(r, s) = (q, p)$ needs more care.

The expression of the 'current group' generators as bilinear combinations of the fermionic operators explains once again why B_θ^{pq} is a density rather than a half-density, and why it is 'more singular' than the ψ_θ.

14
THE KAC CHARACTER FORMULA AND THE BERNSTEIN–GELFAND–GELFAND RESOLUTION

This chapter is devoted to the Kac character formula for positive energy representations of loop groups. The formula is an exact analogue of the Weyl character formula for compact groups, which we have described in Section 14.2 as an introduction and motivation. Kac's formula is stated and discussed in Section 14.3, and is proved by his methods in Section 14.4, in a version which we learnt from Macdonald [110].

There seem to be two ways to 'understand' the formula. One is as a 'fixed-point' formula: we have adopted that for our motivation in Section 14.2. The other, which is superficially more complicated but ultimately much more fruitful, is as an expression of the stratification of the complex homogeneous space LG/T into convex pieces indexed by the affine Weyl group. This leads to the so-called Bernstein–Gelfand–Gelfand resolution, which has Kac's formula as an immediate consequence. In Section 14.5 we have given a general account of the resolution—and one or two of its consequences—without quite proving its exactness. We hope that this compromise will prove useful: in the existing literature it seems hard to find an accessible account even of the analogous theorem for compact groups, and the present treatment covers that case completely. We have followed the paper of Kempf [91], but have tried to simplify it as much as possible. In Section 14.6 we have described some simple applications of the resolution. In particular we calculate the cohomology of the Lie algebra $L\mathfrak{g}$.

14.1 General remarks about characters

The character χ_V of a finite dimensional representation V of a group G is the function on G defined by $\chi_V(g) = \mathrm{trace}(T_g)$, where T_g is the transformation of V corresponding to g. It is well known that when G is a compact group the character determines the representation up to equivalence. The character is a *class-function* on G, i.e. $\chi_V(g)$ depends only on the conjugacy class of g. If G is a compact Lie group then each element is conjugate to an element of the maximal torus T, and so χ_V is completely described by giving its restriction to T, which is a function invariant under the conjugation-action of the Weyl group W on T.

For positive energy representations of loop groups it turns out, perhaps surprisingly, that there is a character theory which is as simple and convenient as that for compact groups. Let us consider representations of $\mathbb{T} \tilde{\times} \tilde{L}G$, where $\tilde{L}G$ is an arbitrary central extension of LG by $\mathbb{T}$. This group has the maximal torus $\mathbb{T} \times \tilde{T}$, where $\tilde{T}$ is the inverse-image in $\tilde{L}G$ of a maximal torus T of the subgroup G of constant loops. Of course it is very far from true that all elements of $\mathbb{T} \tilde{\times} \tilde{L}G$ are conjugate to elements of $\mathbb{T} \times \tilde{T}$. Nevertheless, positive energy representations of $\mathbb{T} \tilde{\times} \tilde{L}G$ are determined up to equivalence by their restrictions to $\mathbb{T} \times \tilde{T}$: that follows at once from the fact that each irreducible representation Γ_λ contains a unique minimal weight vector, up to a scalar multiple, transforming according to a character λ of $\mathbb{T} \times \tilde{T}$ which determines Γ_λ completely. If E is a representation of finite type, i.e. each energy level $E(k)$ is finite dimensional, then in the decomposition of E according to the characters of $\mathbb{T} \times \tilde{T}$ each character λ will occur with finite multiplicity d_λ, and so the following definition is very natural.

Definition (14.1.1). *The* character *of E is the formal sum*

$$\chi_E = \sum_\lambda d_\lambda \cdot e^{i\lambda}.$$

In writing this formula we have departed from our usual practice and made a distinction between the linear form $\lambda: \mathbb{R} \oplus \tilde{\mathfrak{t}} \to \mathbb{R}$ and the homomorphism $e^{i\lambda}: \mathbb{T} \times \tilde{T} \to \mathbb{T}$.

The character χ_E so defined determines E up to equivalence.

The formal character actually makes sense analytically in an appropriate framework. One can look at it in either of two ways. The first is along the lines of the theory of characters of locally compact groups. If G is a locally compact group which acts by unitary operators $\{U_g\}$ on a Hilbert space H then it is customary to associate to each smooth function ρ with compact support on G the 'smeared' operator

$$U(\rho) = \int_G \rho(g) U_g \, dg,$$

where the integral is with respect to Haar measure on G. If H is an irreducible representation then $U(\rho)$ is an operator of trace class, and the linear map $\rho \mapsto \text{trace } U(\rho)$ is a distribution on G which is called the distributional character of H.

This distributional definition does not immediately make sense for loop groups, as they have no Haar measure. But one can try to apply it to the restriction of the representation to the maximal torus. One then finds that the operator $U(\rho)$ is not of trace class for an arbitrary smooth smearing function ρ, but is of trace class when ρ is real-analytic. In other words the character exists not as a distribution but as a kind of hyperfunction. A

14.1 GENERAL REMARKS ABOUT CHARACTERS

typical example is the character of the basic representation of $T \tilde\times \tilde L T$ studied in Chapter 10. We saw there that, as a formal sum, the character of the basic representation was given by

$$\chi(z, u, a) = a(1 + u) \prod_{k>0} (1 + uz^k)(1 + u^{-1}z^k)$$

$$= a \prod_{k>0} (1 - z^k)^{-1} \cdot \sum_{n\in\mathbb{Z}} u^n z^{\frac{1}{2}n(n-1)}$$

for (z, u, a) in the torus $\mathbb{T} \times \mathbb{T} \times \mathbb{T}$ (see (10.5.3)). If we integrate χ over the second copy of $\mathbb{T}$, i.e. we extract the coefficient of u^0, then we get $a\pi(z)$, where

$$\pi(z) = \sum \pi_n z^n = \prod (1 - z^k)^{-1}$$

is the partition function, in which the coefficient of z^n is the number of partitions of n. This is holomorphic for $|z| < 1$, but the hyperfunction which is its boundary value on $|z| = 1$ is very badly behaved, with a 'pole' at each rational point of the circle.

A more interesting way to think of the character analytically is to observe that not only does the complexified group $\tilde L G_{\mathbb{C}}$ act on the homogeneous space LG/T and hence on the positive energy representations, but the semigroup $\mathbb{C}^\times_{\leq 1} \tilde\times \tilde L G_{\mathbb{C}}$ does so too. (This has been explained in Sections 7.6 and 8.6). The positive energy unitary representations of $\mathbb{T} \tilde\times \tilde L G$ are therefore 'boundary values' of holomorphic representations of the semigroup $\mathbb{C}^\times_{\leq 1} \tilde\times \tilde L G_{\mathbb{C}}$. The operators representing elements of this semigroup turn out to be of trace class, and so there is a character which is a holomorphic function on $\mathbb{C}^\times_{\leq 1} \tilde\times \tilde L G_{\mathbb{C}}$. The formal trace of Definition (14.1.1) is the boundary value of this holomorphic function. We observe that if $|z| < 1$ then the operator $U_{z,\gamma}$ representing $(z, \gamma) \in \mathbb{C}^\times_{\leq 1} \tilde\times \tilde L G_{\mathbb{C}}$ is obtained by smearing the operators $U_{\zeta,\gamma}$ for $|\zeta| = 1$ with respect to the real-analytic Cauchy measure

$$\frac{1}{2\pi i} \frac{d\zeta}{\zeta - z}$$

on the circle.

We shall not pursue either of the analytical approaches to the character theory. One reason is that the formal character seems adequate for all practical purposes. More fundamentally, at present the existence of the analytical character is known only by inspecting the analytical properties of the formal character. If one could see a priori why the operators $U_{z,\gamma}$ for $|z| < 1$ are of trace class then the analytic theory would become much more interesting.

Before leaving these general remarks we should point out one intriguing and mysterious fact. The partition function π is a *modular function* in the sense that if (cf. Section 14.3)

$$\eta(t) = e^{\pi i t/12}\pi(e^{2\pi i t})$$

then

$$\eta(t) = (-it)^{\frac{1}{2}}\eta(-1/t).$$

It follows from the Kac character formula that the characters of all positive energy representations of loop groups are constructed from modular functions in an appropriate sense. No explanation of this phenomenon seems to be known at present.

14.2 Motivation of the character formula: Weyl's formula for compact groups

The most important result about characters is the formula due to Kac for the character χ_λ of the irreducible representation with lowest weight λ. It is an exact analogue of the Weyl character formula for compact groups.

One way to motivate the result is as a 'fixed-point' formula. Let us think first of the action of a finite group G on the vector space Γ of sections of a line bundle L on a finite set X. We assume G acts compatibly on L and X. The space Γ is of course just the direct sum $\bigoplus_{x \in X} L_x$ of the fibres L_x of L. As $g \in G$ maps L_x to L_{gx} the matrix of the action of g on Γ has non-zero diagonal entries only when $gx = x$, and so its trace is given by

$$\sum_{x \in F} \text{trace}(g \mid L_x),$$

where $F = \{x \in X : gx = x\}$ is the set of fixed points of the action of g on X.

Now let us consider a more general group G which acts by holomorphic transformations of a holomorphic line bundle L on a complex manifold X. It is reasonable to hope that if $g \in G$ acts on X with isolated fixed points then the trace of the action of g on the space Γ of holomorphic sections of L—assuming it can be defined—is related to the sum of the traces of the actions of g on the spaces $\mathcal{O}_x(L)$ of germs of holomorphic sections of L at the fixed points x of the action of g on X. In other words, only the infinitesimal neighbourhoods of the fixed points should contribute to the trace. Now the germ at x of a holomorphic section is simply a 'Taylor series' at x, so $\mathcal{O}_x(L)$ can be identified with $L_x \otimes S(T_x^*)$, where T_x^* is the dual of the tangent space to X at x (with its complex structure), and S denotes the symmetric algebra. (More invariantly, the vector space of germs $\mathcal{O}_x(E)$ is filtered by the order to which the germs vanish at x, and the associated graded vector space is $L_x \otimes S(T_x^*)$; the difference between the two, however, is irrelevant for the purpose of calculating the trace.) If the eigenvalues of the action of g on T_x^* are $\{\alpha_1, \ldots, \alpha_n\}$ the trace of its action on $S(T_x^*)$ is—cf. (10.5.2)—formally equal to $\prod (1 - \alpha_i)^{-1}$, i.e. to the reciprocal of the determinant $\det(1 - g \mid T_x^*)$. The hoped-for formula for the trace of the action of g on Γ is therefore

$$\sum_{x \in F} \frac{\operatorname{trace}(g \mid L_x)}{\det(1 - g \mid T_x^*)}. \quad (14.2.1)$$

The Weyl character formula is the assertion that (14.2.1) holds in the case of the action of a compact Lie group G on the complex homogeneous space $X = G/T = G_\mathbb{C}/B^+$, at least for a positive line bundle L (i.e. one which possesses non-vanishing holomorphic sections). The Kac character formula is the same assertion for the action of the maximal torus of $\mathbb{T} \tilde{\times} \tilde{L}G$ on $Y = LG/T = LG_\mathbb{C}/B^+ G_\mathbb{C}$. (Cf. Section 8.7.) We shall now make these formulae more explicit, beginning with the finite dimensional case.

We want to calculate the trace of the action of a general element g of the torus T, when L is the homogeneous line bundle associated to a character λ of T. We shall assume that g is sufficiently generic for its powers to be dense in T. The set of fixed points F of g can then be identified with the Weyl group $W = N(T)/T$. Indeed if the coset hT is fixed we have $ghT = hT$, and so $h^{-1}gh \in T$; this means that h belongs to the normalizer $N(T)$ of the torus T.

If $n \in N(T)$ then the action of n maps L_H and T_H^* isomorphically to L_{nH} and T_{nH}^* in such a way that the action of g on the latter pair corresponds to the action of $n^{-1}gn$ on the former. The action of T on the fibre L_H of L at the base-point is given by the character λ, and the tangent space T_H can be identified with the subspace $\mathfrak{n}^-$ of $\mathfrak{g}_\mathbb{C}$ spanned by the negative root-vectors. Thus the characters of T which occur in the dual $(\mathfrak{n}^-)^*$ are the positive roots.

Proposition (14.2.2). (First form of the Weyl character formula).

The character χ_λ of the irreducible representation of G with lowest weight λ is given by

$$\chi_\lambda = \sum_{w \in W} w \cdot \left\{ \frac{e^{i\lambda}}{\prod_{\alpha > 0} (1 - e^{i\alpha})} \right\}.$$

Here α runs through the positive roots of G, and we have once again distinguished in our notation between the linear forms $\alpha, \lambda : \mathfrak{t} \to \mathbb{R}$ and the corresponding homomorphisms $e^{i\alpha}, e^{i\lambda} : T \to \mathbb{T}$. For $w \in W$ and any function f on T we denote by $w \cdot f$ the function given by $(w \cdot f)(g) = f(n^{-1}gn)$, where $n \in N(T)$ represents w.

14 THE KAC CHARACTER FORMULA

Example. Let $G = SU_2$, with its maximal torus T identified with $\mathbb{T}$ by $u \mapsto \begin{pmatrix} u & 0 \\ 0 & u^{-1} \end{pmatrix}$. Suppose that $\lambda(u) = u^{-n}$. The homogeneous space $X = G_{\mathbb{C}}/B^+$ is the complex projective line $P^1_{\mathbb{C}} = S^2$ on which T acts by rotation with the poles as fixed points. The element $u \in T$ acts on the tangent space $\mathbb{C}$ at the basepoint by u^{-2}, and the non-trivial element of the Weyl group acts on T by $u \mapsto u^{-1}$. Then

$$\chi_\lambda(u) = \frac{u^{-n}}{1-u^2} + \frac{u^n}{1-u^{-2}} = u^n + u^{n-2} + u^{n-4} + \ldots + u^{-n}.$$

If $n \geq 0$ this is of course the character of the action of G on the homogeneous polynomials of degree n in the homogeneous coordinates on $P^1_{\mathbb{C}}$.

The formula (14.2.2) can be written in a slightly different form. Each element $w \in W$ permutes the roots of G and transforms a certain number $\ell(w)$ of positive roots to negative ones. Because $1 - e^{-i\alpha} = -e^{-i\alpha}(1 - e^{i\alpha})$ we have

$$w \cdot \prod_{\alpha > 0} (1 - e^{i\alpha}) = (-1)^{\ell(w)} e^{-is(w)} \prod_{\alpha > 0} (1 - e^{i\alpha}) \qquad (14.2.3)$$

where $s(w)$ is the sum of all the positive roots α such that $w^{-1}\alpha$ is negative. This gives us

Proposition (14.2.4). *(Second form of the Weyl character formula.)*

$$\chi_\lambda = \prod_{\alpha > 0} (1 - e^{i\alpha})^{-1} \sum_{w \in W} (-1)^{\ell(w)} e^{i(w \cdot \lambda + s(w))}.$$

Remark. If we write as usual ρ for half the sum of the positive roots then we have $s(w) = \rho - w \cdot \rho$, and we can rewrite the formula as

$$\chi_\lambda = \prod_{\alpha > 0} (e^{-\frac{1}{2}i\alpha} - e^{\frac{1}{2}i\alpha})^{-1} \cdot \sum_{w \in W} (-1)^{\ell(w)} e^{iw \cdot (\lambda - \rho)}.$$

Applying this formula to the trivial representation, i.e. when $\lambda = 0$ and $\chi_\lambda = 1$, gives a famous identity called the Weyl denominator formula:

$$\prod_{\alpha > 0} (e^{\frac{1}{2}i\alpha} - e^{-\frac{1}{2}i\alpha}) = \sum_{w \in W} (-1)^{\ell(w)} e^{iw \cdot \rho}. \qquad (14.2.5)$$

Example. If $G = SU_n$ then W is the symmetric group S_n, and $(-1)^{\ell(w)}$ is the sign of w. The character $e^{i\rho}$ takes $u = \mathrm{diag}\{u_1, \ldots, u_n\}$ to $u_1^{n-1} u_2^{n-2} \ldots u_{n-1}$, and the positive roots are $u \mapsto u_i u_j^{-1}$ for $i < j$. The formula (14.2.5) then becomes the expansion of the Vandermonde determinant:

$$\begin{vmatrix} u_1^{n-1} & u_2^{n-1} & \cdots & u_n^{n-1} \\ \vdots & \vdots & & \vdots \\ u_1 & u_2 & \cdots & u_n \\ 1 & 1 & \cdots & 1 \end{vmatrix} = \prod_{i<j} (u_i - u_j).$$

The proof of the Weyl character formula from the present point of view is provided by the Atiyah–Bott–Lefshetz fixed point theorem [6]. That tells us that if $\chi^{(q)}$ is the character of the representation of G on $H^q(X; \mathcal{O}_L)$, where $\mathcal{O}_L$ is the sheaf of holomorphic sections of the line bundle L, then the right-hand-side of (14.2.2) is equal to $\sum (-1)^q \chi^{(q)}$. It is known, however, that if the line bundle L arises from an antidominant weight λ, and we write $\mathcal{O}_\lambda = \mathcal{O}_L$, then $H^q(X; \mathcal{O}_\lambda) = 0$ when $q > 0$, and so the alternating sum reduces to the desired character $\chi^{(0)} = \chi_\lambda$.

Let us mention that there are at least four quite different methods known for proving the vanishing of $H^q(X; \mathcal{O}_\lambda)$ when $q > 0$ and λ is antidominant. One is to observe that L_λ is a 'positive' line bundle, i.e. that it has positive curvature in an appropriate sense, and to use Kodaira's vanishing theorem, which depends on Hodge theory. Another way is to prove that if λ is antidominant then

$$H^q(X; \mathcal{O}_\lambda) \cong H^{q+\ell(w)}(X; \mathcal{O}_{w*\lambda}), \tag{14.2.6}$$

where $w * \lambda = w \cdot \lambda + s(w)$, for all q and all $w \in W$: if w is of maximal length and $q > 0$ then $H^{q+\ell(w)}$ vanishes for dimensional reasons. Bott gives a very simple and direct proof of (14.2.6) in [15] (10.5). Two other proofs of the vanishing will be mentioned in Section 14.5.

The known analytic proofs of the Weyl character formula—the one just sketched and the one that will be described in Section 14.5—are very interesting and illuminating; but it must nevertheless be admitted that they are much longer, and depend on much deeper results, than the simple algebraic argument which we shall give in Section 14.4, and which works equally well in the infinite dimensional case.

14.3 The character formula

In this section we shall assume that the group G is semisimple.

We wish to calculate the character of the irreducible representation of $\mathbb{T} \tilde{\times} \tilde{L}G$ on the space Γ_λ of holomorphic sections of the line bundle L_λ on $Y = LG/T = (\mathbb{T} \tilde{\times} \tilde{L}G)/(\mathbb{T} \times \tilde{T})$. The complex structure of Y comes from its description as $LG_\mathbb{C}/B^+G_\mathbb{C}$ (see Section 8.7). Formally everything is exactly as in the case of compact groups considered in the previous section. The fixed points of the action of the torus $\mathbb{T} \times \tilde{T}$ on Y can be identified with the elements of the affine Weyl group W_{aff}, and the tangent space to Y at the base-point is $N^- \mathfrak{g}_\mathbb{C}$, on which $\mathbb{T} \times \tilde{T}$ acts by the set of negative affine roots. We write down exactly the same formula as (14.2.4).

Theorem (14.3.1) (The Kac character formula).

$$\chi_\lambda = \prod_{\alpha>0} (1 - e^{i\alpha})^{-1} \sum_{w \in W_{\text{aff}}} (-1)^{\ell(w)} e^{i(w.\lambda + s(w))}.$$

Here $s(w)$ denotes the sum of all the positive roots α of LG for which $w^{-1}\alpha$ is negative, and the product is over all the positive roots taken with their appropriate multiplicities. (In the finite dimensional case each root has multiplicity one, but for a loop group $\mathbb{T} \times T$ acts on all of $z^k t_\mathbb{C} \subset L\mathfrak{g}_\mathbb{C}$ by the same character $(k, 0)$.)

As just written the character formula applies to all semisimple groups G. (In fact it applies to all compact groups providing the level of λ is strictly positive.) It assumes a slightly more familiar form if G is simple and simply connected and $\tilde{L}G$ is its universal central extension. For then, although we cannot form half the sum of the positive roots of LG, we can define

$$\boldsymbol{\rho} = (0, \rho, -c) \in \mathbb{Z} \times \hat{T} \times \mathbb{Z}, \qquad (14.3.2)$$

where ρ is half the sum of the positive roots of G and c is the eigenvalue of the Casimir operator of $\mathfrak{g}$ in its adjoint representation. (The integer c is given by $c = \langle \rho, \alpha_0 \rangle + 1$, where α_0 is the highest root of G and the inner product is such that $\langle \alpha_0, \alpha_0 \rangle = 2$. If G is simply laced then c is the *Coxeter number* ([20] Chapter 6 Section 1.11) of G.)

Proposition (14.3.3). *For any $w \in W_{\text{aff}}$ we have*

$$\boldsymbol{\rho} - w \cdot \boldsymbol{\rho} = s(w),$$

where $s(w)$ is the sum of the positive roots α such that $w^{-1}\alpha$ is negative.

Proof. Because W_{aff} is generated by the reflections s_α in the planes of the simple roots α (see (5.1.4)) it is enough to check the formula when $w = s_\alpha$. But then $s(w) = \alpha$, while $\boldsymbol{\rho} - w \cdot \boldsymbol{\rho} = \boldsymbol{\rho}(h_\alpha)\alpha$. The definition of $\boldsymbol{\rho}$, however, was chosen precisely to make $\boldsymbol{\rho}(h_\alpha) = 1$ for each simple affine root α.

We can now rewrite the character formula.

Proposition (14.3.4). *If G is simply connected and simple then*

$$\chi_\lambda = e^{i\boldsymbol{\rho}} \prod_{\alpha>0} (1 - e^{i\alpha})^{-1} \sum_{w \in W_{\text{aff}}} (-1)^{\ell(w)} e^{iw(\lambda - \boldsymbol{\rho})}.$$

Example. If $G = SU_2$ with maximal torus T we can identify $\hat{T}$ with $\mathbb{Z}$ so

14.3 THE CHARACTER FORMULA

that $\lambda \in \mathbb{Z}$ corresponds to

$$\begin{pmatrix} u & 0 \\ 0 & u^{-1} \end{pmatrix} \mapsto u^{\lambda}.$$

Then the lattice of weights of $\mathbb{T} \tilde{\times} \tilde{L}G$ is $\mathbb{Z}^3$, and the roots are the elements $(n, \lambda, 0)$ for $n \in \mathbb{Z}$ and $\lambda = \pm 2$ or 0. (The positive roots are those with $n > 0$, together with $(0, 2, 0)$.) We find $\boldsymbol{\rho} = (0, 1, -2)$. If the translation part $\check{T}$ of W_{aff} is written $\{w_m^+\}_{m \in \mathbb{Z}}$ then the action on the weights is

$$w_m^+ \cdot (n, \lambda, h) = (n - \lambda m + hm^2, \lambda - 2hm, h)$$

(see (9.3.3)). The remaining elements $w_m^- = s \cdot w_m^+$, where s is the non-trivial element of the Weyl group of SU_2, act by

$$w_m^- \cdot (n, \lambda, h) = (n - \lambda m + hm^2, -\lambda + 2hm, h).$$

Applying the formula (14.3.4) for the characters of the basic representation and the other representation of level one, with lowest weights $\omega_0 = (0, 0, 1)$ and $\omega_1 = (0, -1, 1)$ respectively, we obtain

$$\chi_{\omega_0}(z, u, a) = a \cdot \Pi^{-1} \cdot \sum_{m \in \mathbb{Z}} z^{m(3m+1)} D_{6m+1}(u) \qquad (14.3.5)$$

and

$$\chi_{\omega_1}(z, u, a) = a \cdot \Pi^{-1} \cdot \sum_{m \in \mathbb{Z}} z^{m(3m+2)} D_{6m+2}(u) \qquad (14.3.6)$$

where

$$\Pi = \prod_{k > 0} \{(1 - u^2 z^k)(1 - z^k)(1 - u^{-2} z^k)\}$$

and

$$D_n(u) = (u^n - u^{-n})/(u - u^{-1}),$$

which is the character of the n-dimensional irreducible representation of SU_2 if $n \geq 0$ (and $D_{-n} = -D_n$).

In earlier chapters we have found two different explicit constructions of the basic representation of $\tilde{L}SU_2$. In Section 13.3 we realized it on the space E of an irreducible representation of $\tilde{L}T$. This representation, in turn, was induced from a Heisenberg representation of the identity component

$$(\tilde{L}T)^0 = T \times \tilde{V},$$

where V is the vector space of maps $S^1 \to \mathbb{R}$ of integral zero. In fact

$$E = \bigoplus_{n \in \mathbb{Z}} E_n,$$

where the E_n are identical as representations of $\tilde{V}$, but $u \in T$ acts as u^{2n}

on E_n. From this description we find
$$\chi_{\omega_0} = a \cdot \prod_{k>0}(1-z^k)^{-1} \cdot \sum_{m\in\mathbb{Z}} u^{2m}z^{m^2}.$$

Comparing this with (14.3.5) gives us a very unobvious identity which when $u=1$ reduces to
$$\prod_{k>0}(1-z^k)^{-2} = \sum_{m\in\mathbb{Z}}(6m+1)z^{m(3m+1)} \bigg/ \sum_{m\in\mathbb{Z}} z^{m^2}. \qquad (14.3.7)$$

Our other construction, in Section 10.6, realized the basic representation of $\tilde{L}U_2$ as an exterior algebra. The character of this is
$$a \cdot (1+u_1)(1+u_2) \prod_{k>0}(1+u_1 z^k)(1+u_2 z^k)(1+u_1^{-1}z^k)(1+u_2^{-1}z^k),$$
$$(14.3.8)$$
where an element of the torus of U_2 is written as $\text{diag}\{u_1, u_2\}$.

The Kac character formula in the version (14.3.1) applies to the basic representation of $\tilde{L}U_2$, which corresponds to $\lambda = (0, 0, 1) \in \mathbb{Z} \times \hat{T} \times \mathbb{Z}$. The affine Weyl group $W \tilde{\times} \check{T}$ is $S_2 \tilde{\times} \mathbb{Z}^2$. If α is the positive root $(1, -1)$ of U_2 then $w = (p, q) \in \check{T}$ transforms the affine roots by (see (4.9.5))
$$(n, \alpha, 0) \mapsto (n+p-q, \alpha, 0)$$
$$(n, -\alpha, 0) \mapsto (n-p+q, -\alpha, 0)$$
$$(n, 0, 0) \mapsto (n, 0, 0).$$

We find $s(w) = (\tfrac{1}{2}(p-q)(p-q-1), (p-q)\alpha, 0)$, while
$$w \cdot (0, 0, 1) = (\tfrac{1}{2}(p^2+q^2-p-q), (p, q), 1).$$

Assembling all these ingredients gives us the following formula for the character
$$\chi(z; u_1, u_2; a) = a \cdot \Pi^{-1} \cdot \sum_{(p,q)\in\mathbb{Z}^2} z^{p^2+q^2-pq-p} D_{2p-q, 2q-p+1}(u_1, u_2), \qquad (14.3.9)$$
where
$$\Pi = \prod_{k>0}\{(1-z^k)^2(1-u_1 u_2^{-1}z^k)(1-u_1^{-1}u_2 z^k)\},$$
and
$$D_{r,s}(u_1, u_2) = (u_1^s u_2^r - u_1^r u_2^s)/(u_2 - u_1).$$

Once again, the fact that the formulae (14.3.8) and (14.3.9) agree is the reverse of obvious. We now have, essentially, three completely different formulae for the character.

The general form of the character

To understand better what the character formula tells us let us observe that any element w of the affine Weyl group can be expressed uniquely in

14.3 THE GENERAL FORM OF THE CHARACTER

the form $w = w_0 \cdot \xi$, where w_0 belongs to the finite Weyl group W of G, and ξ belongs to the lattice $\check{T}$. In the formula (14.3.1) we can therefore perform the sum over W_{aff} in two stages. We sum first over $w_0 \in W$ by using the Weyl character formula (14.2.4) for G. If $\lambda = (0, \lambda, h)$ we obtain

$$\chi_\lambda(z, t, a) = a^h \cdot z^{-\kappa c_\lambda} \cdot \Pi^{-1} \cdot \sum_\mu \chi_\mu(t) z^{\kappa c_\mu} \tag{14.3.10}$$

where

$$c_\mu = \tfrac{1}{2}(\|\mu - \rho\|^2 - \|\rho\|^2),$$
$$\kappa = (h + c)^{-1},$$
$$\Pi = \prod_{k>0} \left\{ (1 - z^k)^\ell \prod_\alpha (1 - e^{i\alpha} z^k) \right\} \tag{14.3.11}$$

(with $\ell = \text{rank}(G)$ and the product over all roots α of G), χ_μ is given by the Weyl formula (14.2.4), ignoring the fact that μ is not necessarily antidominant, and Σ_μ denotes the sum over all $\mu \in \hat{T}$ of the form $\mu = \lambda + (h + c)\xi$, with $\xi \in \check{T}$ and $\check{T}$ embedded in $\hat{T}$ by the standard inner product.

Let us notice that both sides of (14.3.10) are naturally power series in z whose coefficients are class-functions on G. The expression (14.3.11) can be written more intelligibly as

$$\Pi = \Pi(z, g) = \prod_{k>0} \det(1 - z^k \, \text{ad}(g)), \tag{14.3.12}$$

where $\text{ad}(g)$ denotes the adjoint action of $g \in G$.

For any $\mu \in \hat{T}$ and $w \in W$ we have $\chi_{w*\mu} = (-1)^{\ell(w)} \chi_\mu$, where $w * \mu = w(\mu - \rho) + \rho$. For each μ we can find w so that $w(\mu - \rho) = w * \mu - \rho$ is dominant, and we can therefore rewrite (14.3.10) as follows.

Proposition (14.3.13). *If G is simple and simply connected then*

$$\chi_\lambda(z, t, a) = a^h \cdot z^{-\kappa c_\lambda} \Pi^{-1} \cdot \sum_V N_V z^{\kappa c_V} \chi_V(t),$$

where

the sum is over the irreducible representations V of G,
$\Pi = \Pi(z, t)$ is given by (14.3.12),
c_V is the value of the Casimir operator $\Delta_\mathfrak{g}$ on V,
N_V is an integer,
h is the level of λ, and
$\kappa = (h + c)^{-1}$.

A striking property of the formula (14.3.13) is that the power of z which accompanies each finite dimensional character χ_V does not depend on λ (except for its level), and is proportional to the value of the Casimir operator in V. It would be very interesting to have an explanation of this fact.

Let us now return to the construction of the representations of level 1 of simply laced groups in Section 13.3. As we explained above in the case of SU_2 the construction leads at once to a character formula. For the representation whose lowest weight is $\omega = (0, -\omega, 1)$ it gives

$$\chi_\omega = \alpha \cdot \pi(z)^\ell \cdot \sum_{\xi \in \check{T}} e^{i\langle\xi-\omega\rangle} \cdot z^{\frac{1}{2}\langle\xi,\xi\rangle - \langle\omega,\xi\rangle}, \qquad (14.3.14)$$

where

$$\pi(z) = \prod_{k>0} (1-z^k)^{-1}$$

is the partition function. This is quite different from Kac's formula, and in practice seems to be the most useful formula for the character. Equating (14.3.14) and (14.3.13) when $\omega = 0$ gives the remarkable identity

$$\prod_{\alpha,k} (1-e^{i\alpha}z^k) = \sum_{\xi \in \check{T}} \chi_{(c+1)\xi} z^{\frac{1}{2}(c+1)\langle\xi,\xi\rangle} \Big/ \sum_{\xi \in \check{T}} e^{i\xi} z^{\frac{1}{2}\langle\xi,\xi\rangle}, \qquad (14.3.15)$$

which generalizes (14.3.7). (Cf. [83](3.38).)

Macdonald ([131](6.8)) has shown that (14.3.14) can be rewritten as a sum over the irreducible representations of G:

$$\chi_\omega = a \cdot \pi(z)^\ell \cdot \sum_V \chi_V \cdot f_V(z), \qquad (14.3.16)$$

where V runs through the representations whose lowest weights are congruent to ω modulo $\check{T}$, and

$$f_V(z) = z^{\frac{1}{2}(\|\lambda\|^2 - \|\omega\|^2)} \prod_{\alpha > 0} (1 - z^{\langle\alpha,\rho-\lambda\rangle}),$$

where λ is the lowest weight of V.

Macdonald's identity

If we apply the character formula to the trivial representation, i.e. we take $\lambda = 0$, then we know that $\chi_\lambda = 1$, and so we obtain Macdonald's identity [107]

$$\prod_{\alpha > 0} (1 - e^{i\alpha}) = \sum_{w \in W_{\text{aff}}} (-1)^{\ell(w)} e^{is(w)}. \qquad (14.3.17)$$

Assuming that G is simple and simply connected we can once again make this more explicit by writing $w = w_0 \cdot \xi$ as above and summing over $w_0 \in W$. That gives

Proposition (14.3.18).

$$\Pi(z, g) = \sum_{\xi \in \check{T}} z^{a(c\xi)} \chi_{c\xi}(g),$$

14.3 MACDONALD'S IDENTITY

where Π is given by (14.3.12), $c\xi \in \check{T}$ is regarded as a weight by means of the standard inner product on $\mathfrak{t}$ and

$$a(c\xi) = \frac{1}{2c}(\|c\xi - \rho\|^2 - \|\rho\|^2).$$

We have already mentioned in Section 10.5 that when $G = SU_2$ Macdonald's identity is the Jacobi triple product identity. In Section 10.5, however, it arose by comparing representations of $\tilde{L}\mathbb{T}$ which did not extend to $\tilde{L}SU_2$.

Two slightly different ways of writing Macdonald's identity are worth mentioning. The Killing form $(\ ,\ )$ on $\mathfrak{g}$ is $2c$ times the standard inner product $\langle\ ,\ \rangle$ we are using, and Freudenthal's 'strange formula' (cf. [51] p. 243) tells us that

$$(\rho, \rho) = \frac{1}{24} \dim G,$$

so we can rewrite (14.3.18) in the form

$$z^{\frac{1}{24}\dim G}\Pi(z, g) = \sum_{\lambda \in L} \chi_\lambda(g) z^{(\lambda - \rho, \lambda - \rho)}, \quad (14.3.19)$$

where L is the sublattice of $\hat{T}$ which is the image of $\check{T}$ under the map $\check{T} \to \hat{T}$ defined by half the Killing form. If we put $g = 1$ in the left hand side of (14.3.19) we obtain $\eta(z)^{\dim G}$, where

$$\eta(z) = z^{\frac{1}{24}} \prod_{k>0} (1 - z^k)$$

is Dedekind's η-function, a modular form of weight $\frac{1}{2}$ (cf. Serre [135] Chapter 8).

The second reformulation is due to Kostant [93], who observed that (14.3.18) can be written

$$\Pi(z, g) = \sum_V \chi_V(g_0)\chi_V(g) z^{c^{-1}c_V}, \quad (14.3.20)$$

where the sum is over the irreducible representations V of G, and $g_0 \in G$ is an element belonging to a certain distinguished conjugacy class.† This formula is very interesting because of its relation to the 'heat kernel' on the group G, as we shall now indicate.

The heat equation on G is the equation

$$\frac{\partial f}{\partial t} = -\Delta_\mathfrak{g} f,$$

---

† In fact the conjugacy class of $\exp(4\pi\rho^*)$, where $\rho^* \in \mathfrak{t}$ corresponds to ρ under the Killing form.

where f is a real-valued function on $\mathbb{R}_+ \times G$, and $\Delta_{\mathfrak{g}}$ is the Casimir operator of $\mathfrak{g}$, which can be identified with the Laplace operator on G. The solution of the Cauchy problem for the heat equation is given by convolution with the heat kernel H:

$$f(t, g) = \int_G H(t, gh^{-1}) f(0, h) \, dh.$$

Thus H is the solution of the heat equation when the initial datum is a delta-function at $g = 1$. Because the Hilbert space $L^2(G)$ breaks up as

$$L^2(G) \cong \bigoplus_V V^* \otimes V$$

under the action of $G \times G$, where V runs through the irreducible representations of G, it is easy to see that

$$H(t, g) = \sum_V \dim(V) \chi_V(g) e^{-c_V t}.$$

Thus (14.3.20) gives us

$$H(t, g_0) = \Pi(e^{-ct}, 1).$$

More generally, Fegan [42] has pointed out that (14.3.20) asserts that

$$\tilde{H}(t, g) = \Pi(e^{-ct}, g),$$

where $\tilde{H}$ is the solution of the heat equation whose initial datum is a delta-function concentrated on the conjugacy class of g_0. The relation of the heat equation on G to the representations of LG remains, nevertheless, mysterious (but cf. Frenkel [47]).

14.4 The algebraic proof of the character formula

In this section we shall abandon the analytic and global point of view for that of pure algebra. Although our interest is still in the representations of a loop group $\tilde{L}G$ we shall introduce as a tool the much larger class $\mathcal{V}$ of abstract complex vector spaces V equipped with the following additional structure:

(i) an action of the Lie algebra $\tilde{L}_{\text{pol}} \mathfrak{g}_\mathbb{C}$ of polynomial loops, and
(ii) an action of $\mathbb{T}$ which intertwines with the rotation action of $\mathbb{T}$ on $\tilde{L}_{\text{pol}} \mathfrak{g}_\mathbb{C}$,

which we shall assume to satisfy

(a) V is the algebraic direct sum $\bigoplus_k V(k)$, where $V(k)$ is the part of V where $e^{i\theta} \in \mathbb{T}$ acts as $e^{-ik\theta}$, i.e. the 'part of energy k',

14.4 PROOF OF THE CHARACTER FORMULA

(b) V is of positive energy, i.e.
$$V(k) = 0 \quad \text{when} \quad k < 0,$$

(c) the action of $\tilde{\mathfrak{t}}_\mathbb{C}$ on V arises from a diagonalizable action of the torus $\tilde{T}$, and the characters of $\tilde{T}$ occur with finite multiplicity in each $V(k)$.

We know from Chapter 11 that the subspace $\check{E}$ of vectors of finite energy in any irreducible representation E of $\tilde{L}G$ (with positive energy) belongs to the class $\mathscr{V}$. On the other hand most members of $\mathscr{V}$ do *not* arise from representations of $\tilde{L}G$. The typical example for our purposes is the finite energy part of the space $\Gamma(L_\lambda | U)$ of holomorphic sections of the restriction of the line bundle L_λ of Section 11.1 to the dense open stratum U of the homogeneous space Y. Because U is not stable under the action of $\tilde{L}G$ the group does not act on $\Gamma(L_\lambda | U)$; but it still makes sense to differentiate sections along the tangent vectors corresponding to elements of $\tilde{L}\mathfrak{g}_\mathbb{C}$.

It may be worth mentioning, though we shall not need it, that there is a simple criterion for a member V of $\mathscr{V}$ to come from a representation of $\tilde{L}G$: it is that each of the three-dimensional subalgebras $i_\alpha(\mathfrak{sl}_2(\mathbb{C}))$ corresponding to the simple roots of $\tilde{L}G$ acts on the space V by a sum of finite dimensional representations. Kac [86] calls such spaces V integrable.

In this section we shall make passing use of the concept of the *universal enveloping algebra* $\mathcal{U}(\mathfrak{a})$ of a Lie algebra $\mathfrak{a}$ (cf. [20] Chapter 1 Section 2; [37]). To give an action of $\mathfrak{a}$ on V is equivalent to giving an action of $\mathcal{U}(\mathfrak{a})$. The Poincaré–Birkhoff–Witt theorem ([20] Chapter 1 Section 2.7) tells us that $\mathcal{U}(\mathfrak{a})$ is isomorphic to the symmetric algebra $S(\mathfrak{a})$ as a vector space.

Let us write $\mathcal{U} = \mathcal{U}(\tilde{L}_{\text{pol}}\mathfrak{g}_\mathbb{C})$ and, employing unsymmetrical notation, $\mathcal{U}_+ = \mathcal{U}(N^+_{\text{pol}}\mathfrak{g}_\mathbb{C})$ and $\mathcal{U}_- = \mathcal{U}(\tilde{B}^-_{\text{pol}}\mathfrak{g}_\mathbb{C})$. Because $\tilde{L}_{\text{pol}}\mathfrak{g}_\mathbb{C} = N^+_{\text{pol}}\mathfrak{g}_\mathbb{C} \oplus \tilde{B}^-_{\text{pol}}\mathfrak{g}_\mathbb{C}$ the Poincaré–Birkhoff–Witt theorem implies that $\mathcal{U}_+ \cdot \mathcal{U}_- = \mathcal{U}$ and more precisely that the multiplication map
$$\mathcal{U}_+ \otimes \mathcal{U}_- \to \mathcal{U}$$
is an isomorphism of vector spaces.

The crucial idea for the proof of the character formula is that of a *Verma module*.

Definition (14.4.1). *A Verma module V_λ is a member of the class $\mathscr{V}$ which is generated by a vector ξ_λ such that*
  (i) *ξ_λ is annihilated by $N^-_{\text{pol}}\mathfrak{g}_\mathbb{C}$,*
  (ii) *ξ_λ is an eigenvector of $\mathbb{T} \times \tilde{T}$ corresponding to the character λ,*
and
  (iii) *V_λ is as free as possible subject to (i) and (ii), i.e. the map*
$$\mathcal{U} \otimes_{\mathcal{U}_-} \mathbb{C} \to V_\lambda$$

given by $a \otimes x \mapsto xa$. ξ_λ is an isomorphism. (Here $\mathcal{U}_-$ acts on $\mathbb{C}$ by λ.)

Evidently if a member V of $\mathcal{V}$ contains a vector ξ with the properties (i) and (ii) of (14.4.1) then there is a unique homomorphism $V_\lambda \to V$ taking ξ_λ to ξ.

Proposition (14.4.2).
 (i) *If V_λ is a Verma module then the map $\mathcal{U}_+ \to V_\lambda$ given by $a \mapsto a \cdot \xi_\lambda$ is an isomorphism.*
 (ii) *If μ is any weight of V_λ then $\mu - \lambda$ is a sum of positive roots of $\tilde{L}G$, and the lowest weight λ occurs in V_λ with multiplicity one.*

Proof. The first statement follows from (14.4.1)(iii) and the isomorphism $\mathcal{U}_+ \otimes \mathcal{U}_- \cong \mathcal{U}$. The second statement follows at once from the first.

Example. The finite energy part of the antidual of $\Gamma(L_\lambda | U)$ is a Verma module V_λ. The cyclic vector ξ_λ is given by the evaluation $\varepsilon_1 : \Gamma(L_\lambda | U) \to \mathbb{C}$ at the base-point. Indeed ε_1 is cyclic because a section of $L_\lambda | U$ vanishes if all its derivatives at the base-point vanish, and on the other hand (see (11.1.2)) $\Gamma(L_\lambda | U)$ is essentially isomorphic to $S(N_{\text{pol}}^- \mathfrak{g}_\mathbb{C})^*$, whose antidual is $S(N_{\text{pol}}^+ \mathfrak{g}_\mathbb{C})$, which is isomorphic to $\mathcal{U}_+$ by the Poincaré–Birkhoff–Witt theorem.

Let us notice that the canonical map $V_\lambda \to \Gamma_\lambda$, where Γ_λ is the irreducible unitary representation with lowest weight λ, is the antidual of the restriction $\Gamma(L_\lambda) \to \Gamma(L_\lambda | U)$.

Any space in $\mathcal{V}$ has a formal character. The Verma module V_λ is isomorphic to $\mathbb{C}_\lambda \otimes S(N_{\text{pol}}^+ \mathfrak{g}_\mathbb{C})$ as a representation of $\mathbb{T} \times \tilde{T}$, where $\mathbb{C}_\lambda$ denotes $\mathbb{C}$ with the action of $\mathbb{T} \times \tilde{T}$ given by the character λ. So we have (cf. (10.5.2))

Proposition (14.4.3). *The character ϕ_λ of the Verma module V_λ is given by*

$$\phi_\lambda = e^{i\lambda} \cdot \prod_{\alpha > 0} (1 - e^{i\alpha})^{-1},$$

where the product is over all the positive roots of $\tilde{L}G$, taken with their appropriate multiplicities.

Our next step is to relate the characters of the irreducible representations Γ_λ to those of the Verma modules. For the rest of this section we shall assume that the algebra $\mathfrak{g}$ is *simple*. That is only for the sake of clarity: to treat the general case where $\mathfrak{g}$ is a product of k simple factors one ought to replace $\mathcal{V}$ by the class of modules for the universal central extension $\tilde{L}\mathfrak{g}$ of $L\mathfrak{g}$ by $\mathbb{R}^k$ which admit an intertwining action of $\mathbb{T}^k$ (and are of positive energy and finite type); one would then use the inner product (9.4.11) on $\mathbb{R}^k \tilde{\times} \tilde{L}\mathfrak{g}$, and the following discussion would hold without change.

14.4 PROOF OF THE CHARACTER FORMULA

Proposition (14.4.4). *The character χ_λ of the irreducible representation Γ_λ is a countable sum of the form $\sum n_\mu \phi_\mu$, where $n_\mu \in \mathbb{Z}$ and μ runs through weights such that $\mu \geq \lambda$ and $\|\mu - \rho\|^2 = \|\lambda - \rho\|^2$.*

In this statement we write $\mu \geq \lambda$ to mean that $\mu - \lambda$ is a sum of positive roots of $\tilde{L}G$. For the definition of ρ we refer to (14.3.2). The sum converges—in the formal sense—because for any energy-level k there are only finitely many weights $\mu \geq \lambda$ with energy $\leq k$.

Proof of (14.4.4). Let V be a member of $\mathscr{V}$, and let n be the lowest energy which occurs in it. Let us choose weight vectors $\xi_1, \ldots, \xi_r$ in $V(n)$ which form a basis for the space of lowest weight vectors for the action of $\mathfrak{g}$ on $V(n)$. Suppose that ξ_i has weight $\tilde{\lambda}_i$ for $\tilde{L}G$, and weight λ_i for G. Then there is a map $V_{\tilde{\lambda}_i} \to V$ which takes the generator to ξ_i, and so we obtain an exact sequence in $\mathscr{V}$

$$0 \to V' \to \bigoplus V_{\tilde{\lambda}_i} \to V \to V'' \to 0. \qquad (14.4.5)$$

The map in the middle induces an isomorphism in the energy level n—notice that $V_{\tilde{\lambda}_i}(n)$ is the irreducible representation of G with lowest weight λ_i. The kernel V' and cokernel V'' therefore have lowest energy greater than n. As characters behave additively for exact sequences we obtain

$$\chi_V = \sum \phi_{\tilde{\lambda}_i} + \chi_{V''} - \chi_{V'}.$$

Let us apply this first with $V = \Gamma_\lambda$, and then apply it again with V replaced by V' and V'', and so on indefinitely. We clearly obtain an expression for χ_λ of the desired form. From (14.4.2) it follows that if all the weights in V are $\geq \lambda$ then the same holds for V' and V'', and so the ϕ_μ which occur in the expansion of χ_λ all satisfy $\mu \geq \lambda$.

To obtain the statement about $\|\mu - \rho\|^2$ we use the Casimir operator Δ of Section 9.4. From (9.4.10) we know that Δ acts on V_μ as the scalar $c_\mu = \frac{1}{2}(\|\mu - \rho\|^2 - \|\rho\|^2)$. It then follows from (14.4.5) that if Δ acts as a scalar on V it acts as the same scalar on both V' and V''. But it acts as c_λ on Γ_λ, and so $c_\mu = c_\lambda$ for all the weights μ which occur.

It is now very simple to complete the proof of the character formula. We need two lemmas.

Lemma (14.4.6). *For any weight λ and any $w \in W_{\text{aff}}$ we have*

$$w \cdot \phi_\lambda = (-1)^{\ell(w)} \phi_{w * \lambda},$$

where $w * \lambda = w \cdot \lambda + s(w) = w \cdot (\lambda - \rho) + \rho$.

Lemma (14.4.7). *If λ is an antidominant weight, and ν is a weight which satisfies*
 (i) $\nu \geq \lambda$,

(ii) $\nu - \rho$ *is antidominant, and*
(iii) $\|\nu - \rho\|^2 = \|\lambda - \rho\|^2$,
then $\nu = \lambda$.

Granting the lemmas the proof is as follows. Suppose that $\chi_\lambda = \sum n_\mu \phi_\mu$. Then $w \cdot \chi_\lambda = (-1)^{\ell(w)} \sum n_\mu \phi_{w*\mu}$ by (14.4.6). But the character χ_λ must be invariant under W_{aff}, so we must have $n_{w*\mu} = (-1)^{\ell(w)} n_\mu$ for all μ and w, as the formal series ϕ_μ are clearly linearly independent.

Now suppose that $n_\mu \neq 0$. Choose w so that $w \cdot (\mu - \rho)$ is antidominant. Write $w \cdot (\mu - \rho) = \nu - \rho$, i.e. $\nu = w * \mu$. Then $n_\nu \neq 0$, so that $\nu \geqslant \lambda$ and $\|\nu - \rho\| = \|\lambda - \rho\|$. By (14.4.7) we conclude that $\nu = \lambda$. But we know that $n_\lambda = 1$, so we have proved the character formula in the version

$$\chi_\lambda = \sum_w (-1)^{\ell(w)} \phi_{w*\lambda}. \tag{14.4.8}$$

Proof of (14.4.6). This is simply the combination of (14.4.3), (14.2.3), and (14.3.3).

Proof of (14.4.7). We have

$$0 = \|\nu - \rho\|^2 - \|\lambda - \rho\|^2 = \langle \nu - \lambda, \nu + \lambda - 2\rho \rangle.$$

Now $\nu - \lambda$ is a sum of positive roots, and $\nu + \lambda - 2\rho$ is not only antidominant but *strictly* so, in the sense that $\langle \nu + \lambda - 2\rho, \alpha \rangle < 0$ for every positive root α—for $\langle \rho, \alpha_i \rangle = 1$ for every simple root α_i. Thus the only possibility is that $\nu - \lambda = 0$.

It will be noticed that this is the only point in the proof where we used the hypothesis that $\mathfrak{g}$ was semisimple; and it is unnecessary even then if the level of λ is >0.

14.5 The Bernstein–Gelfand–Gelfand resolution

In this section we shall assume that the group G is simply connected and that λ is an antidominant weight.

The character formula (14.4.8)

$$\chi_\lambda = \sum_{w \in W_{\text{aff}}} (-1)^{\ell(w)} \phi_{w*\lambda},$$

where $w * \lambda$ denotes $w \cdot \lambda + s(w)$, would be explained very naturally if we could see the existence of an exact sequence or 'resolution'

$$0 \leftarrow \Gamma_\lambda \leftarrow V_\lambda \leftarrow \bigoplus_{\ell(w)=1} V_{w*\lambda} \leftarrow \bigoplus_{\ell(w)=2} V_{w*\lambda} \leftarrow \ldots \tag{14.5.1}$$

of the irreducible representation Γ_λ by Verma modules. Such a resolution does indeed exist: it is simply the expression of the stratification of the basic homogeneous space $Y = LG/T$ which we studied in Section 8.7. It

14.5 THE BERNSTEIN-GELFAND-GELFAND RESOLUTION

is called the *Bernstein–Gelfand–Gelfand resolution* (cf. [10]), and is important and useful in its own right. In this section, which is a direct continuation of Chapter 11, we shall try to describe and elucidate the resolution and point out one or two applications; but we shall not give a complete proof of its exactness.

Geometrically it is more natural to discuss the sequence antidual to (14.5.1): the two are of course equivalent. Before approaching the BGG resolution proper we shall first, following the original paper [10], describe a more obvious resolution which we shall call the *weak BGG resolution*. This serves as well as the sequence (14.5.1) for explaining the character formula, and also for many other applications. (For loop groups it was first obtained by Garland and Lepowsky in [56].)

We begin with the case $\lambda = 0$. The de Rham complex

$$\Omega^0_{\text{hol}}(U) \xrightarrow{d} \Omega^1_{\text{hol}}(U) \xrightarrow{d} \Omega^2_{\text{hol}}(U) \xrightarrow{d} \ldots \qquad (14.5.2)$$

of holomorphic differential forms on the dense open subset U of $Y = LG_{\mathbb{C}}/B^+$ is acyclic because U is holomorphically contractible. Thus (14.5.2) is a resolution of $\mathbb{C}$ by $\tilde{L}\mathfrak{g}_{\mathbb{C}}$-modules which are of positive energy and finite type. The Casimir operator Δ of $\tilde{L}\mathfrak{g}_{\mathbb{C}}$ acts on (14.5.2). Because Δ commutes with d we can write

$$\Omega^{\cdot}_{\text{hol}}(U) = \Omega^{\cdot}_{\text{hol}}(U)_{(0)} \oplus \Omega^{\cdot}_{\text{hol}}(U)_{(\neq 0)},$$

where the first summand on the right is the 0-supereigenspace of Δ, and the second is the sum of the other supereigenspaces. (In a finite dimensional space the c-supereigenspace of a linear operator T means the set of vectors annihilated by a power of $T - c$. In the present situation Δ preserves the energy decomposition of $\Omega^{\cdot}_{\text{hol}}(U)$, and we define $\Omega^{\cdot}_{\text{hol}}(U)_{(0)}$ as the closure of the 0-supereigenspaces of the action of Δ in each energy level.) Because Δ annihilates the constant functions in $\Omega^0_{\text{hol}}(U)$ it follows that $\Omega^{\cdot}_{\text{hol}}(U)_{(0)}$ is a resolution of $\mathbb{C}$.

To describe $\Omega^p_{\text{hol}}(U)_{(0)}$ more explicitly we observe that $\Omega^p_{\text{hol}}(U)$ is the space of holomorphic sections over U of the homogeneous vector bundle on $Y = LG_{\mathbb{C}}/B^+$ whose fibre is the representation $\Lambda^p(L\mathfrak{g}_{\mathbb{C}}/B^+\mathfrak{g}_{\mathbb{C}})^*$ of B^+. This representation can be filtered

$$\Lambda^p(L\mathfrak{g}_{\mathbb{C}}/B^+\mathfrak{g}_{\mathbb{C}})^* = F_0 \supset F_1 \supset F_2 \supset \ldots$$

by B^+-invariant subspaces in such a way that each quotient F_k/F_{k+1} is one-dimensional and corresponds to a character of B^+ which is the sum of p positive roots of $\tilde{L}G_{\mathbb{C}}$. The filtration of the fibre leads to a filtration of the homogeneous vector bundle, and hence to a composition series for the space $\Omega^p_{\text{hol}}(U)$ in which the successive quotients are of the form $\Gamma(L_\mu | U)$, i.e. the antiduals of Verma modules V_μ, where μ runs through the sums of p positive roots. The summand $\Omega^p_{\text{hol}}(U)_{(0)}$ accordingly has a composition series where the factors are those $\tilde{V}^*_\mu$ such that the Casimir eigenvalue c_μ is zero. It follows from Lemma (14.4.7) and the discussion following it that the only possible μs are $\mu = s(w)$, where w runs through the elements of length p in the affine Weyl group.

The complex $\Omega^{\cdot}_{\text{hol}}(U)_{(0)}$ is the weak BGG resolution of $\mathbb{C}$. It differs from the true BGG resolution by the fact that its terms are not sums of antidual Verma modules, but only possess composition series with antidual Verma modules as quotients.

The weak BGG resolution of a general representation Γ_λ is the c_λ-supereigenspace of the Casimir operator on the holomorphic de Rham complex $\Omega^{\cdot}_{\text{hol}}(U; \Gamma_\lambda)$ of U with coefficients in Γ_λ. It needs little comment. The only point to notice is that holomorphic p-forms on $Y = \hat{L}G_{\mathbb{C}}/\tilde{B}^+$ with values in Γ_λ can be identified with holomorphic sections of the homogeneous vector bundle on Y whose fibre is the representation $\text{Hom}(\Lambda^p(L\mathfrak{g}_{\mathbb{C}}/B^+\mathfrak{g}_{\mathbb{C}}); \Gamma_\lambda)$ of B^+.

We turn now to the true BGG resolution. We have seen that when Γ_λ is realized as the space of holomorphic sections of the line bundle L_λ on the homogeneous space Y we can identify the antidual of the canonical surjection $V_\lambda \to \Gamma_\lambda$ of (14.5.1) with the restriction map of sections

$$\Gamma(L_\lambda) \to \Gamma(L_\lambda \mid U), \tag{14.5.3}$$

where U is the dense open stratum of Y. We shall be working with a fixed line bundle L_λ from now on, and when W is any open subset of Y we shall write $\Gamma(W)$ for $\Gamma(L_\lambda \mid W)$. Thus the map (14.5.3) becomes $\Gamma(Y) \to \Gamma(U)$.

A section of $L_\lambda \mid U$ comes from one of L_λ if and only if it extends over the strata Σ_w of codimension one: for then it extends over the strata of higher codimension by Hartogs's theorem (cf. Section 11.3). Now Σ_w is contained in the open set $U_w = w \cdot U$, which is isomorphic to $\mathbb{C}^{\ell(w)} \times \Sigma_w$. Let us write $U^*_w = U_w - \Sigma_w$. If Σ_w has codimension one, i.e. if $\ell(w) = 1$, then $U \cap U_w = U^*_w$, and so $s \in \Gamma(U)$ restricts to give $s_w \in \Gamma(U^*_w)$. Thus s comes from $\Gamma(Y)$ if and only if each s_w extends to U_w, and we have an exact sequence

$$0 \to \Gamma(Y) \to \Gamma(U) \to \bigoplus_{\ell(w)=1} \Gamma(U^*_w)/\Gamma(U_w). \tag{14.5.4}$$

The quotient $\Gamma(U^*_w)/\Gamma(U_w)$ is the space of principal parts of meromorphic sections of L_λ which have a pole along the stratum Σ_w. The sequence (14.5.4) is the beginning of the desired resolution because we have

Lemma (14.5.5). *If $\ell(w) = 1$ then $\Gamma(U^*_w)/\Gamma(U_w)$ is naturally antidual to the Verma module $V_{w*\lambda}$.*

14.5 THE BERNSTEIN–GELFAND–GELFAND RESOLUTION

Proof. We must show that
$$\Gamma(U_w^*)/\Gamma(U_w) \cong \mathbb{C}_{w*\lambda} \otimes \mathrm{Hol}(N^-)$$
compatibly with the actions of $N^- \mathfrak{g}_{\mathbb{C}}$ and of $\mathbb{T} \times \tilde{T}$, where $\mathrm{Hol}(N^-)$ denotes the space of holomorphic functions on $N^- = N^- G_{\mathbb{C}}$, and $\mathbb{C}_{w*\lambda}$ denotes $\mathbb{C}$ with the action of $\mathbb{T} \times \tilde{T}$ given by the character $w*\lambda$.

The elements of length one in W_{aff} are the reflections in the planes of the simple affine roots. The open set U_w corresponding to the simple root α is the free orbit of y_w under the group $wN^-w^{-1} = N_w^- \cdot A_w$, where $N_w^- = N^- \cap wN^-w^{-1}$ and A_w is the one-parameter subgroup of N^+ generated by e_α (cf. Section 11.3). If we trivialize $L_\lambda | U_w$ compatibly with N_w^- then $\Gamma(U_w^*)/\Gamma(U_w)$ can be identified with the space of holomorphic functions from N_w^- into $\Gamma(C_w^*)/\Gamma(C_w)$, where $C_w = A_w \cdot y_w$ and $C_w^* = C_w - \{y_w\}$. So (because $N^- = N_w^- \cdot \bar{A}_w$) our task is to show that
$$\Gamma(C_w^*)/\Gamma(C_w) \cong \mathbb{C}_{w*\lambda} \otimes \mathrm{Hol}(\bar{A}_w),$$
compatibly with $\mathbb{T} \times \tilde{T}$ and with $\bar{A}_w$.

Let us identify the Riemann sphere $S^2 = \mathbb{C} \cup \{\infty\}$ with its image in Y under the map induced by $i_\alpha: SL_2(\mathbb{C}) \to \tilde{L}G_{\mathbb{C}}$ (cf. (11.3.2)). Then $0 \in S^2$ is the base-point of Y, ∞ is y_w, and C_w is $S^2 - \{0\}$. The element $\exp(ae_{-\alpha})$ of $\bar{A}_w$ acts on S^2 by $t \mapsto t + a$. The bundle $L_\lambda | S^2$ can be trivialized over $S^2 - \{\infty\}$ compatibly with $\bar{A}_w$. Then $\Gamma(C_w^*)$ becomes simply $\mathrm{Hol}(S^2 - \{0, \infty\})$, and $\Gamma(C_w)$ is the subspace of functions which have a pole at ∞ of order $\leq n$, where $n = -\lambda(h_\alpha)$ is the degree of $L_\lambda | S^2$. The quotient $\Gamma(C_w^*)/\Gamma(C_w)$ can accordingly be identified with the subspace of functions in $\mathrm{Hol}(S^2 - \{\infty\})$ which vanish to order $>n$ at 0; and these are identified with $\mathrm{Hol}(\mathbb{C}) = \mathrm{Hol}(\bar{A}_w)$ by differentiating $n+1$ times.

Finally we come back to the action of $\mathbb{T} \times \tilde{T}$. When we trivialized $L_\lambda | (S^2 - \{\infty\})$ we should have identified $\Gamma(C_w^*)$ with $\mathbb{C}_\lambda \otimes \mathrm{Hol}(S^2 - \{0, \infty\})$. But differentiating $n+1$ times brought us into $\mathbb{C}_{\lambda+(n+1)\alpha} \otimes \mathrm{Hol}(\bar{A}_w)$. This is what we want, because $w \cdot \lambda = \lambda - \lambda(h_\alpha)\alpha = \lambda + n\alpha$, and $s(w) = \rho - w\rho = \alpha$.

Remark. The lowest weight vector of the Verma module antidual to $\Gamma(U_w^*)/\Gamma(U_w)$ is the complex conjugate of
$$\Gamma(U_w^*)/\Gamma(U_w) \to \Gamma(C_w^*)/\Gamma(C_w) \to \mathbb{C},$$
where the second map is the *residue* at y_w.

The bare-handed approach we have been following is obviously not practicable when we come to extend the sequence (14.5.4) by studying the strata of higher codimension. We must introduce some more sophisticated machinery. For reasons which we shall explain presently we wish to allow for the possibility of replacing Y by a 'thickened' space Z which contains Y as a dense subspace. To keep everything as general and as clear as possible we shall proceed axiomatically, making the following assumptions about Z.

(I) Z is a complex manifold on which $\mathbb{T} \tilde{\times} \tilde{L}G_{\mathbb{C}}$ acts, and the action of $\tilde{L}G_{\mathbb{C}}$ is holomorphic.

(II) There is a base-point z_1 in Z whose orbit is dense and whose stabilizer is $\mathbb{T} \tilde{\times} \tilde{B}^+ G_{\mathbb{C}}$.

(III) Z is stratified by subsets Z_w indexed by the elements of W_{aff}. The strata are invariant under the action of N^-, and $Z_1 = U$ is open.

(IV) For each w the action of $A_w = N^+ \cap wN^- w^{-1}$ induces an isomorphism of complex manifolds
$$A_w \times Z_w \to U_w = wU.$$

Let us write $z_w = w \cdot z_1$ and $C_w = A_w z_w$. It follows from the preceding assumptions that the map $\gamma \mapsto \gamma \cdot z_w$ embeds $N_w^- = N^- \cap wN^- w^{-1}$ as a dense subspace of Z_w. We shall assume further:

(V) The natural injection $\mathrm{Hol}(Z_w) \to \mathrm{Hol}(N_w^-)$ has dense image.

All of the preceding assumptions hold if $Z = Y$. We shall add one more, asserting roughly that the strata Z_w are holomorphically convex. It is not known whether this is true when $Z = Y$.

(VI) For each w and any finite dimensional Stein manifold P the groups $H^q(Z_w \times P; \mathcal{O})$ vanish for $q > 0$, where $\mathcal{O}$ is the sheaf of holomorphic functions on $Z_w \times P$.

The assumptions I—IV imply that the holomorphic line bundle L_λ on Y extends canonically to Z: in fact L_λ is canonically N_w^--equivariantly trivial on each open set U_w of Z, and its transition functions depend only on the finitely many variables transverse to the strata, which are the same for Y and for Z. We shall write $\mathcal{O}_\lambda$ for the sheaf of holomorphic sections of L_λ on Z, and in the following discussion all the cohomology groups mentioned will be understood to have coefficients in the appropriate restriction of $\mathcal{O}_\lambda$.

For a general stratum Z_w we have to decide what will assume the role which was played by $\Gamma(U_w^*)/\Gamma(U_w)$ when $\ell(w) = 1$. The group we need is $H_{Z_w}^{\ell(w)}(U_w)$, the cohomology of the sheaf $\mathcal{O}_\lambda \mid U_w$ with supports in the closed subset Z_w. We recall that $H_{Z_w}^*(U_w)$ is defined by choosing a flabby resolution [21], [91]
$$0 \to \mathcal{O}_\lambda \mid U_w \to \mathcal{F}^0 \to \mathcal{F}^1 \to \mathcal{F}^2 \to \ldots$$
of the sheaf $\mathcal{O}_\lambda \mid U_w$ and taking the cohomology of the cochain complex $\Gamma_{Z_w}(\mathcal{F})$, where $\Gamma_{Z_w}(\mathcal{F}^q)$ denotes the sections of $\mathcal{F}^q$ with support in Z_w. If

14.5 THE BERNSTEIN-GELFAND-GELFAND RESOLUTION

$\ell(w) = 1$ then $H^1_{Z_w}(U_w) \cong \Gamma(U_w^*)/\Gamma(U_w)$ because of the long exact sequence associated to the short exact sequence of cochain complexes

$$0 \to \Gamma_{Z_w}(\mathcal{F}^\cdot) \to \Gamma(\mathcal{F}^\cdot) \to \Gamma(\mathcal{F}^\cdot | U_w^*) \to 0.$$

In general we have

Proposition (14.5.6). *The group $H^{\ell(w)}_{Z_w}(U_w; \mathcal{O}_\lambda)$ is naturally antidual to the Verma module $V_{w*\lambda}$, and $H^q_{Z_w}(U_w; \mathcal{O}_\lambda) = 0$ if $q \neq \ell(w)$.*

We shall postpone the proof of this proposition for the moment.

To obtain the BGG resolution we consider the question of calculating $H^*(Z; \mathcal{O}_\lambda)$. Let $\mathcal{F}^\cdot$ be a flabby resolution of $\mathcal{O}_\lambda$ on Z, and let Γ^q denote the space of sections of $\mathcal{F}^q$. Then $H^*(Z)$ is by definition the cohomology of the cochain complex $\Gamma^\cdot$. Let Z^p denote the union of the strata of Z of codimension $\geq p$. This is a closed subset of Z. We can filter the cochain complex $\Gamma^\cdot$ by subcomplexes $\Gamma^\cdot_{(p)}$, where $\Gamma^q_{(p)}$ denotes the sections of $\mathcal{F}^q$ which have support in Z^p. The filtration leads in a standard way [62] to a spectral sequence whose termination is $H^*(Z)$ and which begins with

$$E_1^{pq} = H^{p+1}(\Gamma^\cdot_{(p)}/\Gamma^\cdot_{(p+1)}). \tag{14.5.7}$$

Lemma (14.5.8).

$$E_1^{pq} \cong \bigoplus_{\ell(w)=p} H^{p+q}_{Z_w}(U_w).$$

Proof. Because $\mathcal{F}^q$ is a flabby sheaf the quotient $\Gamma^q_{(p)}/\Gamma^q_{(p+1)}$ is simply the space of sections of $\mathcal{F}^q | (Z - Z^{p+1})$ which have support in $Z^p - Z^{p+1}$. But in $Z - Z^{p+1}$ the subset $Z^p - Z^{p+1}$ is the union of the disjoint closed sets Z_w for $\ell(w) = p$. The sections with support in $Z^p - Z^{p+1}$ are therefore sums of sections with supports in the Z_w. As U_w is a neighbourhood of Z_w in $Z - Z^{p+1}$ we find

$$\Gamma^q_{(p)}/\Gamma^q_{(p+1)} \simeq \bigoplus \Gamma_{Z_w}(\mathcal{F}^q | U_w),$$

from which the lemma follows.

The Lemma together with Proposition (14.5.6) gives us

Proposition (14.5.9). *The groups $H^*(Z; \mathcal{O}_\lambda)$ are the cohomology of the cochain complex*

$$C^0 \to C^1 \to C^2 \to \ldots,$$

where

$$C^p = \bigoplus_{\ell(w)=p} H^p_{Z_w}(U_w).$$

Proof. We have shown that in the spectral sequence $E_1^{pq} \Rightarrow H^*(Z)$ we have $E_1^{pq} = 0$ if $q \neq 0$. This means that the spectral sequence collapses to the cochain complex of (14.5.9). Let us check that the spectral sequence really does converge. We observe that if we replace Z by $Z - Z^n$ for some n then the filtration of the cochain complex $\Gamma(\mathscr{F}^{\cdot} | Z - Z^n)$ has only n steps, and hence leads to an obviously convergent spectral sequence. This shows that $H^*(Z - Z^n)$ can be calculated from the truncated complex

$$C^*_{[n]} = C^0 \to C^1 \to \ldots \to C^{n-1}.$$

But Z is the union of the sequence of open sets $Z - Z^n$, and so, because the sheaves $\mathscr{F}^q$ are flabby, we have

$$\Gamma^{\cdot} = \varprojlim_n \Gamma(\mathscr{F}^{\cdot} | Z - Z^n).$$

That leads to a short exact sequence [113]†

$$0 \to R^1 \varprojlim H^{q-1}(Z - Z^n) \to H^q(Z) \to \varprojlim H^q(Z - Z^n) \to 0.$$

In this sequence, however, the $R^1 \varprojlim$ term vanishes because $H^{q-1}(Z - Z^n)$ is independent of n when $n > q$. Hence $H^q(Z) = H^q(Z - Z^n)$ if $n > q + 1$.

The cochain complex $C^{\cdot}$ of (14.5.9) is the desired BGG resolution. We have therefore shown that its exactness is equivalent to the vanishing of $H^q(Z; \mathcal{O}_\lambda)$ for $q > 0$. Unfortunately we do not know a direct proof of this vanishing theorem: neither of the two proofs of the finite dimensional analogue which we mentioned in Section 14.2 applies to the present case. On the other hand the combinatorial construction of the BGG resolution which is given in [10] *does* work for Kac–Moody algebras, and from that we can deduce the vanishing theorem. (In the finite dimensional case this gives a third proof of the vanishing theorem.)

We shall now give the postponed proof of (14.5.6).

Proof of (14.5.6).

We shall prove the following statement by induction on $p = \ell(w)$.

If Z'_w is a holomorphically convex open subset of Z_w, and we define $U'_w = A_w \cdot Z'_w$ and $U' = w^{-1} \cdot U'_w$, then

$$H^q_{Z'_w}(U'_w; \mathcal{O}_\lambda) = 0 \quad \text{if} \quad q \neq p,$$

and

$$H^p_{Z'_w}(U'_w; \mathcal{O}_\lambda) \cong \Gamma(U'; \mathcal{O}_{w*\lambda}),$$

compatibly with the action of $\tilde{L}\mathfrak{g}_\mathbb{C}$.

Here 'holomorphically convex' means that Z'_w has the property VI

---

† One has such a sequence for the cohomology of $\varprojlim C^{\cdot}_{[n]}$ for any *surjective* inverse system $\{C^{\cdot}_{[n]}\}$ of cochain complexes.

14.5 THE THICKENING OF THE HOMOGENEOUS SPACE Y

asserted above for Z_w. Proposition (14.5.6) is obtained by taking $Z'_w = Z_w$ in the statement.

Suppose that $w = s \cdot v$, where s is the reflection associated with a simple root α, and $\ell(v) = p - 1$. Then $U_w = s \cdot U_v$, and $Z_w \subset s \cdot Z_v$: in fact

$$A_s \times Z_w \xrightarrow{\cong} sZ_v.$$

(It is easy to check that $A_w = A_s \cdot sA_v s$, and that $N_v^- = \bar{A}_s \cdot sN_w^- s$.) Let $Z'_v = s \cdot (A_s \cdot Z'_w)$ and $U'_v = A_v \cdot Z'_v = s \cdot U'_w$. Then restriction of sections from U'_w to $U'_w - Z'_w$ leads to an exact sequence

$$\ldots \to H^{q-1}_{sZ'_v}(U'_w) \to H^{q-1}_{sZ'_v - Z'_w}(U'_w - Z'_w) \to H^q_{Z'_w}(U'_w) \to H^q_{sZ'_v}(U'_w) \to \ldots \quad (14.5.10)$$

The first two groups here are isomorphic to

$$H^{q-1}_{Z'_v}(U'_v) \to H^{q-1}_{Z'_v - sZ'_w}(U'_v - sZ'_w), \quad (14.5.11)$$

but with the $\tilde{L}\mathfrak{g}_\mathbb{C}$-action twisted by conjugation by s. Now $Z''_v = Z'_v - sZ'_w = s \cdot (A_s - \{1\})$. Z'_w is a holomorphically convex open set in Z_v (as it is the complement of a non-singular hypersurface in Z'_v), and $U''_v = a_v \cdot Z''_v$ is an open neighbourhood of Z''_v in $U'_v - sZ'_w$. Thus the second group in (14.5.11) is the same as $H^{q-1}_{Z''_v}(U''_w)$, and our inductive hypothesis tells us that both groups in (14.5.11) vanish when $q \neq p$, and that when $q = p$ the homomorphism coincides with the restriction map

$$\Gamma(U'; \mathcal{O}_{v*\lambda}) \to \Gamma(U''; \mathcal{O}_{v*\lambda}).$$

The corresponding homomorphism in (14.5.9) when $q = p$ is therefore

$$\Gamma(U'_s; \mathcal{O}_{v*\lambda}) \to \Gamma(U''_s; \mathcal{O}_{v*\lambda}), \quad (14.5.12)$$

where $U'_s = s \cdot U'$ and $U''_s = s \cdot U''$. Because this is injective it follows that $H^q_{Z'_w}(U'_s) = 0$ when $q \neq p$, and that $H^p_{Z'_w}(U'_w)$ is its cokernel.

But $U'_s = A_s \cdot V_s$, where $V_s = sv^{-1}A_v sZ'_w$ is an open set of the stratum Z_s; and $U''_s = (A_s - \{1\}) \cdot V_s$. The argument of (14.5.5) now implies that the cokernel of (14.5.12) is $\Gamma(\bar{A}_s.s.V_s; \mathcal{O}_{s*v*\lambda})$. Because $A_s.s.V_s = U'$ and $s * v * \lambda = w * \lambda$ that completes the proof of (14.5.6). (In (14.5.5) the weight λ was assumed to be antidominant. But all that was used there was that $\lambda(h_\alpha) < 0$. In the present situation we have $(v*\lambda)(h_\alpha) < 0$, for that is precisely the condition that $\ell(s \cdot v) > \ell(v)$.)

The thickening of the homogeneous space Y

The usual approach to groups like $H^q(Z; \mathcal{O}_\lambda)$ is by means of Dolbeault's theorem [15]. If we knew that the $\bar{\partial}$-Poincaré-lemma ([68] p. 25) were true locally on the complex manifold Z then $H^*(Z; \mathcal{O}_\lambda)$ would be the cohomology of the $\bar{\partial}$-complex $A^{\cdot}$, where A^q is the space of smooth forms of type $(0, q)$ with coefficients in L_λ. We could then have proceeded more concretely in the discussion above. Instead of choosing a flabby resolution of the sheaf $\mathcal{O}_\lambda$ we could simply have filtered the complex $A^{\cdot}$ by the subcomplexes $A_{(p)}^{0\cdot}$ consisting of forms which vanish in a neighbourhood of the union of the cells of dimension $<p$ in the manifold Z.

Unfortunately it is not known at present for what class of infinite dimensional manifolds the $\bar{\partial}$-lemma is true. (It is known to be false for forms of class $C^{(1)}$ on Hilbert space [28].) The most likely conjecture is that it holds for manifolds modelled on spaces which are duals of nuclear Fréchet spaces (cf. [31], and also [36], [30]).

We have also mentioned before that it would be very interesting to have an invariant measure on the homogeneous space Y. Such a measure, however, is most unlikely to exist on Y itself. If we consider the more or less equivalent space $X = LG/G$, which is the smooth loop space of G, then there is the well-known Wiener measure on the *continuous* loop space [81]. This measure is quasi-invariant under LG, but the smooth loops X are of measure zero for it.†

It seems reasonable to conjecture that there is a 'thickening' Z of Y such that

(i) Z is a complex manifold modelled on the dual of the tangent space $N^-\mathfrak{g}_{\mathbb{C}}$ to Y,

(ii) Z has the properties I–VI listed above,

(iii) Z possesses for each antidominant weight λ a measure μ_λ with coefficients in the line bundle $\bar{L}_\lambda \otimes L_\lambda$ which is invariant under LG,

(iv) every holomorphic section of $L_\lambda | Y$ which is of finite energy extends to Z and is square-integrable for μ_λ.

One might hope in addition that the $\bar{\partial}$-lemma should hold locally on Z.

If the conjecture about the thickening is true then the representation theory of loop groups is considerably more perspicuous. Analogues of the conjecture can easily be made for the groups of real-analytic or of polynomial loops. In those cases it is at least clear what the candidate for Z should be. In the real-analytic case we should take $Z = L_{\text{hyp}} G_{\mathbb{C}} / B_{\text{an}}^+ G_{\mathbb{C}}$, where $L_{\text{hyp}} G_{\mathbb{C}}$ is the group of all holomorphic maps from a (variable) annulus $A_\varepsilon = \{z : 1 - \varepsilon < |z| < 1\}$ to $G_{\mathbb{C}}$, i.e.

$$L_{\text{hyp}} G_{\mathbb{C}} = \lim_{\substack{\rightarrow \\ \varepsilon \to 0}} \text{Hol}(A_\varepsilon; G_{\mathbb{C}}),$$

and $B_{\text{an}}^+ G_{\mathbb{C}}$ is the subgroup of maps which extend over the hemisphere $|z| \leq 1$. In the polynomial case $L_{\text{hyp}} G_{\mathbb{C}}$ would be replaced by the group of all 'loops' in $G_{\mathbb{C}}$ whose matrix entries belong to the quotient field of the ring $\mathbb{C}[[z]]$ of formal power series.

---

† The most relevant work on measures on infinite dimensional homogeneous spaces is that of Pickrell [122].

14.6 Applications of the resolution: the cohomology of $\tilde{L}\mathfrak{g}$

A Verma module V_λ for $\tilde{L}\mathfrak{g}_\mathbb{C}$ is a free module over the enveloping algebra $\mathcal{U}_+$ of $N^+_{\text{pol}}\mathfrak{g}_\mathbb{C}$. This enables us to use the BGG resolution for calculating the cohomology of the Lie algebra $N^-\mathfrak{g}_\mathbb{C}$ with coefficients in any positive energy representation E of $\tilde{L}G$. (The cohomology here is defined by the complex of *continuous* Lie algebra cochains.) Let us notice that the torus $\mathbb{T} \times \tilde{T}$ of $\mathbb{T} \,\tilde{\times}\, \tilde{L}G$ acts naturally on $H^*(N^-\mathfrak{g}_\mathbb{C}; E)$.

Proposition (14.6.1). *If Γ_λ is the irreducible representation of $\tilde{L}G$ with lowest weight λ then*

$$H^q(N^-\mathfrak{g}_\mathbb{C}; \Gamma_\lambda) \cong \bigoplus_{\ell(w)=q} \mathbb{C}_{w*\lambda}$$

as a representation of $\mathbb{T} \times \tilde{T}$.

Proof. The cochain complex defining the cohomology consists of vector spaces of positive energy and finite type, so we can consider each energy level separately, and can equally well prove the antidual result

$$H_q(N^+_{\text{pol}}\mathfrak{g}_\mathbb{C}; \check{\Gamma}_\lambda) \cong \bigoplus_{\ell(w)=q} \mathbb{C}_{w*\lambda}.$$

The homology groups here can be calculated from the chain complex $\mathbb{C} \otimes_{\mathcal{U}_+} C.$, where $C.$ is any resolution of $\check{\Gamma}_\lambda$ by free $\mathcal{U}_+$-modules, such as the BGG resolution

$$0 \leftarrow \Gamma_\lambda \leftarrow V_\lambda \leftarrow \bigoplus_{\ell(w)=1} V_{w*\lambda} \leftarrow \ldots \tag{14.6.2}$$

But $\mathbb{C} \otimes_{\mathcal{U}_+} V_\mu \cong \mathbb{C}_\mu$, and the differentials in $\mathbb{C} \otimes_{\mathcal{U}_+} C.$ are zero because of their $\mathbb{T} \times \tilde{T}$-equivariance.

Remark. We have obtained (14.6.1) from the BGG resolution; but in fact all we required of the resolution (14.6.2) was its behaviour under $\mathcal{U}_+$ and under $\mathbb{T} \times \tilde{T}$. For that it is sufficient to know that $\check{\Gamma}_\lambda$ has a resolution $C.$ such that C_q belongs to the class $\mathscr{V}$ and has a composition series $C_q = C_q^{(0)} \supset C_q^{(1)} \supset \ldots$ for which the quotients $C_q^{(i)}/C_q^{(i+1)}$ are the Verma modules $V_{w*\lambda}$ with $\ell(w) = q$. In other words we only need the *weak* BGG resolution described on p. 291.

Proposition (14.6.1) is the crucial step towards calculating the cohomology of the Lie algebra $L\mathfrak{g}$. This cohomology maps to that of the topological space LG, and we wish to prove the following

Theorem (14.6.3). *The natural map*

$$H^*(L\mathfrak{g}; \mathbb{R}) \to H^*(LG; \mathbb{R})$$

is an isomorphism.

We have already proved in Section 4.11 that the map is surjective. In fact it is enough to prove

Proposition (14.6.4). *The natural map*
$$H^*(L\mathfrak{g}, \mathfrak{t}; \mathbb{R}) \to H^*(LG/T; \mathbb{R})$$
is an isomorphism.

The relative cohomology here is that of the complex of LG-invariant differential forms on the homogeneous space LG/T. (The formal definition can be found in [94] or [13] Chapter 1.) For the proof that (14.6.4) implies (14.6.3), which depends only on the fact that the group T is compact and connected, we refer to [94] (15.3). The cohomology of the space LG/T is, of course, known from its explicit stratification and cell-decomposition. It vanishes in odd dimensions, and in dimension $2p$ it is the free abelian group generated by the strata of complex codimension p, which are indexed by the elements of length p in the affine Weyl group.

To prove (14.6.4) we first complexify, observing that $H^*(L\mathfrak{g}, \mathfrak{t}; \mathbb{R}) \otimes_\mathbb{R} \mathbb{C} \cong H^*(L\mathfrak{g}_\mathbb{C}, \mathfrak{t}_\mathbb{C}; \mathbb{C})$. (The latter group is defined by continuous *complex*-multilinear cochains.) Then we resolve $\mathbb{C}$ by the weak BGG resolution $\mathbb{C} \to \Omega^\cdot$, where $\Omega^\cdot = \Omega^\cdot_{\text{hol}}(U)_{(0)}$. This resolution is a split exact sequence of topological vector spaces (because U is holomorphically contractible), and so the double complex
$$A^\cdot(L\mathfrak{g}_\mathbb{C}, \mathfrak{t}_\mathbb{C}; \Omega^\cdot)$$
of cochains of $(L\mathfrak{g}_\mathbb{C}, \mathfrak{t}_\mathbb{C})$ with coefficients in $\Omega^\cdot$ is exact in the $\Omega^\cdot$-direction. It follows that there is a spectral sequence with
$$E_1^{pq} = H^q(L\mathfrak{g}_\mathbb{C}, \mathfrak{t}_\mathbb{C}; \Omega^p)$$
which converges to $H^*(L\mathfrak{g}_\mathbb{C}, \mathfrak{t}_\mathbb{C}; \mathbb{C})$. Now we use (14.6.1) to prove

Lemma (14.6.5). *We have $E_1^{pq} = 0$ if $p \neq q$, and the dimension of E_1^{pp} is equal to the number of elements of length p in W_{aff}.*

Granting the lemma the proof of (14.6.4) is complete. For the spectral sequence collapses, and
$$H^n(L\mathfrak{g}_\mathbb{C}, \mathfrak{t}_\mathbb{C}; \mathbb{C}) \cong E_1^{pp} \quad \text{if} \quad n = 2p,$$
$$= 0 \quad \text{if} \quad n \text{ is odd.}$$

Thus $H^n(L\mathfrak{g}_\mathbb{C}, \mathfrak{t}_\mathbb{C}; \mathbb{C})$ and $H^n(LG/T; \mathbb{C})$ have the same dimension and are therefore isomorphic. (We know from Section 4.11 that the map between them is surjective.)

Proof of (14.6.5). The $L\mathfrak{g}_\mathbb{C}$-module Ω^p has a finite composition series in which the quotients are the antidual Verma modules $\bar{V}^*_{s(w)}$, where w runs

14.6 COHOMOLOGY OF HOLOMORPHIC LINE BUNDLES

through the elements of length p in W_{aff}. It is therefore enough to show that $H^q(L\mathfrak{g}_\mathbb{C}, \mathfrak{t}_\mathbb{C}; \bar{V}^*_{s(w)})$ vanishes if $q \neq \ell(w)$, and has dimension 1 if $q = \ell(w)$. But because $\bar{V}^*_{s(w)}$ is 'induced' from the representation $\mathbb{C}_{s(w)}$ of $B^+\mathfrak{g}_\mathbb{C}$ we shall see that

$$H^q(L\mathfrak{g}_\mathbb{C}, \mathfrak{t}_\mathbb{C}; \bar{V}^*_{s(w)}) \cong H^q(B^+\mathfrak{g}_\mathbb{C}, \mathfrak{t}_\mathbb{C}; \mathbb{C}_{s(w)}). \tag{14.6.6}$$

The last group is (by definition) the same as the T-invariant part of $H^q(N^+\mathfrak{g}_\mathbb{C}; \mathbb{C}) \otimes \mathbb{C}_{s(w)}$, which by (14.6.1)—or, more precisely, by its complex conjugate—has dimension 1 or 0 according as $q = \ell(w)$ or not.

The isomorphism (14.6.6) would be a formal triviality if we were working in a purely algebraic context (cf. [25] Chapter XIII, 4.2.2a). In our situation, where $\bar{V}^*_{s(w)}$ is the space of holomorphic sections of a line bundle on $U \subset Y$, it can be justified as follows. Let $\tilde{U}$ be the inverse-image of U in $LG_\mathbb{C}$. Then $\tilde{U} = N^- \cdot B^+$, and we can identify $\bar{V}^*_{s(w)}$ with the space of holomorphic maps $f: \tilde{U} \to \mathbb{C}_{s(w)}$ satisfying $f(ub^{-1}) = b \cdot f(u)$ for $b \in B^+$. Thus the cochain complex defining $H^*(L\mathfrak{g}_\mathbb{C}, \mathfrak{t}_\mathbb{C}; \bar{V}^*_{s(w)})$ is precisely the B^+-invariant part of the holomorphic de Rham complex of $T_\mathbb{C}\backslash\tilde{U}$ with coefficients in $\mathbb{C}_{s(w)}$. But $T_\mathbb{C}\backslash\tilde{U}$ is B^+-equivariantly holomorphically contractible to $T_\mathbb{C}\backslash B^+$; and the B^+-invariant part of the latter is the complex which defines $H^*(B^+\mathfrak{g}_\mathbb{C}, \mathfrak{t}_\mathbb{C}; \mathbb{C}_{s(w)})$.

Cohomology of holomorphic line bundles on Y

In the finite dimensional case we can calculate $H^*(G/T; \mathcal{O}_\lambda)$ directly from the analogue of (14.6.1) i.e. from

$$H^*(\mathfrak{n}^-; \Gamma_\lambda) \cong \bigoplus_{w \in W} \mathbb{C}_{w*\lambda}. \tag{14.6.7}$$

This is the last of the four proofs of the fundamental vanishing theorem which we mentioned in Section 14.2. It is worth briefly recalling the argument.

The groups $H^*(G/T; \mathcal{O}_\lambda)$ can be calculated from the $\bar{\partial}$-complex $A^\cdot$, where A^q denotes the smooth forms of type $(0, q)$ on G/T with coefficients in L_λ. We can break up $A^\cdot$ according to the irreducible representations Γ_μ of the compact group G:

$$A^\cdot = \bigoplus_\mu A^\cdot_\mu \otimes \Gamma_\mu,$$

where $A^\cdot_\mu = (A^\cdot \otimes \Gamma^*_\mu)^G$ is the G-invariant part of $A^\cdot \otimes \Gamma^*_\mu$. Accordingly we have

$$H^*(G/T; \mathcal{O}_\lambda) \cong \bigoplus_\mu H^*(A^\cdot_\mu) \otimes \Gamma_\mu. \tag{14.6.8}$$

But

$$(A^\cdot \otimes \Gamma^*_\mu)^G \cong \{\Lambda(\mathfrak{n}^+)^* \otimes \mathbb{C}_\lambda \otimes \Gamma^*_\mu\}^T,$$

and so $H^q(A^\cdot_\mu)$ is the $(-\lambda)$-weight-space of the action of T on $H^q(\mathfrak{n}^+; \Gamma^*_\mu)$.

By (14.6.7) this is zero unless $\lambda = w*\mu$ for some $w \in W$ of length q, in which case it is one-dimensional. In particular $H^q(G/T; \mathcal{O}_\lambda) = 0$ when $q > 0$ if λ is antidominant.

The essential reason why we cannot calculate $H^*(Z; \mathcal{O}_\lambda)$ in the loop-group case from (14.6.1) is that the $\bar{\partial}$-complex of Z does not consist of spaces of positive energy. We can, however, turn the argument around to obtain a rather different result.

For any Lie group $\mathcal{G}$ and any representation E of $\mathcal{G}$ we can define the 'smooth cochain' cohomology $H^*_{s.c.}(\mathcal{G}; E)$ as that of the complex of smooth Eilenberg–Maclane cochains of $\mathcal{G}$ with values in E (cf. [77], [13]). Then $H^1_{s.c.}(\mathcal{G}; E)$ is the group of classes of smooth crossed homomorphisms $\mathcal{G} \to E$, and $H^2_{s.c.}(\mathcal{G}; E)$ classifies the extensions of Lie groups

$$E \to \tilde{\mathcal{G}} \to \mathcal{G}.$$

A striking property of compact groups G which is not shared by non-compact semisimple groups such as $SL_2(\mathbb{R})$ is that $H^q_{s.c.}(G; E) = 0$ for any representation E of G when $q > 0$. This holds essentially because one can *average* over G: the possibility of averaging implies that the fixed point functor $E \mapsto E^G$ is exact, which is an equivalent statement to the vanishing of the cohomology. One more instance of the surprising resemblances between compact groups and loop groups is

Theorem (14.6.9). *If E is a positive energy representation of $\tilde{L}G$ then*

$$H^q_{s.c.}(\tilde{L}G; E) = 0$$

when $q > 0$.

Proof. We can suppose that $E = \Gamma_\lambda$, where λ is an antidominant weight. Let us first give a heuristic argument, assuming that the $\bar{\partial}$-lemma holds on $Y = LG/T$, and that $H^q(Y; \mathcal{O}_\lambda) = 0$ for $q > 0$. Then the $\bar{\partial}$-complex

$$0 \to \Gamma_\lambda \to A^\cdot \qquad (14.6.10)$$

is a resolution. On the other hand $H^q_{s.c.}(\tilde{L}G; A^p) = 0$ when $q > 0$. That is because A^p is the T-invariant part—and hence a direct summand—in the vector space F^p of smooth maps $\tilde{L}G \to \wedge^p(\mathfrak{g}_\mathbb{C}/\mathfrak{b}^+)^*$; and for such an 'induced module' we have $H^q_{s.c.}(\tilde{L}G; F^p) = 0$ when $q > 0$ for elementary formal reasons [77]. Thus the objects in the resolution (14.6.10) are acyclic for $\tilde{L}G$, and it follows that the cohomology $H^*_{s.c.}(\tilde{L}G; \Gamma_\lambda)$ is that of the $\tilde{L}G$-invariant part $(A^\cdot)^{\tilde{L}G}$. Just as in (14.6.8), we find that $H^q_{s.c.}(\tilde{L}G; \Gamma_\lambda)$ is the $(-\lambda)$-weight-space of the action of $\tilde{T}$ on $H^q(N^+\mathfrak{g}_\mathbb{C}; \mathbb{C})$, or equivalently the λ-weight-space of $H^q(N^-\mathfrak{g}_\mathbb{C}; \mathbb{C})$. If $q > 0$ this is zero by (14.6.1).

The preceding argument is only heuristic. A genuine proof can be given along rather different lines, using the *van Est spectral sequence* [149], [13]. We shall only sketch it.

Proposition (14.6.11). *For any representation E of $\tilde{L}G$ there is a spectral sequence with*

$$E_2^{pq} = H_{\text{s.c.}}^p(\tilde{L}G; H^q(\tilde{L}G; E))$$

which converges to the continuous Lie algebra cohomology $H^*(\tilde{L}\mathfrak{g}; E)$. Here $H^*(\tilde{L}G; E)$ denotes the cohomology of the de Rham complex of smooth forms on $\tilde{L}G$ with values in E.

Before proving the existence of the spectral sequence let us point out its relevance to (14.6.9). First suppose E is the trivial representation $\mathbb{C}$. Then the vertical edge homomorphism of the spectral sequence is the map

$$H^*(\tilde{L}\mathfrak{g}; \mathbb{C}) \to H^*(\tilde{L}G; \mathbb{C}) \qquad (14.6.12)$$

defined by regarding Lie algebra cochains as left-invariant forms on $\tilde{L}G$. The theory of spectral sequences tells us that (14.6.12) is an isomorphism if and only if $H_{\text{s.c.}}^q(\tilde{L}G; \mathbb{C}) = 0$ when $q > 0$. We already know that (14.6.12) is an isomorphism; and so (14.6.9) is true when $E = \mathbb{C}$.

If on the other hand E is a non-trivial irreducible representation of positive energy then $H_{\text{s.c.}}^*(\tilde{L}G; E)$ vanishes for elementary reasons, for the action on E of any central element of $\tilde{L}G$ must induce the identity on the cohomology, while on the other hand the centre acts on E by a non-trivial character.

Proof of (14.6.11). The spectral sequence is that of the double complex $C^{\cdot\cdot}$, where

$$C^{pq} = C_{\text{s.c.}}^p(\tilde{L}G; \Omega^q(\tilde{L}G; E))$$

denotes the smooth p-cochains of $\tilde{L}G$ with values in the space of E-valued q-forms on $\tilde{L}G$. Because $H_{\text{s.c.}}^p(\tilde{L}G; \Omega^q(\tilde{L}G; E)) = 0$ when $p > 0$ (for the same reason as explained above for $H_{\text{s.c.}}^p(A^q)$) the cohomology of the total complex of $C^{\cdot\cdot}$ is that of the $\tilde{L}G$-invariant part of $\Omega^{\cdot}(\tilde{L}G; E)$, i.e. it is $H^*(\tilde{L}\mathfrak{g}; E)$. On the other hand we have

$$E_1^{pq} = C_{\text{s.c.}}^p(\tilde{L}G; H^q(\tilde{L}G; E)).$$

That is true because the de Rham complex $\Omega^{\cdot}(\tilde{L}G; E)$ is *split*, i.e.

$$\Omega^q(\tilde{L}G; E) \cong (\text{exact } q\text{-forms}) \oplus H^q(\tilde{L}G; E) \oplus (\text{exact } (q+1)\text{-forms}).$$

(This holds for the de Rham complex of any manifold whose homology groups are finitely generated [9].) That completes the proof.

REFERENCES

1. Adams, J. F., *Lectures on Lie groups*. Benjamin, New York, 1969.
2. Atiyah, M. F., *K*-theory and reality, *Quart. J. Math., Oxford*, **17** (1966), 367–86 (reprinted as an appendix to [3]).
3. Atiyah, M. F., *K-theory*. Benjamin, New York, 1967.
4. Atiyah, M. F., Global theory of elliptic operators, *Proc. Int. Conf. on Func. Anal. and related topics*. Tokyo, 1969.
5. Atiyah, M. F., Instantons in two and four dimensions, *Commun. Math. Phys.*, **93** (1984), 437–51.
6. Atiyah, M. F., and Bott, R., A Lefschetz fixed point formula for elliptic complexes: II Applications, *Ann. of Math.*, **88** (1968), 451–91.
7. Atiyah, M. F., and Bott, R., The Yang–Mills equations over Riemann surfaces, *Philos. Trans. R. Soc. Lond.*, **308**A (1982), 523–615.
8. Atiyah, M. F., and Pressley, A. N., Convexity and loop groups. In: *Arithmetic and Geometry: papers dedicated to I. R. Shafarevich on the occasion of his sixtieth birthday, Volume II: Geometry*. Birkhäuser, 1983.
9. Beggs, E., De Rham's theorem for infinite dimensional manifolds, *Quart. J. Math., Oxford* **38** (1987), 131–54.
10. Bernstein, I. N., Gelfand, I. M., and Gelfand, S. I., Differential operators on the base affine space and a study of $\mathfrak{g}$-modules, 21–64. In: *Lie groups and their representations*, Summer School of the Bolyai Janos Math. Soc., edited by I. M. Gelfand. Wiley, New York, 1975.
11. Birkhoff, G. D., Singular points of ordinary differential equations, *Trans. Amer. Math. Soc.*, **10** (1909), 436–70.
12. Birkhoff, G. D., Equivalent singular points of ordinary linear differential equations, *Math. Ann.*, **74** (1913), 134–9.
13. Borel, A., and Wallach, N., Continuous cohomology, discrete subgroups, and representations of reductive groups, *Ann. of Math. Studies*, **94**, Princeton University Press, Princeton 1980.
14. Bott, R., An application of Morse theory to the topology of Lie groups, *Bull. Soc. Math. France*, **84** (1956), 251–81.
15. Bott, R., Homogeneous vector bundles, *Ann. of Math.*, **66** (1957), 203–48.
16. Bott, R., The space of loops on a Lie group, *Michigan Math. J.*, **5** (1958), 35–61.
17. Bott, R., The stable homotopy of the classical groups, *Ann. of Math.*, **70** (1959), 313–37.
18. Bott, R., and Tu, L. W., *Differential forms in algebraic topology*. Springer–Verlag, New York, 1982.
19. Bourbaki, N., *Espaces Vectoriels Topologiques*. Hermann, Paris, 1964.
20. Bourbaki, N., *Groupes et algèbres de Lie*. Ch. 1, Hermann, Paris, 1960; Ch. 2 et 3, Hermann, Paris, 1972; Ch. 4, 5 et 6, Hermann, Paris, 1968; Ch. 7 et 8, Hermann, Paris, 1975; Ch. 9, Masson, Paris, 1982.
21. Bredon, G. E., *Sheaf theory*. McGraw–Hill, New York, 1967.

REFERENCES

22 Brieskorn, E., Singular elements of semi-simple algebraic groups. In: *Actes du Congrès International des Mathématiciens (1970)*. 279-84, Gauthier-Villars, Paris, 1971.
23 Brown, L. G., Douglas, R. G., and Fillmore, P. A., Extensions of C^*-algebras and K-homology, *Ann. of Math.*, **105** (1977), 265-324.
24 Carey, A. L., and Ruijsenaars, S. N. M., On fermion gauge groups, current algebras, and Kac-Moody algebras, Preprint, A.N.U., Canberra, 1985.
25 Cartan, H., and Eilenberg, S., *Homological algebra*. Princeton University Press, 1956.
26 Chau, L. L., Ge, M. L., Sinha, A., and Wu, Y. S., Hidden symmetry algebra for the self-dual Yang-Mills equation, *Phys. Lett.* **121B** (1983), 391-6.
27 Chern, S. S., *Complex manifolds without potential theory*, Second edition. Springer, New York, 1979.
28 Coeuré, G., L'équation $(\bar{\partial}u = f)^*$ en dimension infinie, *Journées Bruxelles-Lille-Mons d'Analyse Fonctionelle et Équations aux Dérivées Partielles*, Université de Lille, Publications Internes, **131** (1978), 6-9.
29 Coleman, S., Quantum sine-Gordon equation as the massive Thirring model, *Phys. Rev.*, **11D** (1975), 2088-97.
30 Colombeau, J.-F., *Differential calculus and holomorphy*. North-Holland, Amsterdam, 1982.
31 Colombeau, J.-F., and Perrot, B., The $\bar{\partial}$-equation in DFN spaces, *J. Math. Anal. Appl.*, **78** (1980), 466-87.
32 Connes, A., Non-commutative differential geometry, *Publ. Math. I.H.E.S.*, **62** (1986), 257-360.
33 Connes, A., Cohomologie cyclique et foncteurs Ext^n, *C.R. Acad. Sci., Sér. A, Paris*, **296** (1983), 953-8.
34 Conway, J. B., *A course in functional analysis*. Springer-Verlag, New York, 1984.
35 Date, E., Jimbo, M., Kashiwara, M., and Miwa, T., Transformation groups for soliton equations. I: *Proc. Japan Acad.*, **57** A (1981), 342-7; II: *Ibid.*, 387-92; III: *J. Phys. Soc. Japan*, **50** (1981), 3806-12; IV: *Physica*, **4D** (1982), 343-65; V: *Publ. RIMS, Kyoto Univ.*, **18** (1982), 1111-19; VI: *J. Phys. Soc. Japan*, **50** (1981), 3813-18; VII: *Publ. RIMS, Kyoto Univ.*, **18** (1982), 1077-110.
36 Dineen, S., *Complex analysis in locally convex spaces*, Mathematics Studies No. 57. North-Holland, Amsterdam, 1981.
37 Dixmier, J., *Enveloping algebras*. North-Holland, Amsterdam, 1974.
38 Dolan, L., Kac-Moody algebra is hidden symmetry of chiral models, *Phys. Rev. Lett.*, **47** (1981), 1371-4.
39 Dold, A., Partitions of unity in the theory of fibrations, *Ann. of Math.*, **78** (1963), 223-55.
40 Douglas, R. G., *Banach algebra techniques in operator theory*. Academic Press, New York, 1972.
41 Faddeev, L., Operator anomaly for the Gauss law, *Phys. Lett.*, **145B** (1984), 81-4.
42 Fegan, H. D., The heat equation and modular forms, *J. Differential Geom.*, **13** (1978), 589-602.
43 Feingold, A. J., and Lepowsky, J., The Weyl-Kac character formula and power series identities, *Adv. in Math.*, **29** (1978), 271-309.

REFERENCES

44 Finkelstein, D., and Rubinstein, J., Connections between spin, statistics and kinks, *J. Math. Phys.*, **9** (1968), 1762–79.
45 Frenkel, I. B., Two constructions of affine Lie algebra representations and boson–fermion correspondence in quantum field theory, *J. Funct. Anal.*, **44** (1981), 259–327.
46 Frenkel, I. B., Representations of affine Lie algebras, Hecke modular forms and Korteweg–de Vries type equations, pp. 71–110. In *Lie algebras and related topics, Proceedings of a Conference held at New Brunswick, New Jersey, May 29–31, 1981*. Lecture Notes in Mathematics Vol. 933, Springer–Verlag, New York, 1982.
47 Frenkel, I. B., Orbital theory for affine Lie algebras, *Invent. Math.*, **77** (1984), 301–52.
48 Frenkel, I. B., Representations of Kac–Moody algebras and dual resonance models, *Lectures in Applied Mathematics*, Vol. 21, 325–53. American Mathematical Society, 1985.
49 Frenkel, I. B., and Kac, V. G., Basic representations of affine Lie algebras and dual resonance models. *Invent. Math.*, **62** (1981), 23–66.
50 Frenkel, I. B., Lepowsky, J., and Meurman, A., A natural representation of the Fischer–Griess Monster with the modular function J as character, *Proc. Nat. Acad. Sci. USA*, **81** (1984), 3256–60.
51 Freudenthal, H., and de Vries, H., *Linear Lie groups*. Academic Press, New York, 1969.
52 Gabber, O., and Kac, V. G., On defining relations of certain infinite dimensional Lie algebras, *Bull. Amer. Math. Soc.*, **5** (1981), 185–9.
53 Gabriel, P., Représentations indécomposables, *Séminaire Bourbaki, 26e année (1973/74), Exp. 444*. Lecture Notes in Mathematics Vol. 431. Springer–Verlag, Berlin, 1975.
54 Garland, H., The arithmetic theory of loop algebras, *J. Algebra*, **53** (1978), 480–551.
55 Garland, H., The arithmetic theory of loop groups, *Pub. Math. I.H.E.S.*, **52** (1980), 5–136.
56 Garland, H., and Lepowsky, J., Lie algebra homology and the Macdonald–Kac formulas, *Invent. Math.*, **34** (1976), 37–76.
57 Garland, H., and Raghunathan, M. S., A Bruhat decomposition for the loop space of a compact group: a new approach to results of Bott, *Proc. Nat. Acad. Sci. USA*, **72** (1975), 4716–17.
58 Gelfand, I. M. (ed.), *Representation theory*. London Mathematical Society Lecture Note Series Vol. 69, Cambridge University Press, Cambridge, 1982.
59 Gelfand, I. M., Graev, M. I., and Vershik, A. M. Representations of the group of smooth mappings of a manifold into a compact Lie group, *Compositio Math.*, **35** (1977), 299–334.
60 Gelfand, I. M., and Vilenkin, N. Ya., *Generalised functions*, Vol. 4. Academic Press, New York, 1964.
61 Goddard, P., and Olive, D., Kac–Moody algebras, conformal symmetry, and critical exponents. *Nuclear Phys.*, **257**B (1985), 226–52.
62 Godement, R., *Théorie des faisceaux*. Hermann, Paris, 1964.
63 Gohberg, I. C., and Krein, M. G., Systems of integral equations on a half-line with kernels depending on the difference of the arguments, *Amer. Math. Soc. Trans.*, **14** (2) (1960), 217–84.
64 Goodman, R., and Wallach, N. R., Structure and unitary cocycle repre-

sentations of loop groups and the group of diffeomorphisms of the circle, *J. Reine Angew. Math.*, **347** (1984), 69–133.
65 Gorenstein, D., *Finite Groups.* Harper and Row, New York, 1968.
66 Grauert, H., Analytische Faserungen über holomorph-vollständigen Räumen, *Math. Ann.*, **135** (1958), 268–73.
67 Gray, B., *Homotopy theory.* Academic Press, New York, 1975.
68 Griffiths, P., and Harris, J., *Principles of algebraic geometry.* Wiley, New York, 1978.
69 Grothendieck, A., Sur la classification des fibrés holomorphes sur la sphère de Riemann, *Am. J. Math.*, **79** (1957), 121–38.
70 Hamilton, R., The inverse function theorem of Nash and Moser, *Bull. Am. Math. Soc.*, **7** (1982), 65–222.
71 Hardy, G. H., and Wright, E. M., *An introduction to the theory of numbers* (4th edition). Oxford University Press, Oxford, 1960.
72 Helgason, S., *Differential geometry, Lie groups and symmetric spaces.* Academic Press, New York, 1978.
73 Helton, J. W., and Howe, R. E. Integral operators: commutators, traces, index and homology. In: *Proceedings of a Conference on Operator Theory* (Dalhousie Univ., Halifax, N.S., 1973), Lecture Notes in Mathematics Vol. 345, 141–209. Springer-Verlag, Berlin, 1973.
74 Herman, M.-R., Simplicité du groupe des difféomorphismes de classe C^∞, isotopes a l'identité, du tore de dimension n, *C.R. Acad. Sci. Paris, Sér. A*, **273** (1971), 232–4.
75 Hervé, M., *Analytic and plurisubharmonic functions in finite and infinite dimensional spaces*, Lecture Notes in Mathematics Vol. 198. Springer-Verlag, Berlin, 1971.
76 Hochschild, G. *The structure of Lie Groups.* Holden–Day, San Francisco, 1965.
77 Hochschild, G., and Mostow, G. D., Cohomology of groups, *Illinois J. Math.*, **6** (1962), 367–401.
78 Itzykson, C., and Zuber, J. B., *Quantum field theory.* McGraw-Hill, New York, 1980.
79 Iwahori, N., and Matsumoto, H., On some Bruhat decompositions and the structure of Hecke rings of p-adic Chevalley groups, *Pub. Math. I.H.E.S.*, **25** (1965), 5–48.
80 Jacob, M. (ed.), *Dual theory.* North-Holland, Amsterdam, 1974.
81 Kac, M., *Probability and Related Topics in the Physical Sciences.* Interscience, New York, 1959.
82 Kac, V. G., Simple irreducible graded Lie algebras of finite growth, *Math. USSR Izvestija*, **2** (1968), 1271–311.
83 Kac, V. G., Infinite-dimensional algebras, Dedekind's η-function, classical Möbius function and the very strange formula, *Advances in Math.*, **30** (1978), 85–136.
84 Kac, V. G., An elucidation of "Infinite dimensional algebras... and the very strange formula." $E_8^{(1)}$ and the cube root of the modular invariant j. *Adv. in Math.*, **35** (1980), 264–73.
85 Kac, V. G., Infinite root systems, representations of graphs and invariant theory, I: *Invent. Math.*, **56** (1980), 57–92; II: *J. Algebra*, **78** (1982), 141–62.
86 Kac, V. G., *Infinite dimensional Lie algebras.* Birkhäuser, Boston, 1983.
87 Kac, V. G., Kazhdan, D. A., Lepowsky, J., and Wilson, R. L., Realisation

of the basic representations of the Euclidean Lie algebras, *Adv. in Math.*, **42** (1981), 83–112.

88 Kac, V. G., and Peterson, D. H., Spin and wedge representations of infinite dimensional Lie algebras and groups, *Proc. Nat. Acad. Sci. USA*, **78** (1981), 3308–12.

89 Kantor, I. L., Infinite dimensional simple graded Lie algebras, *Sov. Math. Doklady*, **9** (1968), 409–12.

90 Kasparov, G. G., The operator K-functor and extensions of C^*-algebras, *Math. USSR Izvestija*, **44** (1981), 513–72.

91 Kempf, G. R., The Grothendieck–Cousin complex of an induced representation, *Adv. in Math.*, **29** (1978), 310–96.

92 Kirillov, A. A., *Elements of the theory of representations*. Springer–Verlag, Berlin, 1976.

93 Kostant, B., Lie algebra cohomology and the generalized Borel–Weil theorem, *Ann. of Math.*, **74** (1961), 329–87.

94 Koszul, J.-L. Homologie et cohomologie des algèbres de Lie, *Bull. Soc. Math. France*, **78** (1950), 65–127.

95 Kraft, H., and Procesi, C., Minimal singularities in GL_n, *Invent. Math.*, **62** (1981), 503–515.

96 Kuiper, N. H., The homotopy type of the unitary group of Hilbert space, *Topology*, **3** (1965), 19–30.

97 Kumar, S., Rational homotopy theory of flag varieties associated to Kac–Moody groups, in *Infinite dimensional groups with applications*, edited by V. Kac, MSRI Publications Vol. 4. Springer, New York, 1985.

98 Kuo, H.-H., *Gaussian measures in Banach spaces*, Lecture Notes in Mathematics Vol. 463. Springer–Verlag, Berlin, 1975.

99 Lax, P. D., and Phillips, R. S., *Scattering theory*. Academic Press, New York, 1967.

100 Lepowsky, J., Generalized Verma modules, loop space cohomology and Macdonald type identities, *Ann. Sci. École Norm. Sup.*, **12** (4) (1979), 169–234.

101 Lepowsky, J., and Milne, C., Lie algebraic approaches to classical partition identities, *Adv. in Math.*, **29** (1978), 15–59.

102 Lepowsky, J., and Wilson, R. L., Construction of the affine Lie algebra $A_1^{(1)}$, *Commun. Math. Phys.*, **62** (1978), 45–53.

103 Lepowsky, J., and Wilson, R. L., A Lie theoretic interpretation and proof of the Rogers–Ramanujan identities, *Adv. in Math.*, **45** (1982), 21–72.

104 Loday, J.-L. and Quillen, D., Cyclic homology and the Lie algebra homology of matrices, *Comment. Math. Helvetici*, **59** (1984), 565–91.

105 Looijenga, E., Root systems and elliptic curves, *Invent. Math.*, **38** (1976), 17–32.

106 Lusztig, G., Green polynomials and singularities of unipotent classes, *Advances in Math.*, **42** (1981), 169–178.

107 Macdonald, I. G., Affine root systems and Dedekind's η-function, *Invent. Math.*, **15** (1972), 91–143.

108 Macdonald, I. G., *Symmetric functions and Hall polynomials*. Oxford University Press, Oxford, 1979.

109 Macdonald, I. G., *Affine Lie algebras and modular forms*, Séminaire Bourbaki Exp. 577, Lecture Notes in Mathematics Vol. 901, 258–65. Springer–Verlag, New York, 1981.

110 Macdonald, I. G., Kac–Moody algebras, in Lie algebras and related topics, R. V. Moody (ed.), *Conference Proceedings of the Canadian Mathematical Society, Vol. 5,* American Mathematical Society, 1986.
111 Mandelstam, S., Soliton operators for the quantized sine-Gordon equation, *Phys. Rev.,* **11D** (1975), 3026–30.
112 Mickelsson, J., On a relation between massive Yang–Mills theories and dual string models, *Lett. Math. Phys.,* **7** (1983), 45–50.
113 Milnor, J. W., Axiomatic homology theory. *Pacific J. Math.,* **12** (1962), 337–41.
114 Milnor, J. W., Morse theory, *Ann. of Math. Studies,* **51**. Princeton University Press, Princeton, 1963.
115 Milnor, J. W., Remarks on infinite dimensional Lie groups. In: *Relativity, Groups and Topology II,* Les Houches Session XL, 1983, edited by B. S. de Witt and R. Stora. North–Holland, Amsterdam, 1984.
116 Milnor, J. W., and Stasheff, J. D., Characteristic classes. *Ann. of Math. Studies,* **76**. Princeton University Press, 1974.
117 Moody, R. V., A new class of Lie algebras, *J. Algebra,* **10** (1968), 211–30.
118 Moody, R. V., Euclidean Lie algebras, *Canadian J. Math.,* **21** (1969), 1432–54.
119 Mumford, D., and Fogarty, J., *Geometric invariant theory,* (Second edition). Springer–Verlag, New York, 1982.
120 Omori, H., On the group of diffeomorphisms of a compact manifold, *Proc. Symp. Pure Math.,* **15,** 167–83, American Mathematical Society, 1970.
121 Palais, R. S., On the homotopy type of certain groups of operators. *Topology,* **3** (1965), 271–9.
122 Pickrell, D., Measures on infinite dimensional Grassmann manifolds, *J. Functional Anal.* **70** (1987), 323–56.
123 Quillen, D. G., The mod 2 cohomology rings of extra special 2-groups and the spinor groups, *Math. Ann.,* **194** (1971), 197–212.
124 Quillen, D. G., Determinants of Cauchy–Riemann operators on a Riemann surface. *Functional Anal. Appl.,* **19** (1) (1985), 37–41 (Russian), 31–4 (English).
125 Reed, M., and Simon, B., *Methods of modern Mathematical Physics, Volume 1: Functional Analysis.* Academic Press, London, 1980.
126 Sato, M., Soliton equations as dynamical systems on infinite dimensional Grassmann manifolds, *RIMS Kokyuroku,* **439** (1981), 30–40.
127 Sato, M., Miwa, J., and Jimbo, M., Holonomic quantum fields II: The Riemann–Hilbert problem, *Publ. RIMS, Kyoto Univ.,* **15** (1979), 201–78.
128 Segal, G. B., Classifying spaces and spectral sequences, *Publ. Math. I.H.E.S.,* **34** (1968), 105–12.
129 Segal, G. B., Cohomology of topological groups, *Symposia Mathematica Vol. II (INDAM, Rome, 1968/69),* 377–387. Academic Press, London, 1970.
130 Segal, G. B., The topology of spaces of rational functions, *Acta Math.,* **143** (1979), 39–72.
131 Segal, G. B., Unitary representations of some infinite dimensional groups, *Commun. Math. Phys.,* **80** (1981), 301–42.
132 Segal, G. B., and Wilson, G., Loop groups and equations of KdV type, *Pub. Math. I.H.E.S.,* **61** (1985), 5–65.
133 Serre, J.-P., Représentations linéaires et espaces homogènes Kählériens

des groupes de Lie compacts, *Séminaire Bourbaki, 6e année 1953/54, Exp. 100.* Paris, 1959.
134 Serre, J.-P., *Algèbres de Lie semi-simples complexes,* Benjamin, New York, 1966.
135 Serre, J.-P., *A course in Arithmetic.* Springer-Verlag, New York, 1973.
136 Shale, D., Linear symmetries of free Boson fields, *Trans. Amer. Math. Soc.,* **103** (1962), 149–67.
137 Simon B., *Trace ideals and their applications,* London Mathematical Society Lecture Notes Vol. 35. Cambridge University Press, Cambridge, 1979.
138 Singer, I. M., Families of Dirac operators with applications to physics, *Elie Cartan et les Mathématiques d'aujourd'hui*; *Astérisque,* Special Issue, June 1986.
139 Skyrme, T. H. R., Kinks and the Dirac equation, *J. Math. Phys.,* **12** (1971), 1735–42.
140 Slodowy, P., *Simple singularities and simple algebraic groups,* Lecture Notes in Mathematics Vol. 815. Springer-Verlag, Berlin, 1980.
141 Slodowy, P., *Habilitationschrift.* Universität Bonn, 1984.
142 Smale, S., Differentiable dynamical systems, *Bull. Am. Math. Soc.,* **73** (1967), 747–817.
143 Spanier, E. H., *Algebraic Topology.* McGraw-Hill, New York, 1966.
144 Taylor, M. E., *Pseudodifferential Operators.* Princeton University Press, Princeton, 1981.
145 Tits, J., *Théorie des groupes.* Annuaire du Collège de France, (resumé des cours et travaux), Paris 1980–81 and 1981–82.
146 Tits, J., Groups and group functors attached to Kac–Moody data, *Arbeitstagung Bonn, 1984.* Lecture Notes in Mathematics, Vol. 1111, Springer, Berlin, 1985.
147 Turnbull, H. W., *The theory of determinants, matrices and invariants.* Blackie and Son Ltd, London and Glasgow, 1929.
148 Turrettin, H. L., Reduction of ordinary differential equations to the Birkhoff canonical form, *Trans. Amer. Math. Soc.,* **107** (1963), 485–507.
149 van Est, W. T., On the algebraic concepts in Lie Groups I, II, *Proc. Koninkl. Ned. Ak. v. Wet. Amsterdam,* **58**A (1955), 225–33, 286–94.
150 Vergne, M., Seconde quantification et groupe symplectique, *C.R. Acad. Sci. Sér. A, Paris,* **285** (1977), 191–4.
151 Vershik, A. M., Gelfand, I. M., and Graev, M. I., Representations of the group $SL(2, R)$ where R is a ring of functions, *Russ. Math. Surveys,* **28** (1973), 87–132.
152 Wang, H.-C., Closed manifolds with homogeneous complex structure, *Am. J. Math.,* **76** (1954), 1–32.
153 Warner, G., *Harmonic analysis on semi-simple Lie groups I.* Springer-Verlag, Berlin, 1972.
154 Weyl, H., *The classical groups.* Princeton University Press, Princeton, 1939.
155 Witten, E., Non-abelian bosonization in two dimensions, *Commun. Math. Phys.,* **92** (1984), 455–72.
156 Zakharov, V. E., and Shabat, A. B., Integration of the nonlinear equations of mathematical physics by the inverse scattering method II, *Functional Anal. Appl.,* **13** (3) (1979), 13–22 (Russian), 166–74 (English).
157 Zumino, B., Cohomology of gauge groups: cocycles, and Schwinger terms, *Nucl. Phys.,* **253**B (1985), 477–493.
158 Zygmund, A., *Trigonometric Series* Second edition. Cambridge University Press, Cambridge, 1977.

INDEX OF NOTATION

Roman letters

$a(\phi, f)$	59		$\check{E}$	171
$\mathfrak{a}(f)$	253		$\bar{E}^*$	168
A_n	14		E_{sm}	168
A_w	144		$E(k)$	171
A_β	256		E_6, E_7, E_8	14
A_θ^h	264		$\hat{f}$	59
b_0^-	173		$f^\pm$	252
B^+	20, 122		F_4	14
B^-	21		$Fl(\mathbb{C}^n)$	19
B_0^+	143		Fl_k	19
$B^-\mathfrak{g}_\mathbb{C}$	173		$Fl^{(n)}$	145
B_n	14		$Fl_0^{(n)}$	145
			Fred (H_+)	81
c	280		$\mathfrak{g}$	11
c_λ	183		$\mathfrak{g}_\alpha$	15
c_λ	186		$\mathfrak{g}_\mathbb{C}$	13
$C(V)$	230		$\mathfrak{g}'(A)$	76
C_n	14		$\mathfrak{g}(A)$	76
C_w	144		G	11
C_w^*	293		$G_\mathbb{C}$	13
$C_\mathbf{a}$	133		G^0	27
$C_{|\mathbf{a}|}$	134		G_2	14
C_S	109		$GL_{res}(H)$	80
C_λ	140		$GL_n(\mathbb{C})$	13
$C_{|\lambda|}$	141		$GL(H^{(n)})$	79
			$GL_\mathcal{J}$	98
$d(\mathbf{a})$	133		$\widetilde{GL_{res}}$	88
d_λ	140		$Gr_k(\mathbb{C}^n)$	19
$\mathcal{D}$	58		$Gr(H)$	101
D_n	14		$Gr_0(H), Gr_1(H),$	
$\text{Diff}(S^1), \text{Diff}^+(S^1)$	5		$Gr_\omega(H), Gr_\infty(H)$	104
$\text{Diff}(X)$	28		$Gr^{(n)}$	125
			$Gr_0^{(n)}, Gr_1^{(n)}, Gr_\omega^{(n)},$	
e_α	16		$Gr_\infty^{(n)}$	127
e_α	75		$Gr_\mathbb{R}^{(n)}$	137
$\mathcal{E}$	87, 118, 148		$Gr^\mathfrak{a}$	139

INDEX OF NOTATION

h_α	16	MG, M_0G	30
$h_{\mathbf{a}}$	75, 177	M_γ	79
H	85		
$H^{(n)}$	79	$\mathcal{N}_-$	107
$H^{(n)}_\pm$	82	$\mathcal{N}_+$	109
H_S	103	$\mathfrak{n}_0^-$	217
$H_{\mathbf{a}}$	131	N^-	130, 140
$H^{\mathfrak{g}}$	139	N^+	132, 140
H_λ	140	$N_0^\pm$	139
H^Δ	94	N_w^-	144
$\mathscr{H}$	202	$_{\mathbf{a}}N^-$	132
$\tilde{\mathscr{H}}$	251		
		O_n	136
i_α	16	$O(V), \mathfrak{o}(V)$	230
$i_{\mathbf{a}}$	75	O_{res}	244
$\mathscr{I}_1(H)$	86	$\mathcal{O}_\lambda$	294
$\mathscr{I}_2(H)$	80		
$\mathscr{I}_2(H_+; H_-)$	81	Pf	242
$\mathscr{I}_p$	98	$Pf_{\bar{W}}$	233
I	184	$Pf(S)$	235
		$\text{Pin}(V)$	239
$\mathscr{J}(V)$	234, 245		
J^i_θ	265	R_u	111
$J(\xi)$	264	R_α	59, 251
		$s(w)$	278
$\mathscr{K}$	98	s_α	18, 253
		S	59, 63, 103
$\ell(S)$	107	$\mathscr{S}$	103
$\ell(w)$	144	$S_{\mathbf{a}}$	130
LG	27	S_λ	140
L_λ	21	$\text{Spin}(V)$	231
L_λ	216	$\text{Spin}(V_{\mathbb{C}})$	237
$L^\pm GL_n(\mathbb{C})$	120	$\text{Spin}^{\mathbb{C}}$	245
L_1^-	121		
$L_{\mathbf{a}}^-$	131	$\mathbb{T}$ the circle group	
$L_{\mathbf{a}}^+$	132	$T, \mathfrak{t}$	15
$L_{\text{an}}G$	32	$\hat{T}$	15
$L_{\text{pol}}G, L_{\text{rat}}G$	33	$\check{T}$	16
$L_{(\alpha)}G$	36	$\hat{T}_-$	19
$\tilde{L}G$	38	$\mathscr{T}$	87
$\tilde{L}_\mathfrak{g}$	39		
$\tilde{L}, \tilde{L}_{\mathbb{C}}$	89	U	287
		$U_n, \mathfrak{u}_n$	11
$m(X, G)$	65	$\tilde{U}_n$	12
		U_w	144

INDEX OF NOTATION

U_w^*	292	Δ	182, 185
U_S	103	Δ_f	59
$U_\mathbf{a}$	133		
U_λ	140	$\varepsilon(\phi)$	252
$U_{\text{res}}(H)$	80		
$U_{\text{res}}^\sim$	89	η_α	16
$U(\phi)$	251		
		λ	178
virt. dim W	103	$\Lambda_\mathbf{a}$	135
$\text{Vect}(X)$	28	ΛG	174
V_λ	287		
		$\Pi(z, g)$	283
W	18	π_S	110
W_{aff}	71		
W^{fin}	106		
		ρ	183
y_w	224	$\boldsymbol{\rho}$	186
Y	216		
Y_{pol}	227	Σ_S	107
		Σ_w	144
$z^\mathbf{a}$	120	$\Sigma_\mathbf{a}$	131
Z	294	$\Sigma_{|\mathbf{a}|}$	134
Z_w	294	Σ_λ	141
Z^p	295	$\Sigma_{|\lambda|}$	141
Greek letters		Φ_S	203–4
α	15		
$\boldsymbol{\alpha}$	75, 177		
		$\psi(p)$	251
Γ	195	ψ_α	251
Γ_d	198		
Γ_λ	21	ω_i	179
Γ_λ	217	ΩG	48

INDEX

adjoint action of LG 64
adjoint representation 14, 15
admissible basis 109
affine Lie algebras 77–8
affine Weyl group 45, 65, 71, 73–4
alcove 72
anomaly 200
automorphism group 31–2

Banach Lie group 26–7, 32, 79, 80
based loops 125–8, 136–9, 146, 147–54
basic central extension 49
  of $LGL_n(\mathbb{C})$ 89–91
  of $L\mathbb{T}$ 59–60
  of LU_n 57–60, 89–91
basic inner product 49
basic representation 180, 181–2
  construction from arbitrary maximal abelian subgroup 271–2
  homogeneous construction 263–8, 271–2
  of LO_{2n} 246–7
  of LSU_n 211–14
  of LU_n 165, 179, 211–14
  principal construction 250–8, 260–3, 271–2
  simply laced case 263–8
  (*see also* fundamental representation)
Bernstein–Gelfand–Gelfand resolution 290–8
  weak 291
bi-invariant functions 225–7
Birkhoff decomposition 120, 140–1, 144
  and differential equations 123–4
  and holomorphic bundles 124–5
  relation to Bruhat decomposition 135, 141
blip, *see* vertex operator
Borel subgroup 20, 122
Borel–Weil representation 217, 224–7
  action of $\text{Diff}^+(S^1)$ 221, 268–71
  finite type 217–18
  irreducibility 220–1
  positive energy 217
  smoothness 217
  unitarity 218–20
Borel–Weil theorem,
  for compact groups 21–3
  for loop groups 216–29

boson–fermion correspondence 5, 166, 215
  analogy with extra-special 2-group 247–8
Bott periodicity 7, 85, 99, 122, 146–7, 247
Bruhat decomposition 20, 121, 133, 140–1, 144
  relation to Birkhoff decomposition 135

C^∞-topology 27
canonical anti-commutation relations 254
canonical basis 107
canonical commutation relations 96–7, 214
Carey–Ruijsenaars representation 215
Cartan matrix 74, 76
  indecomposable 77
  symmetrizable 77
Casimir operator
  of $\mathfrak{g}$ 182–3
  of $L\mathfrak{g}$ 184–6
central extension 38
  action of $\text{Diff}^+(S^1)$ 39, 46, 48, 57–9, 63, 92
  circle action 93
  complexification 39, 90–1
  of loop algebras 40–1
  of $L\mathbb{T}$ 59–60
  of LU_n 57–60
  of $\text{Map}(X; G)$ 42–3, 65–7, 98–100
  of $\text{Vect}(S^1)$ 43
  semisimple case 46–50, 55–7
  simply connected case 46–50
  universal 38–9, 42, 50
chamber 72
character 273–6
  distributional 274
  formal 208, 274, 289
  hyperfunction 274–5
charge 198, 215
charge density 262
Chern class 51, 54, 88, 115, 117
class function 273
Clifford algebra 230–3
coadjoint action 43–6, 65
cocycle 38–43
  action of $\text{Diff}^+(S^1)$ 63
  as Chern class 54
  for basic extension of GL_{res} 88–9
  for basic extension of $L\mathbb{T}$ 59
  for extensions of $\text{Vect}(S^1)$ 43
  for induced extension of LT 63

INDEX

cocycle condition 39
cohomology,
  continuous Lie algebra 303
  of compact groups 67–8
  of holomorphic bundles on LG/T 301–3
  of LG/T 300
  of loop groups 68–9, 299
  of Map$(X;G)$ 42, 68
  representation by left-invariant forms 68, 98
  smooth cochain 302–3
commensurable subspace 101, 104
commutant 79, 84
complete reducibility,
  for compact groups 13
  for loop groups 176, 223
completely integrable system 4
complex manifold 25
complex structures, space of 234, 245
  homotopy type 245
  Plücker embedding 240
connection 39, 50–3, 169, 200
Connes' cohomology 42
continuous loop group 26, 79, 128–9
co-order 108
coroot 16
  affine 75
coweight lattice 71
Coxeter element 36
Coxeter number 280
creation and annihilation operators 232
critical points of energy function 118, 149
current group 1
curvature 51

$\bar{\partial}$-Poincaré lemma 297–8
derivations 64, 76
determinant 86
determinant bundle 22–3, 113–16, 196
  action of GL_{res} 114–15
  $\mathbb{C}^*_{\leq 1}$-action 116
  Quillen's 116
  sections of,
    as exterior algebra 196–200
    as symmetric algebra 205–8
    Hilbert space structure 200–2
diagram of LG 72
Diff(S^1)
  action on representations 186–7, 221, 258–60, 268–71
  central extensions of 92
  embedding in U_{res} 91–2
  simplicity 30
diffeomorphism groups 28–30
differentiable manifold 24
differential form 25
Dirac operator 94, 98–100, 200, 214
dual resonance theory 249

energy function 118, 148
  classical vs. quantum 118–19, 151–2, 173–4
  critical points 118, 149
  gradient flow 150–1
  tilted 149
essential equivalence 167–8, 176
eta function 285
exponential map 11, 26–7
  for Diff(S^1) 28–9
  non-surjectivity 27–8
extra-special 2-group 247–8

factorization
  Birkhoff 120–1, 131–2
  Bruhat 121
  unitary × holomorphic 120, 129, 137, 139
Faddeev, see Mickelsson–Faddeev extension
field operator 264
finite energy 171, 194
finite order 106
flag manifold 19–20
  periodic 145–6
flat bundle 38, 51
Fock space,
  bosonic 195, 207
  fermionic 195–6, 199–200
Fredholm operator 81
Freudenthal's strange formula 285
fundamental homogeneous space 122
  associated to a Riemann surface 154–60
fundamental representation,
  of GL_{res} 195–208
    action of Diff$^+(S^1)$ 208, 258–60
    as Heisenberg representation 206–7
  of O_{res} 244–6, 260–3

Gabriel's theorem 4
gauge group 1, 37, 157, 262
Gaussian measures 169, 170, 190, 205
Gelfand representation 169–70
generators and relations 16–17, 74–6
gradient flow 112, 119, 150–1
grading 171
Grassmannian 19
Grassmannian of Hilbert space 101–13
  as homogeneous space 102
  $\mathbb{C}^\times_{\leq 1}$-action on 111–13, 150
  cell decomposition 108–9, 112
  circle action on 111, 118
  holomorphic functions on 106
  Kähler structure 117
  Plücker embedding 110–11, 115–16
  smooth 105
  stratification 106–8, 112
  symplectic structure 117

Grassmannian model,
  of measurable loop group 162
  of ΩG 138–43
  of ΩO_n 136–7
  of ΩSp_n 138
  of ΩU_n 125–9

half-differentiable loops 84–5, 125, 128–9, 208
half-spin representations 233
Hamiltonian function 117, 149
Hamiltonian vector field 148
heat equation 170, 285–6
Hilbert–Schmidt operator 80
Hilbert transform 83
holes 198
holomorphic bundles,
  and projective embeddings 22
  on a Riemann surface 156
  on S^2, 124–5, 152–4
holonomy 51
homogeneous space 19–21
horizontal vector field 51
hyperfunction 275, 298

implementable transformation 6
index 81
index homomorphism 99
integral cohomology class 46, 47
invariant bilinear form 14
  on Kac–Moody algebras 77
  on $\mathbb{R} \oplus \tilde{L}_\mathfrak{g}$ 64–5
invariant polynomials 67–8
isotropic subspace 21

Jacobi identity 12
Jacobi triple product identity 208–10, 285

K-groups 65, 99
K-homology 99–100
Kac character formula 280–4
  algebraic proof 286–90
  for LSU_2 281–2
  for LU_2 282
  Macdonald's formulation 284
  simply laced case 284
Kac–Moody algebras 76–8
Kac–Moody groups 78
kinks 249
Kirillov principle 44–5

LG/T 143–6
Lax–Phillips scattering theory 160–2
Lefschetz fixed point theorem 277, 279
left-invariant vector field 12
length,
  of element of $\mathscr{S}$ 107
  of element of Weyl group 20

level 174, 177
Lie algebra 11, 26
  complexification 13
  which does not come from a Lie group 29, 47
Lie bracket 11, 12
Lie group,
  complex 13, 28
  finite dimensional 11
  infinite dimensional 26
  second homotopy group of 142
  loop group 27
  lowest weight 19, 178
  lowest weight vector 173

Macdonald identity 210, 284–5
  Kostant's formulation 285
mass 93–4
maximal abelian subgroup 34–6
maximal normal subgroup 32
measurable loop group 79, 86, 162
metaplectic representation 191, 208, 268
Mickelsson–Faddeev extension 66–7
modular form 276, 285
moduli space 156
monodromy matrix 124
monster 5
Morse decomposition 7
multiplication operator 80, 82, 94

normal ordering 252, 262
normality of Schubert varieties 229

one-parameter subgroup 11
operator-valued distribution 214, 251
orbits of coadjoint action 45–6
outer automorphisms 32, 181
outgoing subspace 161–2

partition 202
partition function 210, 275, 284
Pfaffian 235, 240–2, 246
Pfaffian bundle 242–4, 246
Pin group 239
Plücker coordinates 110–12, 115, 198, 201–2, 206
polynomial loop algebra 74–5
polynomial loop group 23–4, 127, 227–9
  homotopy type 142, 146–7
  density in LG 33, 128
positive energy 171–5
principal Heisenberg 36, 85–6, 175
projective multiplier 38
projective representation 38

quantum field theory 1, 4, 5, 67, 93–8, 199–200, 214–15
quivers 4

rank 15, 73
rational loop group 33, 127–8
real-analytic loop group 33, 128
real-analytic vector 113
regular element 18
regular representation 170
representation 167
  action of $\text{Diff}^+(S^1)$ 172, 176, 186–8, 194, 208
  completely reducible 176, 223
  exterior power, 28–3, 166
  finite type 177
  integrable 287
  irreducible 168
  of $\text{Map}(X; G)$ 168–71
  positive energy 171–5, 190, 192–4, 217
  projective 171, 178
  smooth 168, 217, 227–9
  symmetric 171
  weights of 177
residue 293
resolution 290
restricted general linear group 80
  central extension of 86–9, 115
  embedding $\text{Diff}^+(S^1)$ into 91
  embedding LG into 82–5, 94–5
  topology 80–2, 87
restricted orthogonal group 244–5
restricted unitary group 80
Riemann–Hilbert problem 5
root 15, 70
  affine 71, 72, 177
  of Kac–Moody algebra 76
  of U_n 16
  positive and negative 18, 72
  real and imaginary 77
  simple 18

scattering matrix 160
scattering theory 122, 160–2
Schatten ideal 98, 100
Schubert variety 20
  normality 229
Schur functions 202, 204
  and Plücker coordinates 206
second quantization 196
Serre's theorem 74
Shale group (see restricted general linear group)
simply laced group 17–18, 49, 181, 264
sine–Gordan equation 215
singularities 4
smooth loop group 27
  $\mathbb{C}^*_{\leq 1}$-action on 130, 150
  circle action on 71, 150
  complex structure 149–50, 152
  Kähler structure 145–50, 152
  Plücker embedding 151
  symplectic structure 147
  twisted circle action on 175

smooth maps 24
smooth maps, group of 26–8, 30–2
  automorphism group 31–2
  maximal normal subgroups 32
  perfection 30
smooth vector 113, 168, 194
spin group 231
  complex 237
spin representation 230–46
  as holomorphic sections 242–4
  holomorphic 243
  infinite dimensional 244–6
  irreducibility 243, 246, 247
  unitarity 233
stable manifold 112
standard module (see Borel–Weil representation)
string model 1
symmetric functions 202–5
  power sums 204
  Schur functions 202, 204
symmetrizable 77
symplectic group 191

tame symbol 60–3
thickened homogeneous space 293–4, 297–8
trace class 86
twisted loop group 34–7, 77, 175

unstable manifold 112, 130

Vandermonde determinant 279
van Est spectral sequence 303
vanishing theorems (for cohomology) 279, 286
vector field 25
Verma module 287–8, 295
  formal character of 288
vertex operator 5
  action of $\text{Diff}^+(S^1)$ 258–60, 268
  bosonic 264–7
  fermionic 251–60
Virasoro algebra 43, 167
virtual cardinal 103
virtual dimension 103

weight 19, 177
  antidominant 19, 178
  dominant 19
  fundamental 179
weight lattice 19
Weyl chamber 18
Weyl character formula 276–9
Weyl denominator formula 278
Weyl duality 165, 212
Weyl group 18
  of Kac–Moody algebra 77
  of LG (see affine Weyl group)
width 252
Wiener measure 170, 298